Electrochemical
Engineering and Energy

Electrochemical Engineering and Energy

Edited by

F. Lapicque and A. Storck

Laboratoire des Sciences du Génie Chimique
Ecole Nationale Supérieure des Industries Chimiques
Nancy, France

and

A. A. Wragg

School of Engineering
University of Exeter
Exeter, United Kingdom

Springer Science+Business Media, LLC

Library of Congress Cataloging-in-Publication Data

On file

ISBN 978-1-4613-6070-4   ISBN 978-1-4615-2514-1 (eBook)
DOI 10.1007/978-1-4615-2514-1

Proceedings of the Third European symposium on Electrical Engineering, held
March 23–25, 1994, in Nancy, France

© 1994 Springer Science+Business Media New York
Originally published by Plenum Press, New York in 1994

PREFACE

This volume contains the papers presented at the Third European Symposium on electrochemical engineering "Electrochemical Engineering and Energy", held on March 23-25, 1994 in Nancy, France. This meeting was organised by the Laboratoire des Sciences du Génie Chimique, CNRS, and the Centre de Perfectionnement des Industries Chimiques, in conjunction with the Groupe Français de Génie des Procédés, the Society of Chemical Industry, Dechema, and the European Federation of Chemical Engineering.

The organisers would like to thank the members of the Scientific Committee for the careful selection and refereeing of the papers. We are grateful to all contributors for the thorough preparation of their manuscripts. Thanks are due to the International Society of Electrochemistry, District Urbain de Nancy, Région Lorraine, Electricité de France, Institut National Polytechnique de Lorraine, and CNRS for their support.

The meeting was devoted to the role of electrochemical engineering and its relationship to the general issue of energy. The twenty-four papers contained here represent new research results and the potential technological developments in a variety of important and relevant areas of the field. The remaining three papers are expert reviews in areas of current technological concern where electrochemical engineering can clearly be seen to have major relevance to energy policy and environmental protection.

The papers cover a wide range of material from energy conversion, through electrochemical transport processes, electrochemical synthesis and modelling of electrochemical cells and processes. Throughout the book the reader will be aware of the multidisciplinary interaction of electrochemistry with other areas such as materials science, fluid mechanics, mass transfer and systems analysis. It is hoped that the papers present a convincing scientific base but are also relevant to future progress development and innovation in the field.

The final paper, a review paper on "Towards a Cleaner Environment using Electrochemical Techniques" by Klaus Jüttner and Gerhard Kreysa, was given in response to the award by the SCI of its Castner Medal to Professor Kreysa in recognition of its outstanding and distinguished work in electrochemical engineering.

F. Lapicque
A. Storck
A.A. Wragg

CONTENTS

ELECTROCHEMICAL SYNTHESIS

MODELLING OF ELECTROCHEMICAL SYSTEMS

ENVIRONMENTAL ELECTROCHEMICAL ENGINEERING REVIEW

ON THE ENVIRONMENTAL IMPACT OF ENERGY CONVERSION SYSTEMS

Peter Biedermann, Bernd Höhlein,
Bertram Sackmann and Ulrich Stimming
Institute of Energy Process Engineering (IEV)
Research Centre Jülich (KFA)
D-52425 Jülich, Germany

ABSTRACT

An analysis of the energy consumption and the associated emissions for power plants and road traffic is presented for West Germany as a typical highly industrialized country. While the former is the sector of highest energy consumption and CO_2 emission, the latter shows the highest emissions with respect to NO_x, CO and volatile organic compounds (VOC). Both sectors, each being prominent in its own way, are analyzed with respect to the underlying technical principles of the present technology in comparison to a potential future use of fuel cell systems. The energy consumption and emission levels of fuel cells in stationary electricity production in power plants and in vehicle application in road traffic is compared with the corresponding conventional systems. Conclusions for the future technological development can be drawn from the results of the presented analysis.

INTRODUCTION

In the past 20 years the global energy market has been subjected to intensive changes. Limited primary energy supply and the environmental impact associated with energy conversion lead to new legal requirements at both the national and the international level. New energy conversion processes are being developed in order to reduce the specific primary energy consumption and the associated specific emissions.

An analysis of primary energy requirements in the various areas in an industrialized country like West Germany (1990) reveals that power and heating plants as well as households and small businesses are the top primary energy consumers whereas traffic plays only a minor role. Conversion losses related to primary energy demand in power and heating plants amount to 23 %; the fraction of end energy consumption in road traffic amounts to 16 % (total traffic: 18 %). Almost 40 % of the total national primary energy input is used for conversion processes in power and heating plants.

The CO_2 emission balance specifies approximately 35 % emissions from power and heating plants, whereas road traffic with about 18 % (Figure 1) ranks behind industrial firing systems (20 %).

For CO, NO_x, and VOC (volatile organic compounds) emissions, which contribute to the formation of photochemical secondary pollutants in the troposphere near ground level,

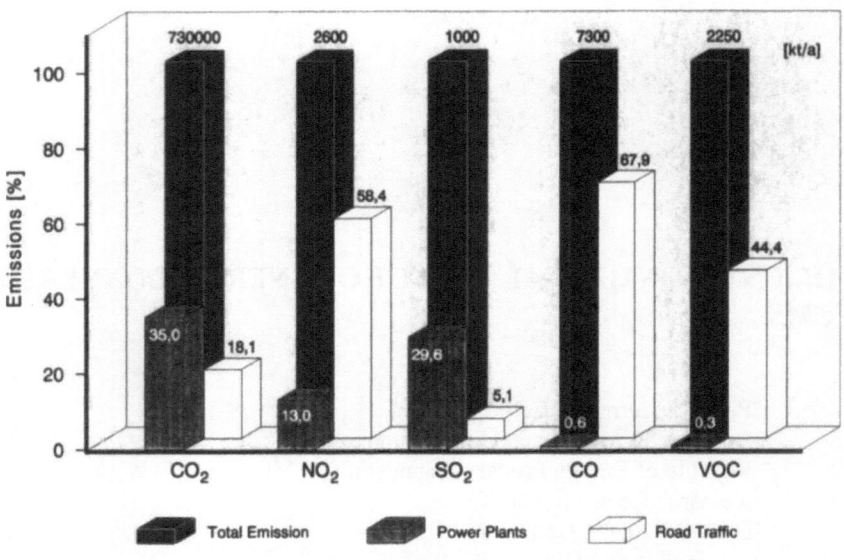

Figure 1. Percentage of Emissions: West Germany 1990[3]

the picture is quite different. Emissions from road traffic are at the top with more than 44 %. Power and heating plants contribute 13 % to NO_x emissions, < 1 % to CO emissions, 30 % to SO_2 emissions and < 1 % to organic compounds (VOC) (Figure 1).

These data show the significance of power and heating plants with respect to the consumption of primary energy and the emissions of carbon dioxide and sulphur dioxide on the one hand and the importance of road traffic with respect to legally restricted emissions of CO, NO_x and VOC on the other hand; such results are typical of many industrial countries. Consequently, a number of national and also international efforts are undertaken aiming at developing solutions for new energy conversion processes in the sector of power and heating plants and in the field of vehicle drive systems, especially for road traffic.

This paper describes the state of the art and the development potential for conventional and new thermal power processes as well as energy conversion processes based on fuel cells for electricity and heat production using fossil fuels. The discussion centres on the future use of high-temperature fuel cells and the resulting potential for lower energy consumption and reduced emissions in comparison to conventional thermal power processes. In addition, conventional drive systems based on internal combustion engines are compared with electric vehicles either with batteries or with fuel cells in terms of their energy consumption and emissions.

CONVERSION SYSTEMS FOR STATIONARY ELECTRICITY PRODUCTION

Primary Energy and Emissions: Worldwide - Germany

The consumption of fossil fuels was negligible until the middle of the 18th century. Energy requirements for urban and rural space heating, for crafts and small industries were covered by regenerative sources of primary energy; man-made CO_2 emissions from fossil fuels were not important. This situation changed significantly with industrialization. The beginning of the industrial era was marked by the invention of the steam engine, the first machine which allowed man to convert the chemical energy contained in fuels into mechanical energy. It was originally applied in mining and thus fired with coal. The mechanization of other industries and manufacturing methods as well as the use of steam engines in railways promoted this form of energy conversion. The second industrial revolution introduced electricity on a broad scale; this was initially generated exclusively on the basis of coal. As motorized traffic developed, another CO_2 source was added, the fossil fuel being oil which, together with natural gas, gradually began to replace coal. The current situation of industrialized countries is as follows:

- The prosperity of an industrial country is closely related to its degree of industrialization and its energy consumption. Apart from the OPEC states and Eastern Europe, today a country's prosperity is deduced from the per-capita energy consumption of its citizens.
- In spite of the introduction of nuclear and other energies, most of our energy requirements are still provided by fossil fuels.

Conventional methods of using the energy contained in fossil fuels is thermal combustion which generates heat from hydrocarbons according to the reaction

$$C_nH_m + \left(n + \frac{m}{4}\right)O_2 \longrightarrow n\,CO_2 + \frac{m}{2}\,H_2O \quad + \Delta H \qquad (1).$$

This leads currently to a carbon consumption from fossil fuels and an associated carbon dioxide emission of 5.4×10^9 t/a C and 19.8×10^9 t/a CO_2, respectively, the error of the estimate being $\pm 9\,\%$[1]. This CO_2 input into the atmosphere has led to an initially slow, but then accelerated increase of the CO_2 concentration in the earth's atmosphere (see Figure 2)[2]. It is being recognized today that this, in conjunction with other gases, can cause the greenhouse effect, i.e. a rise in global temperatures.

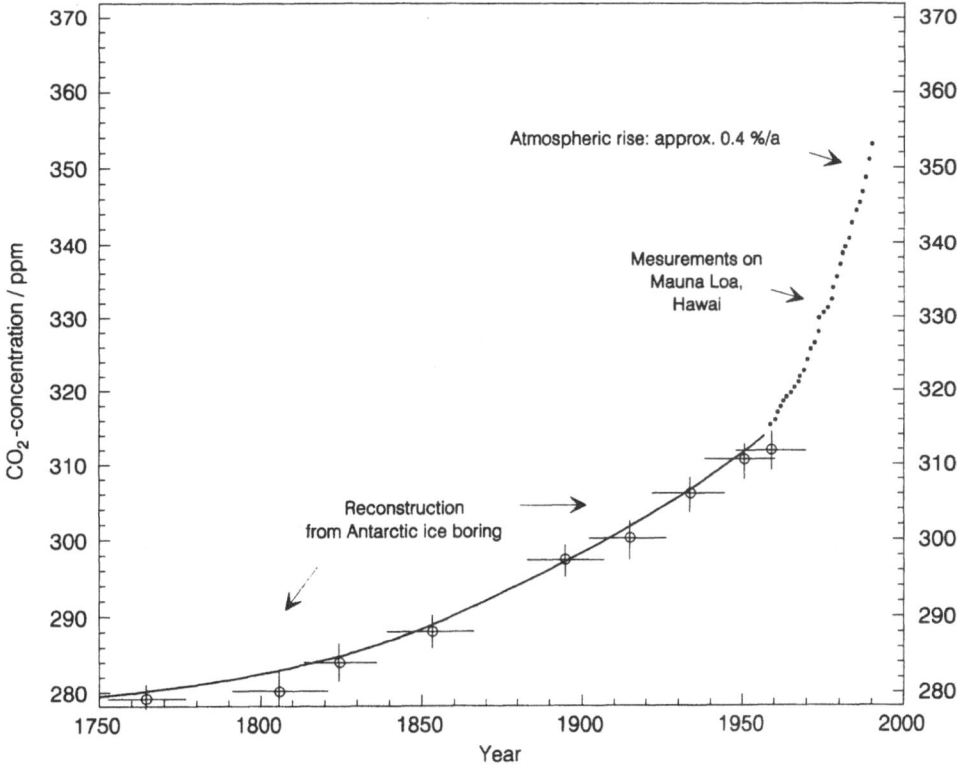

Figure 2. Rise of the CO_2 Concentration in the Earth's Atmosphere [2]

In addition to CO_2 emission, the combustion of carbonaceous energy carriers produce CO due to incomplete reaction, and other pollutants due to impurities in the fuel and nitrogen in air. These emissions are NO and NO_2, both referred to as NO_x, as well as SO_2, hydrocarbons and particles. A detailed list of emissions in West Germany is shown in

Table 1. Various air pollutants according to their sources (West Germany, 1990)[3]

		CO_2	CO	NO_x as NO_2	SO_2	VOC	Particles
Total quantity in million t/a		730	7.3	2.60	1.00	2.25	0.45
Power and heating plants	%	**35.0**	**0.6**	**13.0**	**29.6**	**0.3**	**5.1**
Industrial firing	%	19.8	10.0	9.1	31.3	0.4	4.0
Industrial processes	%	3.0	8.3	0.6	8.5	4.5	27.7
Small businesses	%	6.5	1.5	1.4	5.1	0.2	1.3
Households	%	12.9	7.9	2.8	8.3	1.2	5.3
Road traffic	%	**18.1**	**67.9**	**58.4**	**5.1**	**44.4**	**13.3**
Other traffic	%	4.7	3.7	14.7	12.0	3.4	4.6
Use of solvents	%	-	-	-	-	45.5	-
Transport of bulk goods	%	-	-	-	-	-	38.7

Table 1[3]. It can be seen that electricity generation and traffic contribute significantly to the aforementioned emissions. In the following the emissions produced in the energy consumption sector of power and heating plants will be discussed in more detail.

The most important gaseous impurities in flue gas from electricity generation plants are CO_2, NO_x, SO_2 and particles. In order to prevent SO_2 emission the sulphur can be removed from the fuel or the flue gas is desulphurized. For energy conversion processes requiring almost sulphur-free fuel gas for technical reasons such processes treating fuel can often not be introduced due to high costs. This applies, in particular, if regulatory limits for sulphur emissions permit cheaper desulphurization techniques in the offgas flow.

Particles are of significance only as fly ash in coal-fired power plants; effective filtration systems are available. It should be generally noted that the removal of NO_x, SO_2 and particles from offgases leads to efficiency losses for energy conversion in power plants.

Here, the discussion should be limited to CO_2 and NO_x which are the basic emissions from the thermal combustion of fossil fuels. CO_2 contributes about 50 % to the greenhouse effect and NO_x plays a role in surface water and soil acidification and in the formation of photosmog and ozone.

Energy Conversion Systems

According to Figure 3[4] there are two principle routes for converting the energy contained in fossil fuels into electricity. The first route comprises the thermal conversion processes in heat engines, which are classified into those with external and those with internal combustion. In a conventional steam power plant, the fuel is burned and steam is produced at high pressure and temperature. In gas turbine power plants, the combustion chamber is part of the turbine and combustion takes place inside the heat engine. Turbines and electrodynamic generators convert the enthalpy contained in the steam or gas into mechanical energy and subsequently into electricity. The second conversion route is the fuel

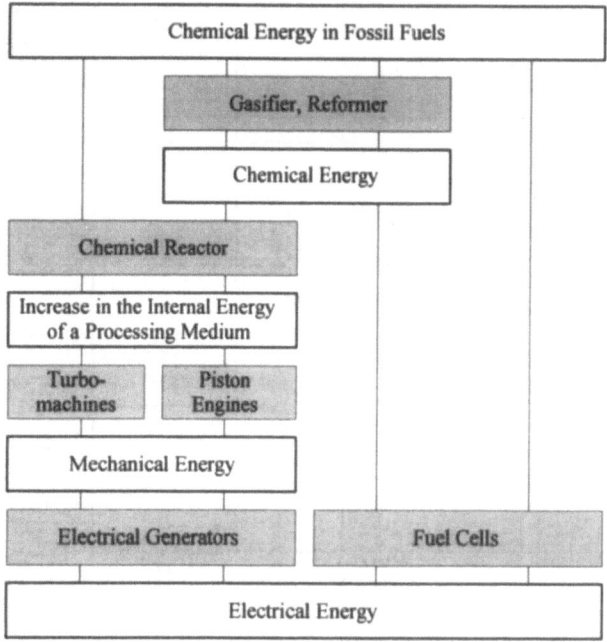

Figure 3. Conversion Routes of Fossil Fuels [4]

cell process. In a fuel cell, chemical energy is directly converted into electrical energy by means of an electrochemical reaction. The fuel is continuously fed to the cell and reaction products are continuously removed.

The maximum theoretical efficiency of a thermal process is given by the Carnot Cycle:

$$\eta_C = 1 - \frac{T_l}{T_u} \qquad (2).$$

T_l and T_u are the lower and upper temperatures during isothermal expansion or compression steps of this cycle.

This is in contrast to the theoretical efficiency of an electrochemical process given by

$$\eta_{th} = \frac{\Delta G}{\Delta H} = \frac{\Delta H - T \cdot \Delta S}{\Delta H} = 1 - \frac{T \cdot \Delta S}{\Delta H} \qquad (3).$$

The fuel cell converts directly the fraction ΔG into electricity. ΔH is the chemical energy stored in the fuel. The entropic energy loss, $T \cdot \Delta S$, is relatively small in most cases. The theoretical efficiency, η_{th}, for fuel cell processes with oxygen as the oxidant is > 90 % under standard conditions at 25°C, if the lower heating value (LHV) of the fuel is used for ΔH. A comparison between thermal conversion process and fuel cell operation in practically applied processes is shown in Table 2. It should be noted that the η_{th}-values are based on

Table 2. Comparison of Heat engines and Fuel Cells in Power Plants

System	Operation Temperature or Range in °C	Carnot Efficiency η_C	Theoretical Efficiency η_{th} based on $H_2 + O_2$ [1]	Technically achieved η_{el} based on **Natural Gas + Air** (Cell Efficiency based on synthesis gas) [1]
Gas engine				0.30 - 0.36
Gas turbine	1100 – 500	0.44		0.32 - 0.36
Steam power plant	535 – 40	0.613		0.42 - 0.45
Combined-cycle power plant (CCP)	1100 – 30	0.78		0.53 [2]
PAFC packaged cogeneration plant	200		0.87	0.40 [3],[4] (0.55) [4]
SOFC cogeneration plant	1000		0.71	0.40 - 0.60 [5] (0.60 - 0.65)
Triple process SOFC + CCP				0.63 - 0.68 [6] (0.60 - 0.65)

[1] relative to lower heat value   [4] given by Wendt et al.[11]
[2] KWU/Ambarli (Turkey)   [5] expected by Riensche et al.[9]
[3] ONSI, with reformer   [6] expected by Reinecke et al.[12]

pure hydrogen and pure oxygen. The technically achieved η_{el}-values are efficiencies of complete systems for which methane was taken as the energy carrier.

Fuel cell types are distinguished according to their electrolytes and vary with respect to their operating temperatures:
- Alkaline Fuel Cell (AFC)   approx. 80°C
- Phosphoric Acid Fuel Cell (PAFC)   approx. 200°C
- Molten Carbonate Fuel Cell (MCFC)   approx. 650°C
- Solid Oxide Fuel Cell (SOFC)   approx. 1000°C

The AFC is mentioned because it is fairly far advanced in its development. It may not be suitable for broader application because it can only be operated with high-purity hydrogen and oxygen. The types of fuel cells under development for stationary electricity generation are the PAFC as a low-temperature fuel cell and MCFC and SOFC as high-temperature fuel cells. Typical materials used in these cells are shown in Table 3.

The fuel used is synthesis gas prepared by reforming natural gas or by coal gasification. The oxidant is air. In practice the fuel is not completely converted; residual gas leaves the cell together with the product gas from electrochemical oxidation, but it can be used for heat production in residual combustion. In Table 2 the cell efficiency based on synthesis gas from methane is shown along with the overall efficiencies η_{el} of fuel cell power plants.

The heat produced due to irreversibilities of the electrochemical process and the heat from residual gas combustion can be used thermally. Heat is, for example, required for reforming natural gas.

Table 3. Examples of Electrode and Electrolyte Materials

	Anode	Electrolyte [mobile ions]	Cathode
PAFC	carbon + Pt	H_3PO_4 $[H^+]$	carbon + Pt
MCFC	carbon + Pt	Li_2CO_3-K_2CO_3 in $LiAlO_2$-matrix $[CO_3^{2-}]$	NiO Li-doped
SOFC	electronic conductor Ni-ZrO_2-Cermet 30 Vol% NiO 70 Vol%$(ZrO_2)_{0.9}$ $/(Y_2O_3)_{0.1}$	ionic conductor $(ZrO_2)_{0.92}$ $/(Y_2O_3)_{0.08}$ $[O^{2-}]$	electronic conductor (perovskite) $La_{0.8}Sr_{0.2}MnO_3$

The current status of the technical development of fuel cells depends on the type of system. Up to 11 MW are reached for the phosphoric acid fuel cell, which is the most advanced type on a technical scale. ONSI Corporation, USA, produces e.g. the PC25 type as a PAFC packaged cogeneration plant with an electrical output of 200 kW in a small-series production[5]. Complete commercilization together with competitive prices are expected by ONSI within the next few years. Material problems encountered for the MCFC due to corrosiveness of the molten carbonate and for the SOFC due to high temperatures still require intensive development work in materials research.

Thermal Conversion Systems

In the following, the terms "system efficiency" and "cell efficiency" will be distinguished. Both denote the ratio of electrical power output, P_{el}, to chemical energy, $\dot{H}$, supplied with the fuel:

$$\eta_{el} = \frac{P_{el}}{\dot{H}} = \frac{P_{el}}{\dot{m}_f \cdot \Delta H} \qquad (4)$$

They differ with respect to the boundaries for which energy changes are taken into account:
- the system efficiency compares the electricity output with the fuel supply for the entire plant
- the cell efficiency compares the electric current at the cell terminals with the fuel input at the gas inlet of the cell stack.

The chemical energy input, $\dot{H}$, is calculated from the mass flow $\dot{m}_f$ and the lower heating value (LHV) of the fuel.

Large Power Plants. The technologies currently used for thermal energy conversion in power plants are the steam- and the gas-turbine process or a combination of both. Their system efficiencies are mainly determined by gas or steam pressure and temperature.

Irreversibilities, friction and other losses cause practical efficiencies, which are substantially lower than the Carnot efficiency. Improvements have been done by increasing pressure and temperature at the turbine inlet, but limits are given by potential material problems. Other measures for an efficiency increase are improvements in heat supply and removal (e.g. multiple reheating, regenerative combustion air and feed water preheating, fluidized-bed combustion, wet cooling etc.).

Distinct thermodynamic improvements and thus a considerable increase in system efficiency are obtained by installing a gas turbine before the steam process. This so-called combined cycle is easy to realize for gaseous fuels and leads to low investment costs. State-of-the-art gas turbines derived from aeroplane engines are available with high power rating and high efficiency. Currently operated combined-cycle power plants with maximum efficiency can achieve a system efficiency of 53 % (KWU/Ambarli, Turkey, see Table 2).

If coal is used in combined cycle power plants, it must first be converted into a suitable gas. The two basic techniques are (i) combustion of the coal under pressure and expansion of the firing gases in the gas turbine and (ii) gasification of the coal into a combustible gas which, similarly to natural gas, is fed into the combustion chamber of the gas turbine. Problems are encountered in the first process with hot gas cleaning, since particles and other impurities from coal flue gases are detrimental to the gas turbine. In the gasification process, the synthesis product gas is cooled to temperatures permitting highly efficient and proven cleaning techniques to be used. The pollutants are extracted to such an extent that they do not represent any problem for the gas turbine. This, however, reduces the system efficiency and causes high investment costs. In this respect, coal gasification is at a disadvantage unless stricter requirements of emission reduction, i.e. SO_2 content in the offgas from coal-fired power plants, are being imposed.

The nitrogen oxides NO and NO_2, summarized as NO_x, arise during fossil fuel combustion in the flame and the surrounding high-temperature zones. Reduction measures distinguish between those concerning combustion technology and those involving separation techniques. Combustion technologies such as premix burners, flue gas recycling, staged combustion or fluidized bed combustion aim at preventing NO_x by keeping peak temperatures as low as possible and/or having combustion partially taking place in a reducing atmosphere. For the separation of nitrogen oxides the flue gas is passed through a reactor. A solid catalyst is used to reduce NO_x to nitrogen and water by ammonia injection.

Table 4. Emissions from Different Types of Fossil-Fuel-Fired 600-MW Power Plants[6]

Power plant	η_{el} %	CO_2 $\frac{g}{kWh}$	NO_x $\frac{mg}{kWh}$	SO_2 $\frac{mg}{kWh}$	CO $\frac{mg}{kWh}$	VOC $\frac{mg}{kWh}$
Pulverized coal fired steam power plant	42	830	600	600	75	< 5
Combined cycle power plant with integrated coal gasification	45	770	305	155	65	< 5
Natural gas fired steam power plant	45	440	200	< 1	18	< 5
Natural gas fired combined cycle power plant	53	375	125	< 1	33	18

System efficiencies, emissions and their reduction in different conventional power plant types can be seen from Table 4[6]. Although the CO_2 emission is linked to the conversion efficiency, Table 4 clearly demonstrate the advantage of natural gas over coal because of the low specific carbon content of natural gas.

Cogeneration Plants. Cogeneration plants are energy conversion systems which provide useful heat in addition to electricity. If heat removal takes place at low temperatures, the Carnot efficiency is high (see equation 2), but this heat can not be used otherwise. A rise in temperature at the cold end lowers the yield of electrical power in steam power plants, but the extracted heat can be efficiently used e.g. as process steam. This operating procedure is referred to as cogeneration. Depending on customer requirements, the overall efficiency of cogeneration can be greater than in case of pure electricity generation.

In principle, cogeneration can be used for any type of power plant, but the influence of the equation 2 on steam turbines is great; they are very sensitive to an increase in cooling temperature. Water-cooled combustion engines supply waste heat at elevated temperatures without efficiency losses. Using the waste heat both of cooling water and of offgas up to 150°C cogeneration plants with combustion engines are only suitable for the heating of buildings and low-grade processes[7]. In this case packaged cogeneration plants with gas combustion engines located close to the heat consumers are a preferred application.

A cogeneration plant consists of one or several modules, equipped with the necessary heat exchangers for the use of cooling-water and offgas heat. Electricity and heat are simultaneously produced in plant operation. An additional boiler system is used when more heat is required than supplied by the cogeneration process.

For emission reduction in combustion engines, three techniques can be distinguished: lean-mix combustion, exhaust gas treatment in a three-way catalyst with a stoichiometric air/fuel ratio ($\lambda = 1$) and exhaust gas treatment with ammonia injection and post-oxidation in an oxidation catalyst. These measures involve costs, but the effectiveness is only sufficient in terms of the emission limits: the pollutant concentration in the released offgas is still in the order of 100 mg/m^3 (see Figure 7).

Conversion Systems with High-Temperature Fuel Cells

Equation 3 gives the theoretical efficiency of energy conversion in fuel cells being very high at low temperatures. This suggests that the phosphoric acid fuel cell operating at temperatures of approx. 200°C seems best suited for stationary systems. Fossil fuels must, however, first be converted into a hydrogen-rich synthesis gas by reforming or gasification. The required heat cannot be covered by the waste heat of this cell type, because reforming takes place at approx. 850°C. In addition, the CO content in the synthesis gas must be reduced in a second process step. Due to these characteristics the system efficiency of a cogeneration plant with phosphoric acid fuel cell using methane only reaches approx. 40 % (see Table 2).

The situation is different for high-temperature fuel cells. Their high operating temperature increases the fraction of entropic energy loss $T \cdot \Delta S$ in electrochemical conversion. For hydrogen as fuel the theoretical efficiency is $\eta_{th} = 71.1$ % at standard pressure and 1000°C compared to 94.5 % at 25°C. If the fuel of the high-temperature system is methane, the amount and temperature of the heat loss from the electrochemical reaction is sufficient to supply the heat for reforming. An efficiency of $\eta_{th} = 87.7$ % is then calculated at 1000°C[8].

The great advantages of high-temperature fuel cells of the MCFC and SOFC type using methane as the fuel are:
  - The theoretical efficiency, η_{th}, of the overall system of a "natural gas fed fuel cell" is high in spite of high operating temperatures.
  - The irreversible heat losses in electrochemical conversion supply enough heat required for natural gas reforming.
  - The steam produced during the reaction in the fuel cell is also available at a high temperature level.
  - CO can be oxidized electrochemically.
  - Reforming can take place partially or entirely in the cell (internal reforming).

In the case of internal heat utilization, the reformer is integrated in the cell or stack. This permits particularly high system efficiencies to be achieved with natural gas as fuel. Various possibilities of optimizing a SOFC cogeneration plant have been presented[9]. The maximum calculated cell efficiency is given as 65 % for 100 % internal reforming and optimistic cell parameters. For an overall plant with SOFC as sole energy converter an electrical efficiency of about 55 % can be derived. In the case of external reforming, the heat and steam must be extracted or removed from the cell and fed to a separate apparatus, the reformer. This causes lower efficiencies.

Of particular importance are the low NO_x emissions from fuel cells. This is due to the low working temperatures in comparison to thermal combustion. Even the SOFC is still below the 1300°C limit above which pronounced formation of thermal NO_x sets in.

The SOFC operating principle is very simple due to the solid ceramic electrolyte involved, but material problems result from the high operating temperature. Three different cell designs have been basically tested to date: the "tubular concept", the "honeycomb concept" and the "planar cell concept". The tubular concept has been advanced by Westinghouse[10]. The planar cell concept involves a cell composed of plate-type elements (anode, electrolyte, cathode as a unit or separated, gas distribution and current conductor plates). Several cells form a stack. This design offers the advantage that its components are well suited for mass production, but problems of sealing the gas-containing spaces still exist. Up to now stacks of the kW class have been successfully tested.

In addition to corrosion problems resulting from the high chemical aggressiveness of molten carbonate, the MCFC process has the disadvantage, that carbon dioxide must always be present in the air at the cathode due to the electrode reactions. This requirement necessitates the separation of this component from the anode gas flow and its admixture to the cathode gas, a process step which makes the periphery more expensive[11]. External and internal reforming processes are currently discussed for gas processing in MCFC.

In principle, systems with high-temperature fuel cells can be used both in large power plants and in decentralized electricity and heat production. A combination of fuel cells and thermal power plants can be an optimum solution in power plant design. The high temperature fuel cell is here installed as a so-called topping cycle before a thermal power process and contributes significantly to electricity production. However, high-temperature fuel cells are also very suitable as the sole energy converters in cogeneration plants.

Power Plants. As described above SOFC systems can be combined with thermal cycles to utilize the heat and the unused part of the fuel leaving the stack in a thermal process. A combination with a gas and steam cycle allows for maximum system efficiency but is most useful only in the > 10 MW class. Figure 4 shows a fuel cell, here an SOFC, as a topping cycle integrated into a combined cycle power plant. The essential process steps are briefly as follows:

Natural gas is reformed and the hydrogen-rich synthesis gas is cleaned and fed into the fuel cell. The electrochemical process produces electricity; the by-products heat and steam are partially recycled into the reformer unit. Since the fuel is not completely converted in the SOFC, the offgas undergoes combustion and enters a gas turbine. A conventional steam process is arranged after the turbine. The electrical energy produced by the SOFC system and by the generator coupled to the gas and steam turbines is fed into the grid.

Instead of natural gas it is also possible to use coal as the primary energy source, which must then be subjected to a gasification process. High demands are made on the purity of the product gas from coal gasification: substances like sulphur, chlorine and fluorine compounds acting as catalyst poisons must be separated from the fuel gas. The exact requirements are not clear as yet but are probably for the sulphur content < 1 ppm, for chlorine and fluorine < 0.1 ppm. Similarly to the use of coal gas in gas turbines, these requirements may be an obstacle to the introduction of fuel cell technology based on coal in the power plant market.

Coal gas contains a maximum of 4 - 15 % methane. The use of cell heat losses for reforming is thus lower in proportion to this content. This explains the smaller efficiency increase in the application of high-temperature fuel cells in coal-fired power plants.

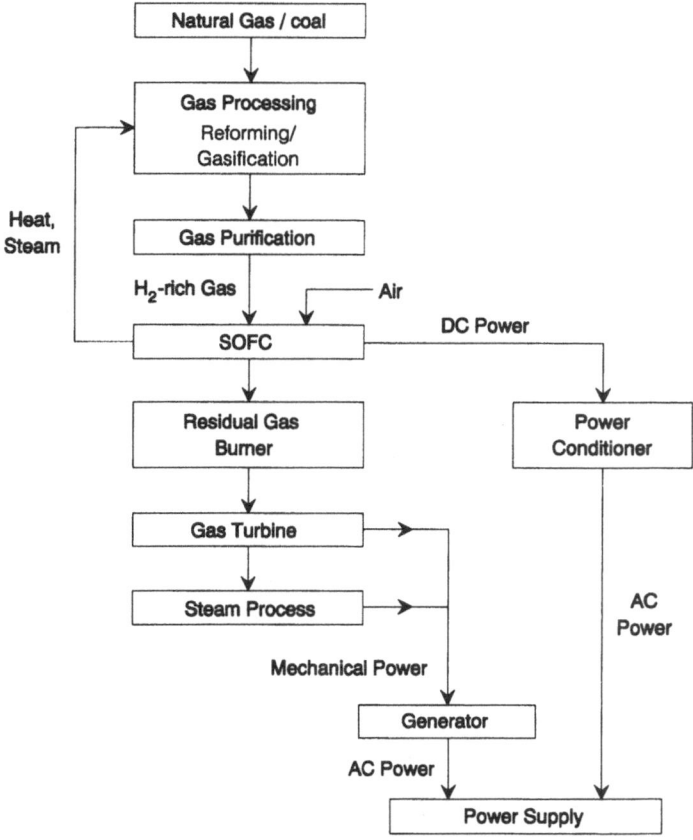

Figure 4. SOFC as the Topping Cycle in Combined Power Plants

Simple calculations show the advantage of a power plant with a fuel cell topping cycle, if compared to the combined cycle used today. The fuel may be natural gas. A conventional power plant having an system efficiency of 50 % releases approximately 390 g CO_2/kW. The emissions of a progressive fuel cell power plant with an system efficiency of 65 % are approximately 300 g CO_2/kW. That means a CO_2-emission reduction by 23 % for natural gas. Similar calculations give an CO_2-saving of 13 % if coal is used as fuel[12].

The results of systematic studies have been summarized for a future power plant with MCFC or SOFC fuel cells as the topping cycle[12]. It was assumed that the cells can be operated under pressure and that cell efficiency reaches 80 % of the theoretical electrochemical efficiency. The calculations for a 40 MW gas-fired power plant gave the following system efficiencies:

with external reforming	MCFC: 47 %	
	SOFC: 63 %	and
with internal reforming	MCFC: 60 %	
	SOFC: 68 %.	

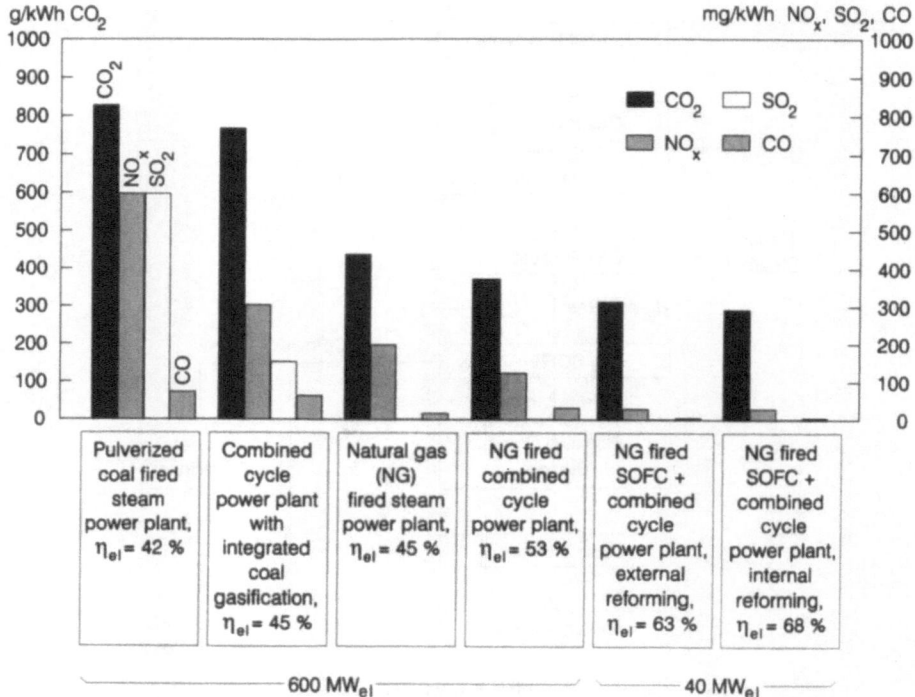

Figure 5. Comparison of Power Plant Emissions [6,12]

Figure 5 shows a diagram of emissions calculated for the 40 MW gas-fired power plant with SOFC and the state-of-the-art gas-fired 600 MW conventional power plant shown in Table 4. The CO_2 emissions from SOFC power plants are calculated from the system efficiencies. The few data available for NO_x emissions from SOFC fuel cell power plants vary between 50[12] and 13 mg NO_x/kW$_{el}$[13]. In the diagram the value of 30 mg NO_x/kW$_{el}$ is used specified by Wendt et al.[11] for different fuel cell power plant types. Values from Reinecke et al.[12] were used for the other emissions. It should be emphasized that a direct comparison of a 40 MW with a 600 MW power plant is difficult, but data are sparse and the principles become clear.

Cogeneration Plants. In a cogeneration plant with an SOFC as the sole energy converter, electricity is generated and the waste heat is used e.g. for heating purposes. Figure 6 shows the process engineering section of an SOFC used in such a cogeneration plant. For reasons

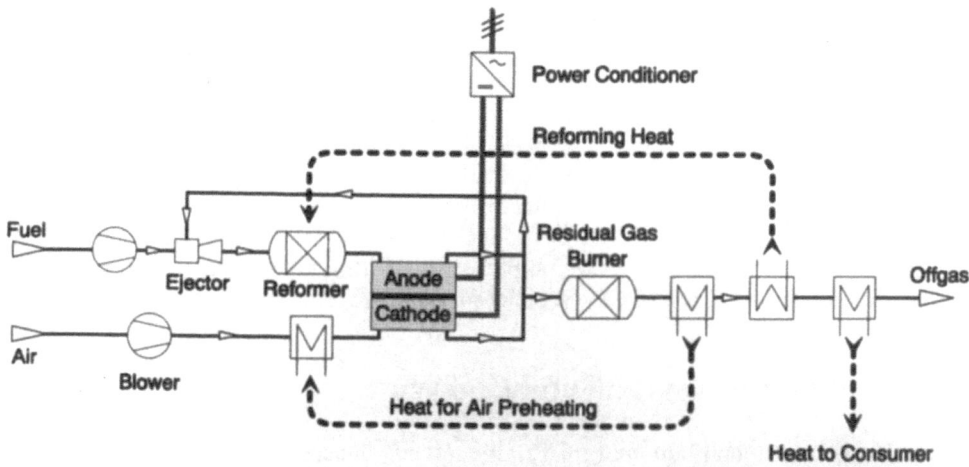

Figure 6. SOFC System as Decentralized Cogeneration Plant

of process control, the natural gas is not reformed internally, but subjected to partial prereforming. Preheated air flows into the cathode part of the cell stack. Natural gas is mixed with anode offgas in an injector and passed into a prereformer. Partially reformed natural gas enters the cell stack and is electrochemically converted up to 85 %. Part of the converted anode gas (approx. 1000°C) flows back into the injector and the other part is passed to the catalytic burner. The heat is used to heat the prereformer and the air preheating system and is then available for other heating purposes.

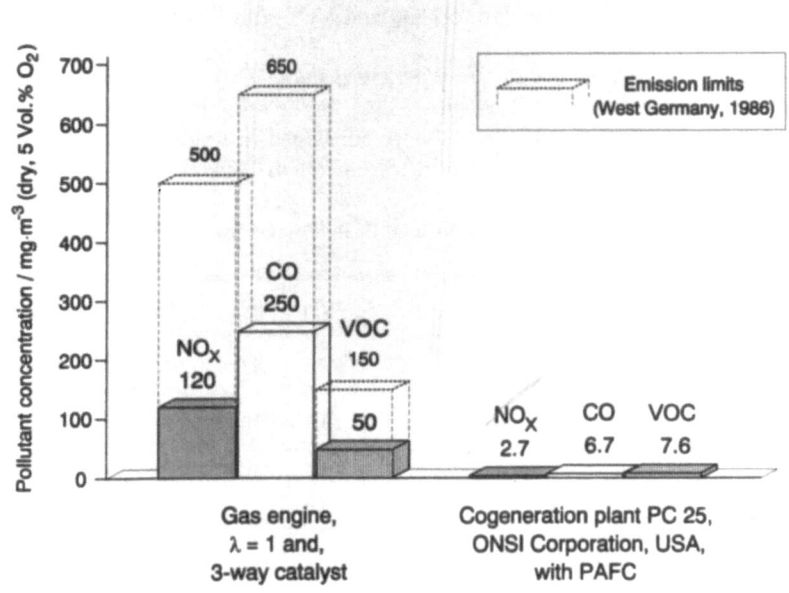

Figure 7. Comparison of Emission: Cogeneration Plant with Gas Engine and Fuel Cell[5,15]

The potential competitors of cogeneration plants with high-temperature fuel cell are gas engine plants. These have two disadvantages. On the one hand, its waste heat is available only at a maximum temperature of 150°C and, on the other, its efficiency strongly decreases at small power ratings. Currently available cogeneration plants with gas engines have a maximum electrical system efficiency of 30 % at 100 kW_{el} or 36 % at 2000 kW_{el} rated power. In comparison, 50 % and more are anticipated for a 200 kW_{el} cogeneration plant with a SOFC[14]. Waste heat is extracted at temperatures of up to 1000°C and can be fed to corresponding production processes or heating systems. Figure 7 shows NO_x, CO and VOC emissions for two existing cogeneration plants, one with a gas motor, the other being the above mentioned PAFC plant PC25, produced by ONSI Corporation, USA[5,15].

CONVERSION SYSTEMS FOR ROAD TRAFFIC

The possible contribution of road traffic to environmental, climate and health changes is increasingly becoming recognized. Pollutant emissions, noise, the consumption of surface area, the water contamination as well as local and global climate change are caused by traffic. Polluters are ships, airplanes, railways and, predominantly, road vehicles.

Emissions from Road Traffic

The most important pollutants from road traffic include carbon monoxide (CO), carbon dioxide (CO_2), nitrogen oxides (NO_x), sulphur dioxide (SO_2), volatile organic compounds (VOC) and particles. In most industrialized countries the emission levels of these compounds are legally restricted.

In addition, other substances are emitted which are not yet subject to legal restrictions. This group includes benzene, which is carcinogenic, and other organic compounds. Emissions may have a local and global impact on man, environment and climate. Examples are the toxicity of CO, NO and NO_2, the acidification of soils by nitrogen compounds, ozone formation near ground level due to NO_x and VOC and the greenhouse effect caused by CO_2.

The primary energy consumption in West Germany (1990) amounted to 11,495 PJ[1], in which road traffic had a share of 1,818 PJ (~ 16 %). In the same period 1,809 PJ in the form of gasoline (64 %) and Diesel fuel (36 %) were consumed in road traffic[3]. A listing of the various pollutants together with their quantities are given in Table 5.

Table 5. Emissions from road traffic in West Germany (1990) [3]

Pollutant	CO_2	CO	NO_2	SO_2	VOC	Particles
Emissions / 10^6 t	132	4.95	1.50	0.05	1.00	0.06
Passenger cars (gasoline) / %	62.5	92.9	55.1	10.3		4.4
Passenger cars (Diesel) / %	12.0	2.1	4.9	29.2	100 %	22.3
Commercial vehicles / %	24.8	3.1	39.9	60.4		73.2

This shows that the passenger car with internal combustion (Otto) engine is by far the largest polluter especially for CO and NO_x emissions.

When energy consumption and emissions due to road traffic are evaluated it is of great importance to consider all contributions. In the energy conversion sequences discussed next the calculation starts with the primary energy carrier and its exploration and ends at the friction losses of the wheel on the road. In a life cycle approach all energy consumption and emissions associated with making, running, maintaining and disposing of the vehicle are taken into account.

Comparison of Drive Systems for Passenger Cars

In order to better understand the energy conversion sequences presented later, the principle of the internal combustion engine will be compared with that of an electric motor powered by an accumulator or a fuel cell. In the evaluation of efficiencies a distinction is made between the efficiency of the energy converter itself, the efficiency of the energy conversion system (energy converter and auxiliaries) and the efficiency of the whole energy conversion sequence. In addition, the theoretical efficiency must be distinguished from the technically achieved efficiency of an energy converter.

Internal Combustion Engine. In an internal combustion engine, a fuel-air mixture is burned within the engine so as to produce mechanical energy. This involves a thermodynamic cycle whose theoretical energy efficiency is determined by the maximum and minimum absolute temperatures of the working medium. Technical processes are realized as Otto and Diesel engines which exhibit poor partial load performances. These engines are sensitive with respect to gasoline which is a mixture of different hydrocarbons and additives.

The emissions of internal combustion engines are composed of hydrocarbons due to incomplete combustion, thermal cracking products, combustion by-products from fuel additives and impurities and combustion by-products from atmospheric oxygen. In addition, oxidants from exhaust components are produced by photochemical reactions.

Electric Motor with Battery. Electrical energy can be stored in accumulators in the form of chemical energy. During charging and discharging electrochemical reactions take place at the electrodes of the cells.

There is a large variety of different battery systems for traction purposes. New systems such as Ni/MeH, Na/S and Na/NiCl can be introduced to the market only after 5 to 10 years while the lead/acid and nickel/cadmium batteries are already available[16].

The efficiencies of the systems range between 70 % and 85 %, the energy densities between 30 and 140 Wh/kg and the power densities between 70 and 200 W/kg. The associated operating temperatures range from ambient temperatures to 350°C. Principal problems are the low energy and power densities.

The battery supplies electricity to the electric motor of the vehicle and thus results in a zero emission vehicle at the location where it is operated. Since in practically all cases the battery is charged with electricity from the grid the emissions originating from the power plants have to be taken into account.

Electric Motor with Fuel Cells. Fuel cells for traction require catalyst electrodes (e.g. Pt, Pd, Ru) to ensure a sufficient rate of the electrochemical reactions.

The main components of a hydrogen-oxygen fuel cell are the anode, at which hydrogen is oxidized, the cathode, at which oxygen is reduced, and the electrolyte. The electrodes have good electronic conductivity in contrast to the ion-conducting electrolyte. The electrodes are permeable to gas, the electrolyte has to be gas tight. The electric circuit is closed by an external ohmic load.

In the overall reaction, the chemical energy stored in the fuel gas is directly converted into electrical energy and entropic loss; the reaction product is water. The reaction is thus not dependent on the Carnot factor! This is the basic distinction between the electrochemical fuel cell system and a conventional heat engine.

The efficiency of a fuel cell for a given fuel cell reaction (e.g. H_2/O_2) is only determined by the ratio of the Gibbs free energy to the lower heating value at a defined working temperature. A maximum efficiency of 83 % can be ideally achieved for the H_2/O_2

reaction under standard conditions at a temperature of 25°C. Efficiencies of up to 69 % were technically achieved for the same reaction with a stack at a temperature of 80°C[17]; this will be discussed later.

The following low-temperature fuel cell types are principally suitable for application in the mobile sector: Alkaline Fuel Cell (AFC) 80 - 120°C, Proton Exchange Membrane Fuel Cell (PEMFC) 50 - 110°C and Phosphoric Acid Fuel Cell (PAFC) 160 - 210°C. The PEMFC is favoured due to its advantageous properties such as a solid electrolyte, a good cold start possibility, a high power density and a high efficiency. A disadvantage is, however, the CO sensitivity of the Pt catalyst (< 10 ppm). This is of importance if the fuel is a hydrogen-rich synthesis gas instead of pure hydrogen. Compared to the AFC the PEMFC has the advantage of a certain tolerance with respect to CO_2; compared to the PAFC it operates at

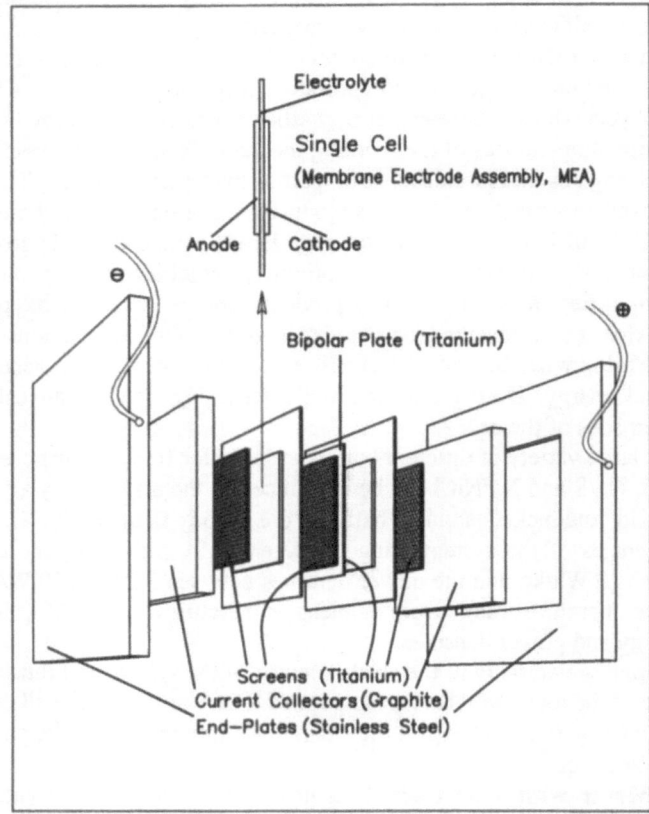

Figure 8. Principle of a PEM Fuel Cell Stack

lower temperatures. In a combination of reformer and fuel cell the PAFC has advantages due to the higher temperature level of the reformer, higher reaction rate and lower CO sensitivity.

The basic components of a PEMFC are the electrolyte, the electrodes and the catalyst. A single cell has a planar sandwich geometry. Porous gas diffusion electrodes with a carbon-bound platinum catalyst are pressed onto both sides of an ion exchange membrane. Titanium meshes serve as distributors of the reaction gases on the electrode surfaces. The individual membrane units are interconnected by bipolar plates and held in place by end plates. Figure 8 schematically shows a design with two single cells in series connection[18]. Hydrogen or a hydrogen-rich fuel gas is fed to the anode, oxygen or air to the cathode. Proton-conducting polymer membranes such as NAFION (DuPont) or Dow XUS (DOW) are used as electrolyte.

For transportation two basic fuel concepts are currently being developed for application with a PEMFC in an electric vehicle: hydrogen/air and methanol reforming (H_2-rich reformer gas)/air. The performance data of the methanol reformer must be adapted to the cell considering parameters such as pressure, temperature and water regime as well as dynamics and long-term stability. Another great problem is the sensitivity of the platinum catalyst of the anode to CO contained in the reformer gas.

In vehicles with fuel cells and electric motors hydrogen or methanol as fuel must be stored and transported on-board. In case of methanol as fuel a hydrogen-rich gas must be synthesized on-board by reforming. The fuel gas is electrochemically converted into electricity in the fuel cell. An electric motor converts it into mechanical energy. The limiting element for energy storage is the gas storage or the methanol tank.

The efficiency of a fuel cell is only dependent on free reaction enthalpy, operating temperature and lower heating value of the fuel gas, while an engine depends on the Carnot factor. Efficiencies of 59 to 69 % were technically achieved for two 30- to 40-kW PEMFC stacks and H_2/O_2 reaction at an operating temperature of 80°C, depending on the operating conditions. Stack efficiency increased from 59 % at full load to a maximum of 69 % at 20 % partial load[17]. Internal combustion engines exhibit the opposite behaviour. Using the example of a 1.6 l Otto and a 1.6 l Diesel engine, the efficiency decreases from 29 % (Otto) and 33 % (Diesel) at optimum load to 8 % (Otto) and 17 % (Diesel), respectively, in partial load operation as is typical for driving in urban areas[19].

PEMFC and battery produce few or no pollutant emissions even in the partial load range at their location of use. The fuel cell vehicle has the advantage of lower weight due to the lack of traction batteries. Transported energy reserves are much greater and thus lead to an extended range. The fuel cell vehicle is much more flexible due to faster refuelling and an existing network of gas stations suitable for liquid fuels.

Vehicles in current road traffic are largely operated at partial load. Even the best internal combustion engines exhibit high fuel consumption which, in connection with currently used engine techniques, leads to the known high emission levels[19]. The exhaust problems and, in particular, nitrogen emissions cannot be coped with by these techniques.

Options for Energy and Emission Reduction

A potential for distinct improvement of the partial load performance of internal combustion engines is the development of new engines operated by mixture stratification. Considerable consumption and emission reductions can be achieved with novel combustion techniques, especially in two-stroke engines[19].

A considerable problem is the storage and transport of electrical energy in the vehicle. Low specific storage capacity (kWh/g) and low specific power density (kW/kg) result in short vehicle ranges. Except for the lead acid battery and the nickel-cadmium battery, all other battery concepts have not yet reached the state of a commercial product.

Hybrid systems are a combination of at least two basically different drive systems. In electric hybrid vehicles, for example, the electric motor is operated in the partial load range with significantly better efficiency, while the internal combustion engine drives the vehicle at higher load. This leads to minimum local emissions combined with long ranges. Such systems are currently being tested for series qualification[19]. A different electric hybrid concept is realized in the Environmental Concept Car by VOLVO[20,21]. A gas turbine operated with Diesel fuel drives a high-speed electric generator. The electric motor can be supplied from the generator, a Ni/Cd battery set, or a combination of both. The emissions from the gas turbine are in compliance with the ULEV (ultra low emission vehicle) standard, anticipated for California, USA, in 1998.

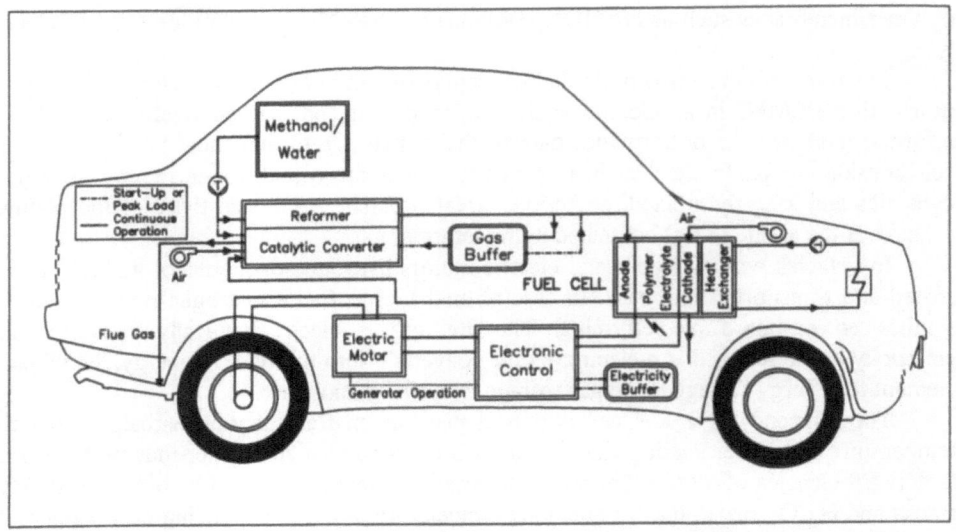

Figure 9. Passenger Car with Electric Motor and PEMFC based on Methanol

A different drive system for passenger cars uses an electric motor together with a fuel cell and hydrogen based on natural gas or methanol as fuel. Figure 9 shows a vehicle in which a PEMFC is used as energy converter[22]. The main components of this system are the methanol reformer to convert methanol and water into a hydrogen-rich synthesis gas, a gas/energy storage covering the energy demand for peak loads, the PEMFC as the electrochemical energy converter, energy storage to supply the vehicle's electric system, and an electric motor driving the vehicle, as well as electronics. In order to cover the heat requirements of the reformer, unconverted anode gas or methanol is catalytically burned. The characteristic feature of this system is the PEMFC-governed long-range driving cycle with minimum emissions and good full and partial load performance.

Energy Conversion Sequences

Energy conversion sequences will be analysed beginning with the primary energy carrier and ending with the conversion of the end energy in a given drive system under defined performance requirements. This will be briefly explained by the example of the crude oil - internal combustion engine - driving cycle sequence (Figure 10). Crude oil as the primary energy carrier is extracted, transported to a refinery and processed into gasoline as the on-board energy carrier. Gasoline is transported to a service station where a vehicle is fuelled. The gasoline is then used to operate a passenger car with internal combustion engine. Power output at the wheel follows the requirements of (e.g.) an ECE driving cycle. All the above parts of the energy conversion sequence involve emissions or due to the use of energy for the production, refining or storage of gasoline.

The balance boundary was assumed to be the border of Germany. A peculiarity is the balancing of locally emission-free battery operated electric vehicles. Their overall emission is affected by locally different emissions depending on the type of power plant. This leads to the situation that such vehicles may be heavily burdened with emissions whereas they are almost emission-free when electricity production is not based on fossil fuels. As an example, while locally operating with zero emission in all cases the total emissions can be quite different if one compares Germany with a high percentage of coal fired power plants (Table 1), France with 70 % nuclear power plants and California, U.S.A., with a high percentage of combined cycle power plants based on natural gas.

For the energy conversion sequences shown in Figure 10 a first estimate was made for the energy and emission balances[23,24]; the results are compiled in Table 6. Data are specified for Otto and Diesel engines, an electric motor exclusively powered by accumulators on the basis of the electricity mix in Germany and a PEM fuel cell with electric motor using methanol based on natural gas. The specific emissions are compared with the values of the Californian ULEV standard for passenger cars to be adopted in 1998.

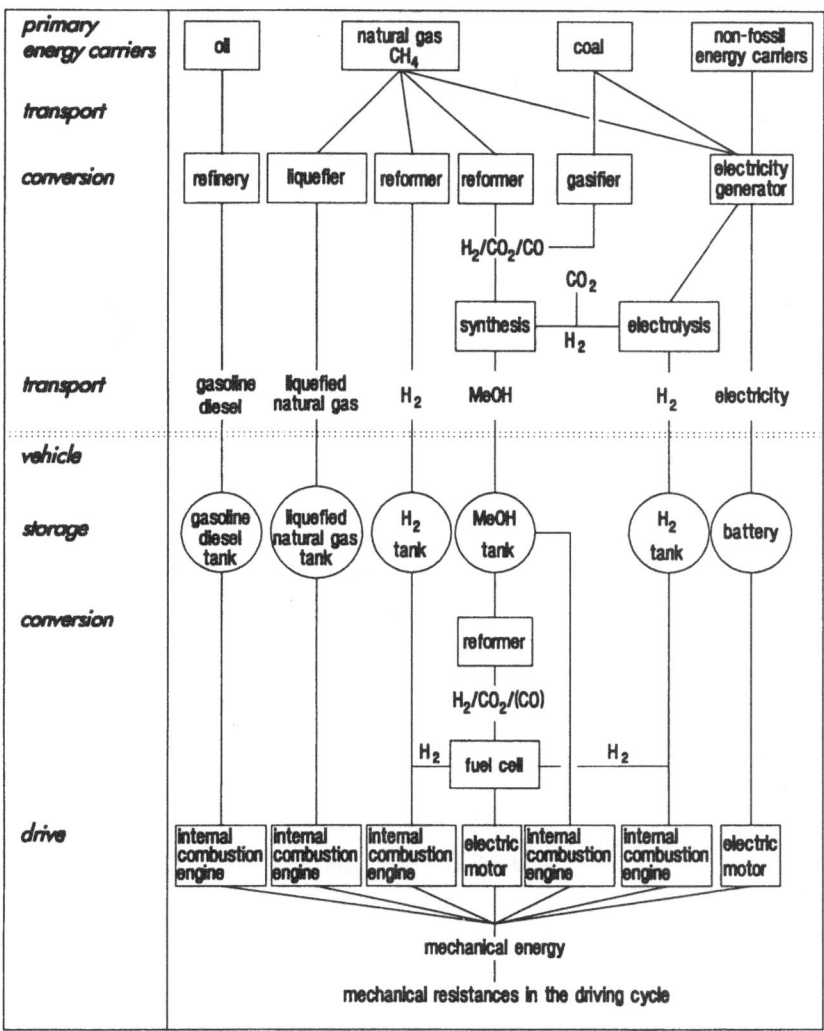

Figure 10. Energy Conversion Sequences for Road Traffic

The estimates for the fuel cell vehicle are based on the following assumptions: Taking a small (1100 kg) passenger car operating in the European ECE-R-15 driving cycle as the basis, 9 kWh/100 km is required at the wheel. Taking into consideration the energy transformation at the motor shaft, electric motor and converter, internal consumption and the recovery of braking energy results in an energy demand of 12 kWh/100 km to be covered by the fuel cell. Assuming a fuel cell efficiency of 62 %, an overall system efficiency of 52 % can be achieved including methanol reformer and catalytic combustion. This translates into a methanol requirement of 22.8 kWh/100 km or 5.2 l MeOH/100 km or 34 kWh/100 km if the energy conversion efficiency of transforming natural gas into methanol is considered. Compared to a conventional drive system, such a passenger car offers a reduction in primary energy of approximately 55 %.

With respect to emissions one should differentiate between greenhouse gases such as CO_2 and CH_4 and emissions such as CO, NO_x and volatile organic compounds (VOC) or NMOC. All the above-mentioned gases can be reduced considerably if a fuel cell with a methanol reformer is used as shown in Table 6. It can be seen that CO_2 can be reduced to almost one third. It can also be seen that the emission of the other pollutants is reduced by a factor of about 10 to 100 and that the ULEV standard is easily fulfilled.

Table 6. Energy conversion sequence for passenger cars [1]

Propulsion [2]	Energy kWh$_{PEC}$ [4] / 100 km	CO_2 kg / 100 km	NO_2 g / 100 km	SO_2 g / 100 km	CO g / 100 km	VOC [3] g / 100 km
Gasoline engine	76	20	65	8	620	210
Diesel engine	63	17	95	19	103	50
Electric motor with accumulator based on electricity mix [5]	79 [6]	20	13	14	4	2 [1]
Electric motor with PEM fuel cell based on Methanol from natural gas	34	7	1	< 1	< 1	< 1
ULEV-standard for passenger cars			12.5		106	2.5

[1] *Without* coal mining, oil and natural gas extraction and PEC transport to German border
[2] ECE cycle of VW Golf
[3] VOC: Volatile **O**rganic Compounds *without* Methane
[4] PEC: **P**rimary **E**nergy **C**arrier
[5] Emissions due to electricity production for West Germany (1990)
[6] corresponding to 29 kWh$_{el}$ / 100 km

CONCLUSIONS

An analysis of the energy consumption and the associated emissions for power plants and road traffic is presented for West Germany. In particular the integration of fuel cells in power plants for stationary electricity production and in vehicle application in road traffic has been discussed. The energy consumption and emission balances for West Germany as typical for other industrial countries show the significance of power and heating plants with respect to the emissions of CO_2 and SO_2 as well as the importance of road traffic with respect to legally restricted emissions of CO, NO_x and VOC.

In terms of energy consumption and emissions, the paper describes

- the state of the art and the development potential for conventional and new thermal power processes as well as energy conversion processes based on fuel cells for stationary electricity and heat production and
- the comparison of energy conversion sequences for road traffic with three different drive systems: internal combustion engine, electric motor with accumulator based on electricity mix from power plants and electric motor with PEM fuel cell based on methanol from natural gas.

The advantage of high temperature fuel cells in power plants can be summarized as follows:
- higher efficiency and correspondingly lower CO_2-emissions,
- significantly lower emissions of non-CO_2 pollutants,
- useful waste heat at a high temperature level and
- good partial load performance.

The chances for fuel cell technology to be introduced in the power plant market depend primarily on its ability to be commercially competitive. In a future power plant market, those technologies will be at an advantage which make it possible to stay within regulatory limits at minimum costs.

New drive systems for road traffic will reach a significant potential for application if they exhibit a higher system efficiency than internal combustion engines and considerably lower emissions. At present, the dominant technology is internal combustion power using oil as a energy source. Battery-powered electric vehicles can only cover a niche market demand. Among the vehicles with energy converters and storage systems, fuel cells in combination with electric motors have a clearly higher efficiency than conventional internal combustion engines. Specific CO, NO_x and VOC emissions from vehicles with fuel cells could be significantly reduced even with respect to the ULEV standard in California.

REFERENCES

1. Bundesministerium für Umwelt, Naturschutz und Reaktorsicherheit, "Umweltpolitik, Klimaschutz in Deutschland", Bonn (1993)
2. C. D. Schönwiese, B. Diekmann, "Der Treibhauseffekt - Der Mensch ändert das Klima", Rowohlt, Reinbek (1991)
3. Bundesminister für Verkehr (Ed.), "Verkehr in Zahlen 1993", DIW, Berlin (1993)
4. D. Bohn, Fortschrittliche Energieumwandlungskonzepte im Hinblick auf CO_2- und NO_x-Minderung und Vermeidung, Lecture WS 1992/1993, RWTH Aachen (1993)
5. H. Knappstein, H. Nymoen, Blockheizkraftwerke mit Brennstoffzellen - Energieerzeugung der Zukunft, *GASWÄRME Int.* 42 (1993) p. 139-144
6. G. Haupt, J. S. Joyce, GUD-Technik nutzt Gas, Öl und Kohle extrem schadstoffarm, *SIEMENS Power Journal* 3 (1993) p. 31-35
7. W. Suttor (Ed.), "Praxis Kraft-Wärme-Kopplung", C.F. Müller, Karlsruhe (1993)
8. Ch. Rechenauer, E. Achenbach, "Dreidimensionale mathematische Modellierung des stationären und instationären Verhaltens oxidkeramischer Hochtemperaturbrennstoffzellen", Jül-2752, Jülich (1993)
9. E. Riensche, H. Fedders, "Modellrechnungen zu Systemvarianten von 200 kW-SOFC-Anlagen", Jül-2787, Jülich (1993)
10. S. C. Singhal, Solid Oxide Fuel Cell Development at Westinghouse, Second International Symposium on Solid Oxide Fuel Cells, Athens, Greece (1991)
11. H. Wendt, V. Plzak, Brennstoffzellen – eine Einführung, in: VDI Berichte Nr. 912, VDI Verlag, Düsseldorf, (1992) p. 89-102
12. J. Reinecke, G. Huppmann, W. Drenkhahn, Brennstoffzellenkraftwerke zur Verstromung von Wasserstoff aus fossilen Rohstoffen und anderen Quellen, VDI Berichte 912 (1992) p. 103-124
13. R. F. Singer, Solid Oxide Fuel Cell Development, Internal Report, Asea Brown Boveri AG, Corporate Rersearch Heidelberg, Germany (1990)
14. H. Fedders, E. Riensche, Process Engineering Problems of Peripheral Components of Natural-Gas-Fuelled SOFC Plants, Second International Symposium on Solid Oxide Fuel Cells, Athens, Greece (1991)

15. W. Drenckhahn, K. Hassmann, Brennstoffzellen als Energiewandler, *Energiewirtschaftliche Tagesfragen* 43 (1993) p. 382-389

16. K. Ledjeff, Batterien für Elektrofahrzeuge im Vergleich, Aspekte alternativer Energieträger für Fahrzeugantriebe, in: "VDI-Berichte 1020", VDI-Verlag, Düsseldorf (1992) p. 315-340

17. K. Strasser, Development of PEM Fuel Cells at Siemens: Status and Potential Applications, International Society of Electrochemistry, Berlin (1993)

18. U. Stimming, J. Divisek, B. Höhlein, V.M. Schmidt, J. Stumper, Neuere Entwicklungen bei Brennstoffzellen, in: "Deutsche Physikalische Gesellschaft, Arbeitskreis Energie", H. Unger (Ed.), Bad Honneff (1993) p. 141-170

19. H. Heitland, G. Rinne und K. Wislocki, K., Chancen hybrider Antriebssysteme im zukünftigen Straßenverkehr, *Motortechnische Zeitschrift* 55 (1994) p. 94-101

20. Volvo, Weltpremiere: Das Zukunftsauto von Volvo, Press Release, Volvo Deutschland GmbH, Dietzenbach-Steinberg (1993)

21. O. von Fersen, Concept Car präsentiert sich als Saubermann, *VDI-Nachrichten*, Nr. 11 (1993) p. 27

22. B. Ganser, "Verfahrensanalyse: Wasserstoff aus Methanol und dessen Einsatz in Brennstoffzellen für Fahrzeugantriebe", Jül-2748, Jülich (1993)

23. B. Höhlein, B. Ganser, R. Juffernbruch, R. Kolke, (Jülich), S. Birkle, H. Voigt, (Erlangen), Energieumwandlungsketten für den Straßenverkehr im Vergleich, *Energiewirtschaftliche Tagesfragen* 43 (1993) p. 828-835

24. J.B. Hansen, and K. Aasberg-Petersen, (Haldor-Topsoe A/S, Lyngby, Denmark), B. Höhlein (Forschungszentrum Jülich, Jülich, Germany), Fuel processing for mobile fuel cell application, in: "Fuel Cells For Tractionary Applications", Symposium at the Royal Swedish Academy of Engineering Sciences, Stockholm, Sweden (1994)

ELECTROCHEMICAL ENERGY CONVERSION :
POTENTIAL APPLICATIONS AND CONDITIONS FOR POSSIBLE FUTURE INTRODUCTION IN ELECTRICAL SYSTEMS

André Marquet

Electricité de France (EDF)
Direction des Etudes et Recherches, Service Matériel Electrique
1, avenue du Général de Gaulle - F 92141 Clamart Cedex

INTRODUCTION

The large scale production of electricity from fuels by traditional thermomechanical means is subject to the limitations of the Carnot cycle; furthermore, such electricity is economically difficult to store. However, the continuous improvement of electrochemical generators and some evolution in electrical networks planning may offer new opportunities to introduce these technologies into electrical energy systems.

International pressure towards clean-up of emissions and clean electric vehicles has aroused new interest in more efficient and long-lived storage batteries. The slow but continued development of fuel cells also offers prospects for the production and conversion of electricity, with or without combined production of heat. Several factors contribute to a renewed interest from the electrical industry.

Autonomy from electricity sources, especially for mobile applications, is very attractive for the development of electric services. Also load levelling by means of batteries enables a better optimisation of electricity production mixes, allowing for reduced environmental impact.

Electricity production is now increasingly diversified world-wide. Increased interest exists for high efficiency, very low pollution modular systems, able to use diverse fuels, adapted to co-generation of heat and electricity close to the customer. Although their maintenance conditions remain poorly assessed, fuel cells are concerned. Their direct electric performances as well as their co-generation facilities raise a growing interest from different utilities.

The will of electrical utilities to improve the quality of the delivered electricity, and to minimise environmental problems raised in the future from the development of the networks may bring some new pay-off from the grid side for decentralised electrochemical sources.

An additional factor relevant to electrochemical technology is the renewal of research and development on systems using hydrogen as an energy vector. Fuel cells, reversible or not, are an essential part of any hydrogen economy scenario.

While R&D consortia are preparing a wider introduction of electrochemical conversion systems, their operating cost will come under scrutiny, particularly in comparison to classical competitor systems: peaking gas turbines in the case of short term storage batteries; combined cycles and/or clean coal technologies like fluidised bed power plants, in the case of fuel cells.

ECONOMIC CONTEXT

Generally speaking, the production systems and the devices used in electrical network substations must have a moderate cost per installed kW and/or stored kWh, a moderate space occupancy, and a very long life combined with an excellent reliability. The modularity of electrochemical systems, which can be a drawback at the investment level, avoiding large scale savings, can bring some advantages for reliability/availability.

Production side

Figure 1 below shows the load duration curve corresponding to an average projected scenario for electricity demand in France in year 2010. This curve represents the cumulative time during which a certain power is demanded through a year.

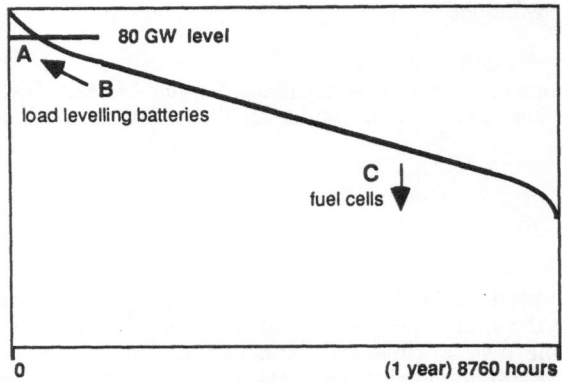

Figure 1. Load duration curve for electricity demand in France (average forecast for year 2010)

Batteries. In the case of batteries, storage allows transfer of energy produced in area B (lower power demand and lower cost), to fulfil the demand in area A (peak power demand and high cost), thus giving an economic advantage. This is usually called load levelling, or peak shaving.

A recent study performed at EDF/DER (Network Studies Department) based on a 2005 to 2010 marginal cost analysis shows that the yearly saving could theoretically reach roughly 50 FF per kWh at that time for a storage battery charged during low demand hours (at night for instance) and discharged for 2 hours at 0.5 kW rate during peak demand, approximately 300 times per year, with a minimum overall efficiency of 70%. Such figures suppose that the battery is optimally operated by someone having previous knowledge of the best time for taking the decision to charge or discharge it; a real operation would be less efficient. For instance a decision to keep the battery as charged as possible immediately after discharging would lower the yearly benefit to around 15 FF per kWh. It is assumed in the following that an optimised operation process makes it possible to reach 38 FF/kWh.year. Discounting such a saving at 8% (official discounting rate in France) for a 30 years operation period gives a total saving of 432 FF, which is rather low.

Additionally, such 0.5 kW load levelling equipment delays the investment in half a kW of peaking production means; considering an average installation cost of 2500 FF/kW for large gas turbine sets, this represents around 1250 FF. **Table 1** gives, according to the minimum cycle number attainable by one battery, the maximum cost allowed per kWh of battery, assuming that (i) the direct investment saving neutralises the investment of the battery system (including the electronic converter) -figures in italics- and (ii) the financial equilibrium is just attained on a 30 years operation basis. The cost of the converter is taken

24

as a parameter representing 1200 FF/kW (today's situation) or 600FF/kW (future situation for large sets of several MW). The other parameter is the cyclability of batteries which determines the frequency of battery replacements during the whole operation period of the system (30 years).

Table 1. Maximum cost per kWh versus life duration for a load levelling battery

Battery system operation: 30 years; 300cycles per year; total 9000 cycles; saving 38 FF/year; discount rate 8%; discounted total saving 432 FF; avoided peak power investment 2500 FF/kW (30 years life)						
Nb of batteries used in 30 years operation	1	2	3	4	5	6
Max number of cycles attainable per battery	9000	4500	3000	2250	1800	1500
Maximum cost per kWh of battery FF — converter 1200 FF/kW	*650*	*427*	*417*	*322*	*266*	*228*
	1082	859	849	754	698	660
Maximum cost per kWh of battery FF — converter 600 FF/kW	*950*	*624*	*610*	*471*	*389*	*333*
	1382	1056	1042	903	821	765

Figures in italics stand for neutralisation of investment/de-investment
Normal cast figures suppose global financial equilibrium, taking yearly benefit into account

The available investment cost per kWh of battery, especially for 1500 cycle batteries which correspond more or less to the standard state of the art for large stationary units, looks limited. But the case is not hopeless, although the growth rate of allowed investment per stored kWh is rather slow when cyclability increases, which makes it even more difficult to improve. Such figures explain why, from the point of view of production economy alone, and considering the present technologies, the load levelling battery investment is generally difficult to justify in a context similar to that of the French network, as to the production mixes and the load profile. For detailed studies however, additional pay-off like the reduction of needed 'spinning reserve' due to the added standby power of the batteries should be taken into account.

Fuel cells (FC). While batteries come under the peak power area, fuel cells would be adapted to the semi-continuous or continuous service area, shown on **figure 1** as zone C. They can be considered as a production means aiming at a very high efficiency, justifying a high investment cost; thus, they would hardly be used as peaking devices. Accordingly, their life duration must be considerably improved, from less than ten thousand hours at present (in general), to several ten thousand hours, keeping in mind that the FC's core could be replaced a few times during a 20 or 30 years operating period.

Presented on **table 2** are the potential production costs of a FC with a 55 or 60% electrical efficiency following two levels of investment cost:12 000 FF/kW, a value with good chances of commercial attainability within the next ten years, and 5000 FF/kW, a longer term goal given as a realistic objective for several FC techniques. The fixed operating charges, although much uncertainty remains about their real value, are taken as comparable to those of a gas combined cycle. The fuel prices are considered both in a depressed situation (0.06 FF/kWh) or a hard pricing situation (0.12 FF/kWh). The production costs of FCs are compared with those of conventional techniques[1].

Table 2. Cost per kWh produced through fuel cells and various competing devices

Direct cost of electricity without heat sale (in centimes of FF per kWh).
30 years, 6000 hrs/year operation; 100 % assumed availability for each device;
discounting rate 8%.

Fuel cells (FC) : fixed operating charges: 165 FF/kW/year ; efficiency η

fuel price	0.06 FF/kWh		0.12 FF/kWh	
η fuel cells				
	55%	60%	55%	60%
FC 12000 FF/installed kW	31.3	30.3	42.2	40.3
FC 5000 FF/installed kW	21	20.1	31.9	30.1
Gas combined cycle (efficiency 46.7%)	21.9		34.7	
Coal circulating fluidised bed (as reference - efficiency 36.4%, cost of coal 0.05 FF/kWh)	32.3			
Gas turbine (for information) (efficiency 30.8%)	23.7		43.2	

It can be seen that the variation in FC efficiency slightly influences the results, especially in the depressed fuel price context, where high cost FCs seem to be competitive with coal fluidised beds, and low cost FCs with gas combined cycles. In the hard price fuel context, the lower cost FC remains competitive with the combined cycle and the coal fluidised bed, even assuming that the price of coal remains unchanged. Although they are based on hypothetic costs, such results encourage deeper analysis.

New pay-off from grid side

Not only do economic and environmental issues from the production side have to be taken into account, but in a 10 to 15 year perspective, the change in conditions for transmission and distribution of electric energy will contribute to enhancement in the profitability of devices able to improve:

(i) the grid delivery **quality** ; this may be the case from the ability of storage batteries to ensure a standby help in case of interruption of delivery from upstream grid faults. This 'quality effect' can be related to the cost of 'not supplied energy' -a concept used in general studies on electrical system reliability- which, in a country like France, should increase from a few FF per kWh at present, to more than one hundred FF/kWh within 10 to 15 years. This is taken into account in the study mentioned above. A positive effect supposes excellent system reliability. Additionally, with the aid of the associated power electronic drives, storage plants might contribute to the improvement of the dynamic control of active and reactive power fluxes, power factors, etc., without specific investment.

(ii) the general **environmental** acceptance of the transmission and distribution works[2,3,4]. Here, in some cases, load levelling batteries may be used to avoid or delay transmission lines enhancement for peak power transits. FCs bring more than an order of magnitude reduction of pollutant emissions like NOx or CO; thus, they can be installed close to the site of consumption and deliver heat at short distance, which enhances its economic interest. FCs, installed, for instance, at distribution medium voltage sources level, may make the development of the transmission network from remote sources unnecessary. The associated avoided costs should at least be in the order of 1500 to 2000 FF/kW; this amount can be cut from the investment cost. Such consideration could greatly help the market introduction of FCs or even batteries, especially at the industrial pilot stage, where the investment level for new techniques generally remains unoptimised. But the questions of investment level and life duration will remain crucial.

Hydrogen scenarii

Several organisations are currently supporting studies on large world-wide hydrogen energy systems : WE-NET in Japan, Euro-Québec and Hydrogen World Council in Canada...**Figure 2** illustrates a simplified 'idealised' scheme for interconnecting the production and use of hydrogen as an energy vector with other energy systems through electro-chemical conversion devices: fuel cells (FCs), batteries and electrolysers.

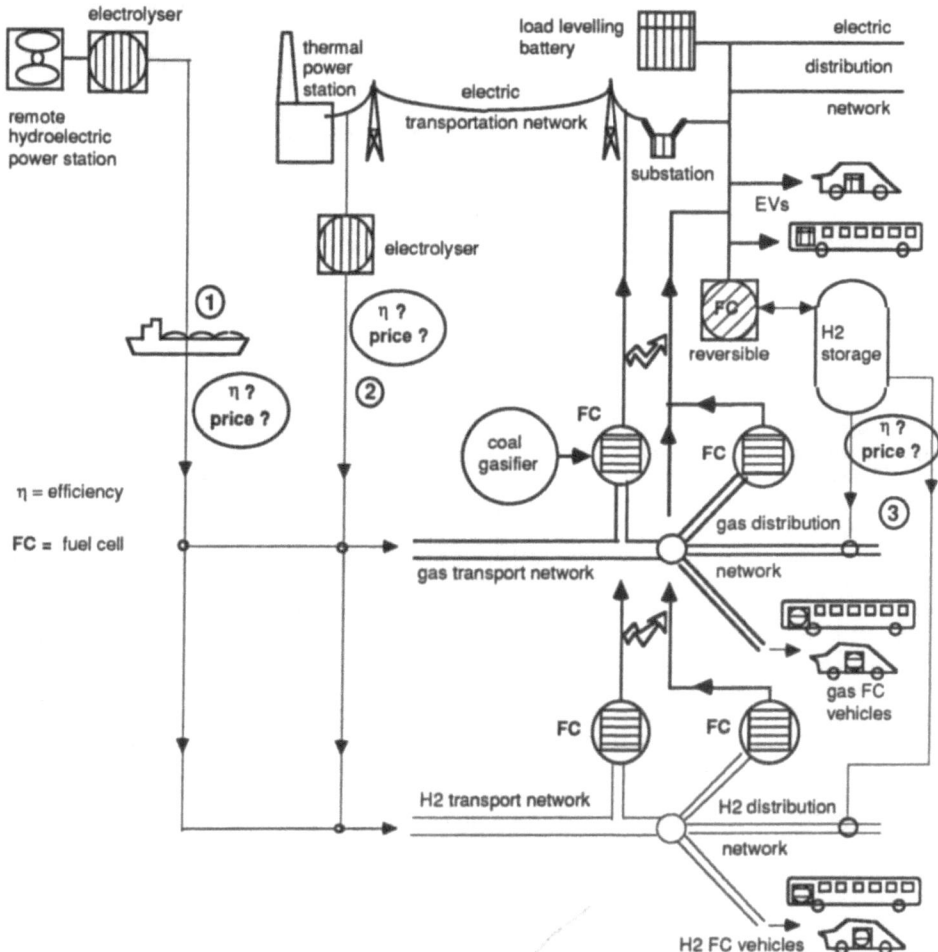

Figure 2. General scheme of hydrogen vector possible connections

The attractiveness of such a scheme is affected on the hydrogen side by two factors:
- the investment cost of new equipment and transport infrastructure, moderated by the environmental benefits from the clean combustion or direct use of hydrogen through FCs;
- the cost of hydrogen energy production through the cost of electricity and the efficiency of the electrolysers or reversible FCs, as compared to those of other fuels. Efficiencies have progressed from 40% with non-catalysed technologies to 80% with solid

polymer membrane techniques and possibly 90% with solid oxide (reversible) high temperature technology.

The three paths depicted on the scheme are different: path 1 should provide competitive hydrogen especially from very low cost electricity of large hydraulic renewable power stations and high efficiency electrolysers. But the cost of plant and infrastructure (electrolysers, liquefiers or hydrogenation/dehydrogenation systems, tankers and transport network) will be high; hydrogenation or dehydrogenation can be done using various 'supporting' compounds / support return agents: methanol/CO_2, methanol/CO, methanol/methylformate, ammonia, cyclohexane/benzene. Paths 2 and 3 start from a much higher cost of electricity from thermal power stations. Some electricity from nuclear stations may be available at lower demand periods at a low marginal cost close to the nuclear fuel cost; but in that case a large part of the infrastructure (electrolyser, network, etc.) will be used only during such periods, thus increasing the final cost. It is clear that the prices of hydrogen in consuming areas will hardly be competitive with natural gas, for instance, in a situation of depressed pricing like the present one. Such projects address longer term change in the fuel market where prices may increase again due to supply/demand unbalance, and also to increased pressure of global environmental problems.

TECHNICAL SITUATION - MAIN INVOLVED PARTIES

Converting these possibilities into realities will require significant technical progress on the electrochemical generator side, especially as investment cost is concerned. Programmes for energy conversion should focus increasingly on improving manufacturing process productivity.

General trends - Stationary / EV synergies

The considerable expansion of R&D prompted by the idea of electrical traction in modern cities will have positive spin-off on stationary applications of electrochemical generators. In turn, these will help to maintain the general effort if some change or delay happens on EV development.

The specific energy criterion may be less important for stationary uses than for mobile ones. In fact it has been shown above that the stress on investment cost is high enough as to impose minimisation of the quantity and the cost of materials used. Similarly, the installation constraints, e.g. volume, effluent management, noise, operating temperature, are very important for stationary devices, and can make energy or power density a very important aspect, as well as for traction. Also reduced maintenance is a growing necessity to address, for instance, large scale installation in urban or suburban substations. In addition, longer life duration/cyclability remains a very strong requirement. Finally, there exists a large synergy between stationary and mobile generators objectives, and it will not be so surprising to see one family of solutions studied for both. A good example is the LIBES program in Japan -Lithium Battery Energy Storage Research Association managed by NEDO and MITI- where lithium secondary batteries are studied for both objectives; though developed for different needs, the technologies will obviously fertilise one another.

The desire of EV supporters to bring about a range of solutions adapted to different types of vehicle has led to the formation of research combines for electrochemical generators, both accumulators and fuel cells. Of course these correspond to the opposed but compatible concepts of close electric link or remote fuel autonomy. Hydrogen-fuel produced through electricity may be of interest for transportation purposes, but, in addition to the unfavourable economic context mentioned above, there is lack of realistic solutions for mobile storage. Every solution comes up against, not only the direct performances of the generators, but mainly the prospective total costs and the robustness of the total system involved.

Batteries

The most well known combine formed for traction is the Advanced Batteries Consortium (USABC) which gathers in the USA the 'big three' car manufacturers, the DOE, EPRI (Electric Power Research Institute), and miscellaneous electric utilities. Los

Alamos, Berkeley, Sandia and Argonne Labs bring in their expertise on battery testing and electrochemical science. This consortium was evolved for managing a provisional 260 M$ fund over five years starting from 1992. It has been decided by USABC to exclude research on conventional technologies like lead-acid or alkaline NiCd. The panel of selected technologies at present involves nickel-metal hydride, lithium polymer electrolyte, high temperature (450°C) lithium iron mono or disulphide, and possibly ß" alumina electrolyte sodium high temperature (350°C).

Lead-acid battery manufacturers, linked by the corresponding mining sector (International Lead Zinc Research Organisation - ILZRO), decided to show that, from the cost point of view, their product remains the most adapted to medium term EV markets and retains a good potential for improvement. They created the Advanced Lead Acid Battery Consortium (ALABC) gathering together most of the large manufacturers in the world. The supported researches (16 M$ - 4 years) involve mainly specific energy and cyclability improvements for valve regulated lead-acid batteries but officially excluding bipolar systems.

In Japan, MITI has also moved, creating LIBES under NEDO administration., with a potential budget of 14 billion ¥ ($\approx$ 120 M$) for 10 years. LIBES intends to develop two main families of lithium high performance large capacity secondary batteries: (i) high energy density (150 Wh/kg) for traction with Li-C anode insertion or Li/polymer techniques or a combination of both, and (ii) very long life (3500 cycles) for load levelling at housing level with lithium-carbon rocking chair systems. Ten companies are involved in the programme.

Funding in Europe remain less important. CEC concentrated for about ten years on lithium polymer systems with a budget of some 7 MECU shared between two research groups. CEC is now directing its efforts towards lithium and/or lithium-carbon systems. In Germany, an agreement has been announced between RWE, bringing the sodium/sulphur technology from SILENT POWER and VARTA, with technologies like lead-acid, nickel metal hydride and lithium or lithium-carbon which they develop both with liquid option for portable devices or polymer electrolyte option for traction. In France, in the field of transportation, the PREDIT program which associates public authorities and car manufacturers for the automobile sector is involved in research for EVs. Lithium batteries development, with both carbon anode/organic liquid electrolyte and Li metal/polymer systems is supported for that purpose.

Lithium. A general research move towards high reactivity systems like Li-insertion with an ambient or moderate working temperature has occurred. Moderating apparent current densities, or even reducing them, is necessary to sustain interfacial properties and improve cyclability; together with the need of high power densities, this leads to use of much larger surfaces and much thinner complexes. At the same time, as the voltages increase, there is a need for electrolytes with improved chemical stability at relatively high temperatures induced by power demands; the move towards pseudo-solids or preferably solid systems where the mobility of chemical species is reduced, these being produced as more or less thin film structures, is a foreseeable trend. The lithium-carbon 'rocking chair' system may be an interesting challenger for load levelling provided sufficient sources of lithium are found; as the power demands are much more continuous than for EVs, the temperature rise would be easier to control and the high cyclability would be preserved; possibly, a liquid organic electrolyte could be used over several thousand cycles as considered in Japan by LIBES.

Lead-acid. For load levelling, only lead-acid is at present cheap enough to be economically considered. Some large prototype installations exist -like BEWAG in Berlin (17 MW, 14.4 MWh stored), CHINO in California (10 MW, 40 MWh stored) as the result of a collaboration between EPRI, Southern California Edison Company and ILZRO) TATSUMI in Japan (1 MW, 4 MWh stored) operated by Kansaï Electric C°- which intend to demonstrate their ability for long life operation at reasonable maintenance cost. With 2000 expected deep discharge cycles and a price evaluated for 'mature' markets in the range of 750 FF/kWh, present prototype large elements should fit the future conditions reported in table 1. But their ergonomics and maintenance conditions must change notably to reach the required low operating costs. Here experience can also be gained from EV developments.

Sodium-sulphur. Being constituted from basic low cost common materials, the ß" alumina sodium sulphur systems were studied originally for load levelling; an experimental plant (1 MW, 8 MWh) has been built in TATSUMI, Japan, for Kansaï Electric C°. Although the cells would be sealed ones, it appears that their level of reliability and sensitivity to standby chemical corrosion presently prevents progress towards large scale production.

Nickel-hydrogen. The only system allowing for sufficient cyclability to withstand a 20 or 30 years service, which represents about 10 000 deep discharge cycles, is the nickel-pressurised hydrogen one. This has been developed for space applications and nowadays reaches more than 30 000 cycles at 40% d.o.d. Its price is very high due to the sophisticated technologies and quality insurance level required. Terrestrial versions have been studied in the USA in particular for load levelling applications. It seems however that their cost could hardly fall under 3000 to 4000 FF/kWh due to the limitation in thickness of the positive electrode, which makes it a two dimensional system, difficult to make cheaper through scale-up.

Fuel cells

The context is different. Research programmes have been continued in United States and -at a lower level- in Japan for 30 years with the help of DOE and MITI, each spending about 500 MFF yearly with a fairly large degree of international exchange.

Europe has practically stopped its efforts towards FCs as energy production systems around 1975, having transferred the responsibility of developing FCs for space programs to Germany. Research programs started again around 1985 under the impulse of CEC with JOULE and BRITE-EURAM programs. The yearly budget nowadays is about 300 MFF. National programs exist in Germany, Italy, the Netherlands, Denmark, United Kingdom, Norway and Switzerland. Recently France started a new program on protonic membrane FCs with the help of the PREDIT program for transportation use.

Presently, FCs programs involving different technologies are increasing on a world-wide scale, for stationary as well as for mobile applications.

PAFCs. Phosphoric acid technology reaches commercialisation at several hundred kW scale with an expectable price of about 20 000 FF/kW. Pilot installations are being tested in many places; TEPCO in Japan operates in Goi the largest installation (11 MW). This technology is suited to co-generation, with a working temperature of about 150 to 200°C, but needs precious metal catalysts; its direct electrical efficiency is limited to under 45%; a fuel or gas reformer is required for hydrogen fed into the stack. The life duration of stacks remains in the order of 5 000 hrs.

MCFCs. Molten carbonate FCs are designed to use carbonated fuels or gases (from gasified coal, methanisation of refuse, etc.) at large scale, with direct reforming on electrodes or indirect reforming. The operating temperature is around 650°C, which avoids the need for any special catalyst and produces reusable heat for industrial or district heating. The efficiency reaches 55-60% in combined cycles, using the vapour produced to drive a complementary turbine. Thermal cycling and corrosion are the two main difficulties. The pilot scale is reached at several hundred kW scale, with life duration reaching 10 000 hrs for a cell. The largest plant (2 MW scale) is now under construction in Santa Clara - California under APPA and PG&E, DOE and EPRI participation. National programs exist in the USA, Japan, the Netherlands, Italy, and Germany.

SOFCs. Solid oxide fuel cells work between 800 and 1000°C to ensure ionic O_2^{--} conductivity in yttrium stabilised zirconia ceramic electrolytes. At such high temperature, no catalyst is necessary. The cells are reversible and able to produce H_2 from electricity with a high efficiency (90%). As FCs their combined cycle electric efficiency should reach 60 to 65%. They are able to use a wide variety of fuels with direct or indirect reforming options. The first evolved technology has a tubular design and now reaches a 25 kW level. with several thousand hours operation, while individual cells have performed well up to 100 000 hrs. The planar bipolar technology is now progressing and seems to be potentially less expensive and more compact, although extrapolation of dimensions and high temperature multi-functional assembly problems make it difficult to meet scale effects. This new technology presently stands at the kW level. Pilot systems in the 10-100 kW range should appear within 5 years. SOFCs are studied very actively world-wide. CEC efforts have resulted in a good position to develop this technology.

PEMFCs. Polymer electrolyte membrane fuel cells were originally developed for space applications. Using a fluorinated protonic membrane on catalysed air electrodes, they have good compactness through a high current density (≥ 2 A/cm²). This feature has made them attractive for possible vehicle application. The dispersion of Pt catalyst on electrodes has made considerable progress, and such stacks now work at 80-100°C with only 0,1 mg/cm² catalyst loading. The present range is 5-50 kW. Larger projects are being discussed. As this technique presently uses only pure hydrogen, it is relevant to hydrogen schemes described above. Some efforts are dedicated to decrease the sensitivity to poisoning by traces of CO. A high direct electric efficiency of 60% is frequently claimed, but it seems that the energy consumption of auxiliary engines may reduce this to the mid 40's range, which would be less motivating. Research programs are currently active mainly in North America, and in Europe (Germany, France, Italy) especially for transportation[5].

Metal / air systems

Both aluminium/air and zinc/air alkaline systems have been considered for traction. Primary systems with a mono-functional air electrode (with too short a life at present) have the advantage of rechargeability through directly 'refuelling' the metallic compound fuel used; secondary systems (in the case of zinc only) use bi-functional (reversible) electrodes whose durability is unsatisfactory at present. The necessary recycling of reaction products (alumina, zinc oxide or zincates) requires either an expensive and cumbersome system of transport to regeneration plants, or a low efficiency on site regeneration. Such systems result in a cost of delivered energy in the range of 2.5 to 5 FF/kWh for aluminium, and 5 to 7 FF/kWh for zinc, far in excess of other systems. In spite of attempts to use zinc-air systems on EVs, the use of metal/air systems looks limited to the field of standby energy supply where the high specific power and compactness of aluminium-air could be very helpful. This supposes an improvement in the operation and maintenance conditions: such a generator must be able to start quickly, and to re-start several times without maintenance, which does not seem to be the case for present systems.

R&D AT ELECTRICITE DE FRANCE

Considering the general potential of these techniques, the R&D Division at EDF decided in 1990 to build a test bench, for EV batteries at first, described on **table 3**.

Table 3. EDF battery test bench in Les Renardières laboratories.

BENCH	working voltage (V)	discharge current (A)	charging current (A)	number of test stands per bench
24 V	0 - 24	0 -350	0 - 200	5
36 V	0 - 36	0 - 350	0 - 200	5
120 V	0 - 120	0 - 350	0 - 200	6

This bench, situated in 'Les Renardières' laboratories permits the testing of batteries intended for EV use and the associated survey and management systems. The 16 stands are now operational, each being able to simulate EV cycles, power demands and braking recovery: On test now are valve regulated lead acid batteries, then nickel-cadmium and, further, lithium-polymer will be tested. Also in 1990, a systematic technological survey activity was started.

In order to prepare a better future market situation for EVs and also stationary applications, EDF is a partner in a consortium with BOLLORE TECHNOLOGIES and CEA (Atomic Energy Commission), with the help of different CNRS public research laboratories, to develop lithium-polymer all solid thin film batteries. A JOULE program on cathode insertion materials associates them with SEDEMA in Belgium under the co-ordination of CEA. The emphasis is on the relation of the technology to the fabrication processes and their productivity in order to keep the price as low as possible. The objective for the next two years is to produce a 100 Wh cell, before starting a manufacturing pilot system necessary to produce prototype batteries.

EDF also has experience in testing the service ability of stationary standby batteries for emergency control in power stations or substations, for telecommunications and photovoltaïc signalling or supply (it is fair to mention here the case of using electrochemical generators when users remain isolated from an electrical grid -islands, mountain site, nascent electrification- and some intermittent renewable source of energy is available: even at apparently high production cost, it may reveal cheaper to use solar, wind or hydraulic sets to feed consumers far enough from the grid; but efficient, reliable and low cost storage through low maintenance batteries is a must; also, in a similar context, fuel cell modules might replace advantageously in the future some diesel generating units).

Currently, the laboratory is preparing to evaluate more accurately the status and progress of SOFC technologies. A technical follow-up is also planned for other technologies like MCFCs and PEMFCs.

REFERENCES

1. Coûts de référence 1992 - EDF-DEPS (internal document)
2. P.Caseau - CIGRE facing several organizations of the electricity supply industry - Electra n°152, Feb. 94
3. H.N. Scherer Jr. - Growing public opposition to overhead transmission lines - ibid
4. Y. Sekine - Pressure to increase utilisation of existing transmission systems, opposition to system expansion plans - ibid
5. P. Zegers - Electric vehicle fuel cells - Clean Vehicles Symposium SEE/CFE/SIA - La Rochelle, France - 15-19 Nov 93

LABORATORY PRODUCTION OF COMPONENTS FOR RECHARGEABLE BIPOLAR MANGANESE DIOXIDE ZINC BATTERIES

Leo Binder[1],Mehdi Ghaemi[1], and Karl V. Kordesch[2]

[1] Institute for Inorganic Chemical Technology, Technical University of Graz, Stremayrgasse 16, A-8010 Graz, Austria
[2] Battery Technologies Inc., 30 Pollard Street, Richmond Hills, Ont., Canada L4B 1A8

ABSTRACT

The most important parameter with respect to the active material utilization and high current capability of manganese dioxide electrodes is the thickness in current flow direction.

Thin electrodes equally providing a sufficient amount of active material and high rate capability may be realized in a battery either as a monopolar, spirally wound system, or in form of a bipolar, flat plate arrangement. For future high energy applications most experts favour the bipolar, flat plate concept.

A special laboratory test cell equipment should permit zinc plating on one surface of a cooled substrate (conductive foil) and - after turning the substrate - deposition of EMD or even an EMD/carbon composite on the opposite face.

INTRODUCTION

The cathodes of standard cylindrical "bobbin type" alkaline manganese dioxide zinc batteries are made by arranging pre-formed rings to a hollow cylinder contacting the cell case on one side and taking up the gelled zinc electrode in the inner cavity. They vary in size according to the actual cell type but in any case the currents on charge as well as on discharge flow radially from the inner current collector ("nail") to the positive terminal ("can") and vice versa.

The most important parameter with respect to the active material utilisation and high current capability (= electrode performance) is thus the wall thickness of the cathode. The idea of thin electrodes equally providing a sufficient amount of active material may be realised following two possible design concepts:

  a) the monopolar, spirally wound system,
  b) the bipolar, flat plate system.

Both designs have been used for other battery systems (a: Ni/Cd, Ni/MeH, lead acid;

b: Zn/MnO$_2$ primary, lithium batteries) but only a limited number of working groups regard them as feasible for the rechargeable alkaline Zn/MnO$_2$ battery[1,2]. Tests at an early state of development indicated a significant advantage of the monopolar system especially from the standpoint of high rate capability.

The bipolar prototypes suffered from inadequate adhesion of the electrodes to the bipolar substrate (carbon-plastic composite) and from corrosion problems of the zinc electrode in contact with carbon materials of any type. A negative aspect of the monopolar, spirally wound test cells was the insufficient energy density. The wire mesh serving as current collector, as well as a mechanical support of the electrodes, the required separators, and the matrices to take up the electrolyte, allowed only a few windings, performed on a simple lab-test facility.

For future high energy applications most experts favour the bipolar, flat plate concept. The key to this technology is a low resistance bipolar plate giving good and reproducible contact to the electrodes and avoiding the usual corrosion problems.

A special laboratory test cell equipment was designed to permit zinc plating on one surface of an inert substrate (conductive foil). After turning the substrate using the previously deposited zinc as current collector either the electrolyte has to be changed or the zinc plated substrate may be transferred to another plating unit. Now, the opposite face can be covered with EMD or even an EMD/carbon composite following a method published for other applications[3] while the foil is cooled from the backside by a special device.

Cooling of the working electrode is necessary to prevent shrinking or warping of the conductive polymer material when EMD is deposited using a 94 to 96 °C hot electrolyte. The electrodes prepared by this method make it possible to study EMD in an "as deposited" form without the influences of grinding, blending, pressing or rolling processes.

The final product is a unit (cathode - bipolar plate - anode) which may be easily assembled to bipolar cell stacks in a periodic arrangement with electrolyte-soaked matrices, separators, and finally the endplates (terminals).

EXPERIMENTAL

Materials and construction

The cell body (or counter electrode housing) was machined from a rectangular prismatic block of DEKADUR-C™, a highly chlorinated type of PVC. This block (see Fig. 1) was equipped with six drill holes (∅ 8 mm) to take up the assembling bolts and with two drill holes (∅ 16 mm) to be fitted with the electrolyte inlet and outlet tubes. The inner (hexagonal) area indicated in Fig. 1 may be divided into a rectangular (dotted) section and two triangles (hatched). In the dotted section the material was removed to a depth of 10 mm by milling, in the triangular fields 2 mm of material were cut off. The rectangular cavity formed the seat of the porous 8 mm thick carbon counter electrode equipped with a central lead-in wire (copper) from the backside.

The counter electrode housing was attached to a vertical wooden rack by means of six threaded assembling bolts which were long enough to hold the additional parts forming the complete test cell as shown in Fig. 2. Those additional components were: a pre-formed flat plate seal (made from ethylene/propylene-diene-terpolymer, EPDM), the conductive substrate (carbon-filled polyethylene), and the working electrode terminal and cooling device, a hollow structure made from an aluminium frame and aluminium plates flooded with water.

Contacting the substrate was done over the whole backside area because of the relatively poor specific conductivity of the material (0.13 to 0.17 $\Omega^{-1}.cm^{-1}$). Cooling was found to be useful to protect the material against the hot EMD plating electrolyte.

The electrolyte pipe system connecting the boiling vessel (three-neck flask) with the test

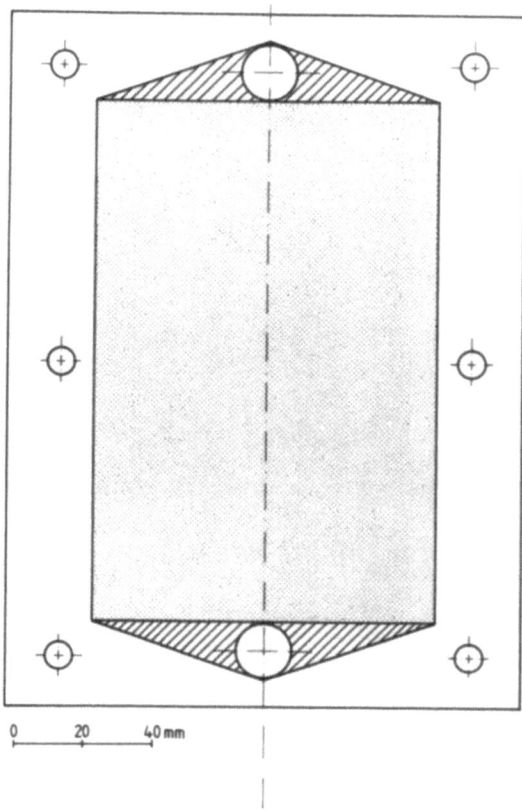

Figure 1. Front view of the counter electrode housing.

cell was built by sticking together straight tubes of appropriate length, pre-formed elbows, and stopcocks all made from DEKADUR-C™ with a PVC-C glue.

The plating electrolytes were circulated by a magnetic drive cencentrifugal pump. All parts of it contacting corrosive liquids were made from either ceramics or glass-filled polypropylene. Even the impeller magnet was encased in polypropylene.

The electrical supply for the required electrolysis current was provided by a controlled conventional DC source or by a pulse-plating device (for alloy plating experiments). Fig.3 shows the whole test set-up with electrolyte loop and electrical circuits.

Deposition experiments

The preliminary deposition experiments were done with substrates made from copper foil instead of the carbon-polyethylene composite material to study the tightness and all the different functions of the test equipment. Fortunately, the test for leaks was negative and all the installations worked without any problem when zinc was deposited at room temperature. When the possibility to deposit manganese dioxide on the opposite side of the substrate was examined the constant (and relatively high) rotation speed of the pump impeller caused negative effects. The electrolyte for depositing EMD is usually applied in a temperature range close to the boiling point of the liquid. In the case of a more complicated electrolyte distribution system the heat loss has to be considered and thus the solution is heated

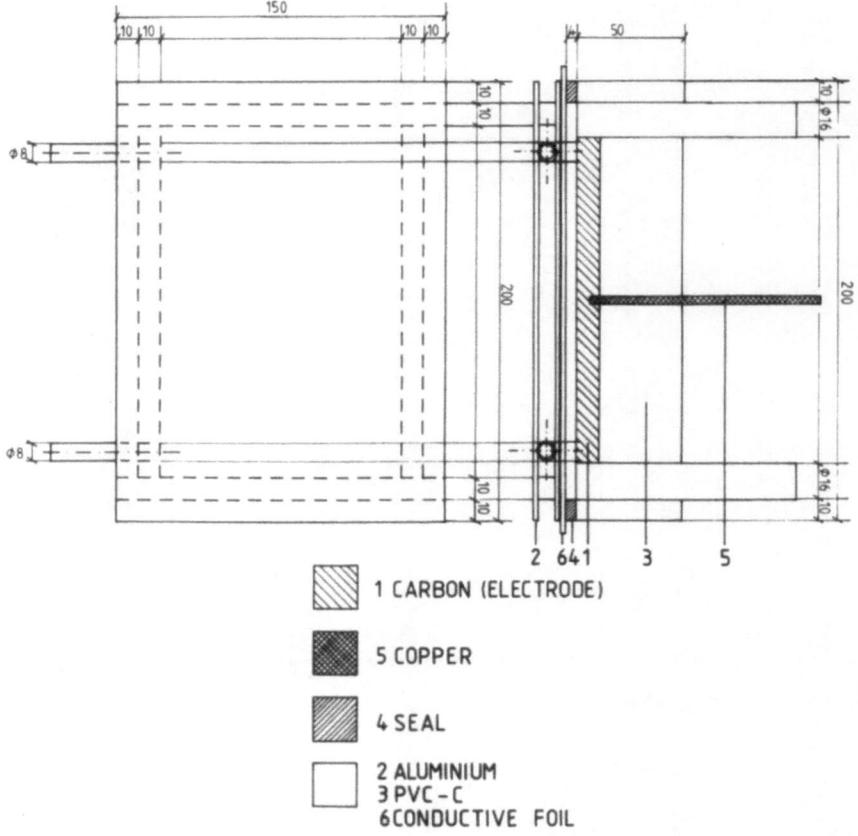

1 CARBON (ELECTRODE)

5 COPPER

4 SEAL

2 ALUMINIUM
3 PVC – C
6 CONDUCTIVE FOIL

Figure 2. Test cell arrangement.

over the boiling point in a separate vessel (see Fig. 3) and fed to the cell, arriving there with the desired temperature. If a liquid is boiling vigorously it evolves large gas bubbles which may enter the intake manifold of the pump and the gas may accumulate in the pump housing, replace the liquid, and cause the impeller to run dry. This possible malfunction depends on the electrolyte velocity determined by the rotation speed of the impeller inside the pump.

The electrolyte flow rate was optimised by changing the rotation speed of the driving motor, but this was impossible with the original one-phase AC motor. The replacement by an easy to control DC motor was the next step to improve the cell function. Depositions of good adherence and uniformity could be obtained maintaining a flow rate of 83 ml.s^{-1} (= 75 cm.s^{-1}). The optimal amount of zinc to be deposited was 5 g (= equivalent to 30 mg.cm^{-2}), while up to 12 g MnO_2 (EMD) could be obtained at the electrode surface (= 75 mg.cm^{-2}). The electrochemical balance with respect to the later utilisation as battery electrodes was: ~4.2 Ah at the anode (Zn, theoretical value) versus ~6.6 Ah at the cathode (EMD, theoretical value).

Subsequently to the described tests depositions on 1 mm thick carbon-polyethylene composite plates were started. According to the final application, a conductive foil of 0.2 mm thickness replaced this material after some plating experiments.

The production process was studied starting with the zinc deposition at room temperature after mechanical or chemical roughening of the surface. Mechanical treatment was done with abrading paper of different grades, the chemical pre-treatment was an anodic oxidation in sulphuric acid. The zinc plating electrolyte consisted of

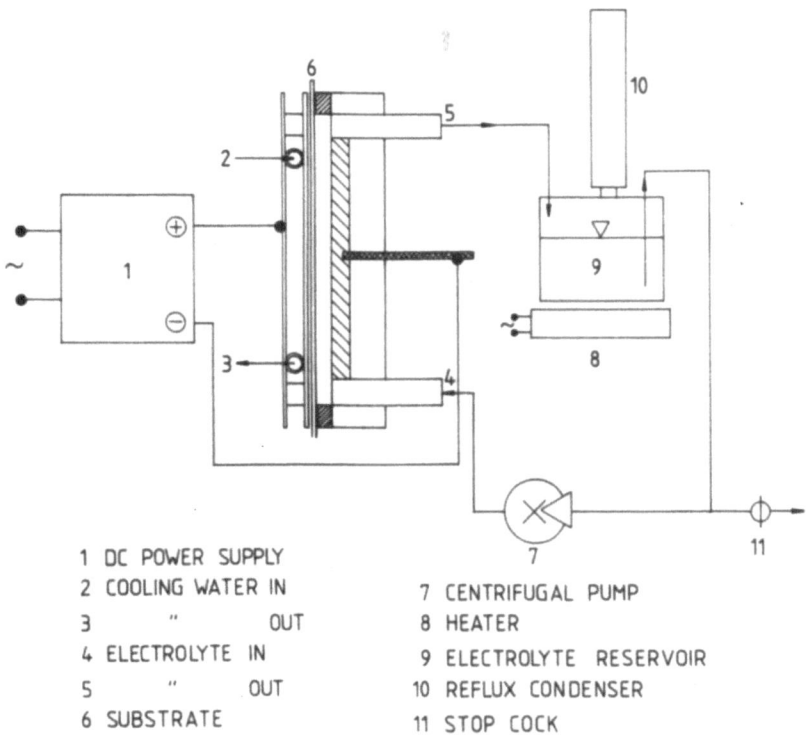

1 DC POWER SUPPLY
2 COOLING WATER IN
3   "   OUT
4 ELECTROLYTE IN
5   "   OUT
6 SUBSTRATE

7 CENTRIFUGAL PUMP
8 HEATER
9 ELECTROLYTE RESERVOIR
10 REFLUX CONDENSER
11 STOP COCK

Figure 3. Electrical connections and electrolyte flow system.

$ZnSO_4.7 H_2O$	380 g/l
$Na_2SO_4.10 H_2O$	72 g/l
$MgSO_4.7 H_2O$	61 g/l
H_2SO_4 conc.	1 ml/l

following literature suggestions 016387 [4]. The applied current density was 6 A.dm^{-2}.

After depositing the required amount of zinc the electrolytic process was interrupted, the test set-up partly disassembled, the substrate rinsed with distilled water, and the deposited layer air-dried. Then the substrate was mounted again in reversed orientation, the electrolyte was changed, and the deposition of EMD started.

For the deposition of EMD a conventional electrolyte containing

$MnSO_4.H_2O$	112 g/l
H_2SO_4 conc.	98 g/l

was used. The current density was maintained at 0.8 A.dm^{-2} and the working temperature was kept at 90 to 93 °C. Only a few experiments were necessary to find out that this electrolyte temperature determined the upper limit of the substrate stability under these conditions (acidic solution, high oxidation potential). On the other hand, a high temperature is required to obtain adherent EMD-deposits and only a small quantity of manganese dioxide in suspension. Tests performed by other groups showed a shift from adherent layer deposition to complete suspension-EMD formation when the electrolysis temperature was lowered from more than 90 °C to room temperature[5]. As a compromise, the electrolyte temperature was fixed at 70 °C in the following experiments. When the projected EMD-deposition was obtained the disassembling procedure was repeated.

The substrate, now covered with an active layer on each side, was thoroughly rinsed with distilled water, air-dried, and kept in a dry place until use.

The complete two-step process from the pure substrate sheet to the double-layered unit for bipolar arrangements is presented in Fig. 4.

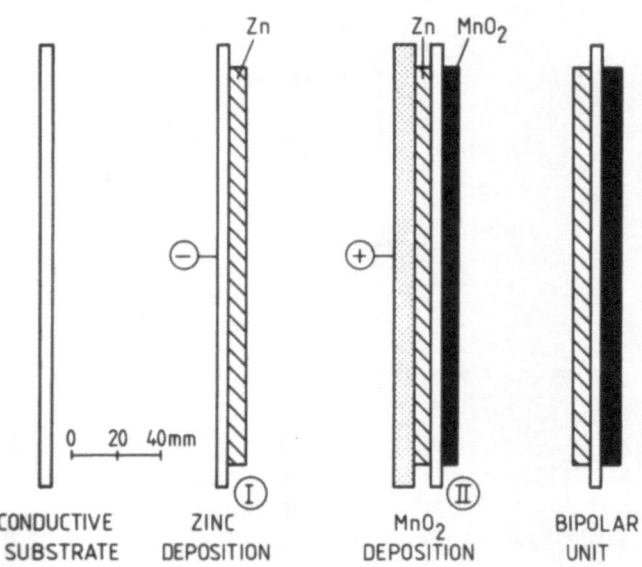

Figure 4. Preparation of bipolar electrodes in a two-step deposition process (deposited layers not in scale).

Stack tests

To obtain information concerning the usability of the produced units a test stack was formed. The zinc layer on one side of the bipolar unit faced the EMD-layer of the next one. The zinc layers were separated by a sheet of fibrous material serving as matrix for the electrolyte and by a sheet of a separator which is impermeable to zinc dendrites which may be formed during the recharging operation. This assembly procedure was repeated according to the projected battery voltage. Each anode/separator/(matrix)electrolyte/cathode-segment contributes 1.5 V to the (nominal) open circuit voltage (OCV).

The two endplates of the stack were slightly different in design. They were equipped with an active layer on one (the inner) side and a galvanic copper coating on the outside, acting as current collector and positive or negative battery terminal.

The horizontally oriented stack was compressed by mechanical means (constant force) to obtain reproducible results. The most important test parameters are OCV, voltage under standardised load, short circuit current, the limiting current density, and rechargeability.

It was no surprise that the first experimental stacks could not fulfil the expectations. Stack No.1 was assembled by arranging bipolar units (three electrode pairs) and intermediate layers formatted to 100 x 160 mm. The stack was compressed between two plates of polymethylmetacrylate ("Plexiglas") of slightly wider dimensions. The upper plate was charged with a load of 20 kg (about 13 kN.m^{-2}). For the initial (informal) tests the edges of the stack were left open.

The first remarkable observation checking stack No.1 was an OCV of 5.3 V, exceeding the nominal value by 0.8 V. This effect may be explained by the high electrochemical activity of the deposited electrode layers, especially of the EMD, and by the absence of additives (graphite, lamp black, binders) in the cathodes. On the other hand, the lack of conductive particles in the cathode layers and the practical impossibility of providing uniform contact across the relatively wide area caused a sudden drop of the cell voltage to 2.3 V under a 10Ω-load. Probably for the same reasons, the short circuit current of 400 mA was lower than expected (current density about 0.27 A.dm^{-2}). The discharge (10 Ω-load) was finished after two hours at a final stack voltage below 0.5 V. An attempt was made to recharge this stack potentiostatically at 5.5 V. The (recorded) charge current dropped from 40 mA at the beginning to 25 mA after 20 minutes and to 10 mA after one and a half hours. The charging period was limited to seven hours by the fact that the stack did not take up any more additional charge. The second discharge failed because the matrices became dry as a result of water evaporation and by part-transformation of the KOH to carbonate by air access.

For the assembly of experimental stack No.2, the bipolar units were modified in composition and size. The anode side - now based on an ABS-foil of 0.2 mm thickness - was equipped with an intermediate (electroless plated) copper layer to separate the carbon containing substrate from the zinc and to improve the adhesion of the zinc electrode. To avoid contact corrosion of the copper/zinc couple, the surface of the very thin copper layer was tin plated prior to zinc anode deposition. The thickness of this layer was 1.4 μm, requiring 100 mg.dm^{-2} tin. The EMD-deposition was done as before.

In order to get better contact, the bipolar units delivered from the final EMD-deposition were cut into two pieces with a size of 60 x 85 mm. To adjust these new units correctly, to prevent air access, and to achieve the required pressure a new battery case had to be built. It consisted of two rectangular shells made from polymethylmetacrylate screwed together with threaded bolts and nuts (Fig.5).

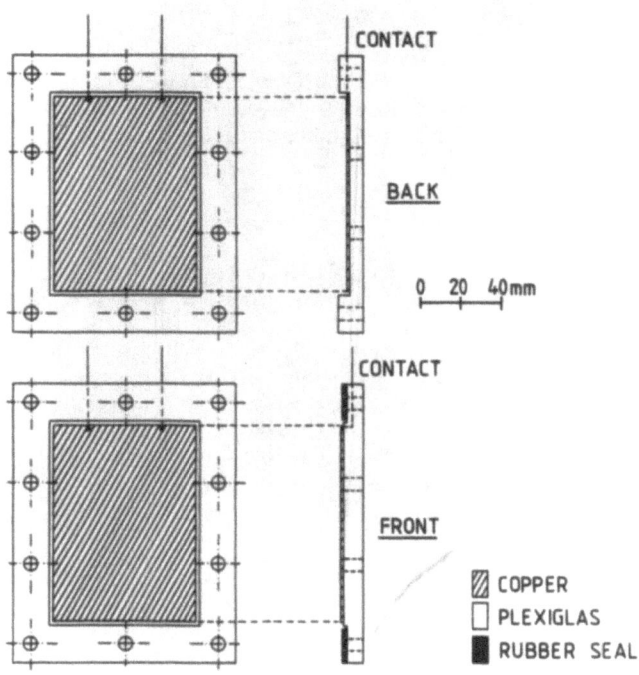

Figure 5. Experimental arrangement for stack tests (subsequent to stack No. 2).

Stack No.2 was found to have an OCV of 5.0 V and could be discharged over a 10Ω-load within one hour. The subsequent charging procedure was finished after restoring about 30 % of the one-electron capacity of manganese dioxide. The second discharge showed an improved performance (double discharge time using the same load resistor) but the recharge ability was faded badly.

RESULTS AND DISCUSSION

Deposition experiments followed by mechanical and electrical tests of the produced coatings (zinc, EMD) showed the principal practicability of a two-step electrolytic process carried out in a special laboratory device fed with circulating electrolyte.

Even with backside cooling of the substrate (carbon/PE composite) and significantly lowered operation temperature, the conditions of EMD deposition are close to the limits of material resistance. While prolonged exposition to the high temperature may destroy the polymer matrix, the conductive carbon particles suffer from oxidation by the hot acidic electrolyte under anodic polarisation. It is possible that future experiments will lead back to metal foil substrates.

If the use of conductive carbon/polymer-foils is requested, considering other than technical motives, the development of a cold EMD-coating process may be essential. In this case, the advantage of in-situ inspection and testing of the EMD will be lost.

For the reasons mentioned above, the two goals of substrate testing and experimental EMD deposition should be separated to achieve one or the other by performing a series of particularly designed experiments.

ACKNOWLEDGEMENTS

The authors wish to thank the Austrian Research Foundation for financial support, C.Conradty GmbH, Nürnberg, for supplying the carbon electrodes, and Deutsche Kapillar-Plastik GmbH, Dautphetal, for the PVC-C tubes and other materials.

REFERENCES

1. L.Binder, J.Daniel-Ivad, K.Kordesch, Rechargeable high energy batteries based on the alkaline manganese dioxide - zinc system, in: ICHEME Symposium Series No.127, Institution of Chemical Engineers, Rugby (1992).
2. L.Binder, K.Kordesch, Wiederaufladbare Hochenergie-Batterien abgeleitet vom alkalischen Zink-Braunstein System, in: DECHEMA-Monographien, Bd.128, VCH Verlagsgesellschaft, Weinheim (1993).
3. M.Matsuki, M.Sugawara, A.Kozawa, Codeposition of carbon materials in electrolytic manganese dioxide, *Progress in Batteries & Battery Materials*, 11:25 (1992).
4. Dettner/Elze, Handbuch der Galvanotechnik, Bd.2, C.Hanser Verlag, München (1966).
5. W.Harer, Die Herstellung von elektrolytischem Braunstein nach dem Prinzip der Umlaufelektrolyse, Dissertation,TU Graz,1985.

IMPROVEMENT OF SECONDARY ZINC ELECTRODES

Elzbieta Frackowiak and Krzysztof Jurewicz

Institute of Chemistry and Applied Electrochemistry
Technical University of Poznan
60-965 Poznan, ul. Piotrowo 3, Poland

ABSTRACT

Various attempts at improving the performance of zinc electrodes were undertaken in the present work. Technological changes connected with the modification of active material composition and application of a new supporting carrier for the active mass are shown to bring about improvements. High purity acetylene black inserted into the active material was the main modifier of the zinc electrodes. Metallized polypropylene fabric with a favourable morphology was a suitable matrix for zinc material. Cycle life enhancement of the improved Zn electrodes was confirmed for Ag-Zn and Ni-Zn accumulators.

INTRODUCTION

During the past few decades much attention has been given to the electrochemical behaviour and improvement of the secondary zinc electrode[1-9]. The main disadvantage of zinc electrodes is their limited cycle life due to the degradation of the zinc material. Shape change, dendrite formation, passivation and densification of the zinc electrode have been identified as the principal causes for the poor cycle life. Shape change is associated with the reduction of the electrochemically active surface area of the zinc electrode due to material movement from the electrode edges to the plate bottom and center. The density gradient model and volume variations of the electrolyte can describe these shape change patterns[1,2]. Suppression of dendritic growth was achieved by use of pulsed current during charging[10,11].

Among the different methods of cycle life enhancement, application of special additives to the active mass or to the electrolyte is relatively simple. This common method is based on adding certain metals eg. Bi, In, Pb, Sn or their oxides[1,3,5] as well as organic compounds[6-8]. Many of the additives exert a useful effect on the zinc morphology owing to inhibition of the initiation time and the rate of dendritic growth. They can often modify the reaction kinetics[6,8]. Much attention has been devoted to the restriction of the solubility of the reaction products through the use of a lower KOH concentration and addition of carbonates or fluorides[4]. It has also been shown that the addition of super fine graphite[12] to the zinc material acts as an adsorptive net for the oxidation products and causes an increase in the electrode durability.

In the present work acetylene black of high porosity was applied as a modifying agent. Constructional changes associated with the application of a new support for the zinc material were also used to enhance electrode performance. Metallized polypropylene fabric was found to be a useful carrier as well as a current collector of significant surface area. The important aim of this matrix was also protection against the zinc shape change.

EXPERIMENTAL

Rotating disc technique

Potentiostatic RDE experiments were performed with pellets made of zinc and high purity acetylene black AB-1402 from Germany. The speed of rotation ranged from 500 rpm to 3000 rpm. For potentiodynamic experiments on pellet electrodes (0.07 cm^2), a potentiostat, a function generator and an integrator, manufactured by Wenking, were used.

Preparation of pressed electrodes

Several electrode formulations were fabricated from zinc oxide, polyvinyl alcohol and acetylene black (AB). 1%, 2%, 5% and 10% of AB were used as additives to a 2,5 Ah zinc electrode. The upper limit for mechanical stability of the electrode was found to be 5% of AB. Electrodes were produced by pressing the active material with the silver collector. Two electrodes were wrapped with a few layers of dendritostatic cellophane separator and underwent cycling in 10 M KOH in a 5 Ah Ag-Zn cell.

Metallization of PP fabric

Polypropylene fibre material was manufactured by a pneumothermic method: its characteristics were porosity of 93%, thickness of 1,5 mm and elementary fibre diameter of 3 µm. The metallizing of PP fabric involved two steps: chemical plating and electrodeposition. Preliminary treatment i.e. degreasing and etching, was achieved in alkaline solution of hydrogen peroxide with uv radiation. This resulted in microcracks on the plastic surface. Electroless copper plating was performed in a stabilized sulfate bath reduced with formalin. Cadmium electrodeposition was then carried out from a typical galvanic bath. The thickness of the copper layer was 1,2 µm and that of cadmium was 3 µm.

Construction of fibre electrode

The general principles of electrode construction based on a PP/Cd matrix are shown in Fig. 1. The main current collector i.e. metallic mesh (a) with current lead (b) was heat sealed with both side claddings (c) made of metallized fibre. A few horizontal heat sealings (f) formed the electrode sectors which held the active mass. The active material was zinc powder inserted into the carrier from suspension in methyl alcohol. The claddings contained a thin barrier layer of dense PP fabric filled with CdO (d) and CaO (e). These oxides amounted to 10% of the electrode weight.

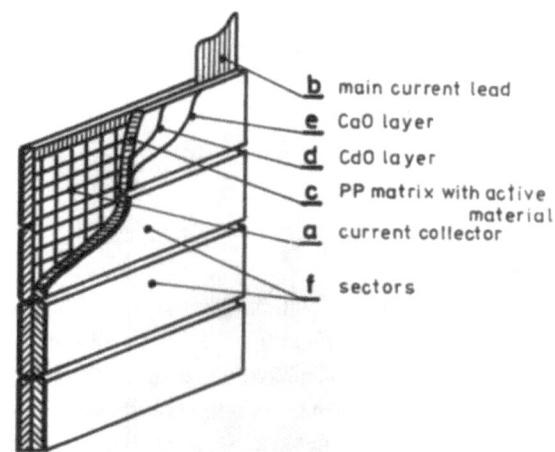

Figure 1. Construction of Zn electrode with PP metallized matrix

The nominal capacity of the fibre electrode was 3 Ah. Electrical cycling was carried out in 7 M KOH saturated with zincates using a Ni-Zn system.

RESULTS AND DISCUSSION

Investigation of active mass

Experiments performed on the pellet electrodes using a potentiostatic RDE technique allowed evaluation of the electrochemical properties of the zinc material with different amounts of the additive AB ranging from 0,1% to 5%. The Levich slopes decreased significantly when the content of AB increased[13]. Active dissolution of zinc was notably decreased in the case of 5% AB content.

Potentiodynamic experiments performed at 2 mV/s on pellet electrodes confirmed the favourable action of acetylene black. In Fig. 2 the improved durability of modified zinc material with acetylene black is shown. The porous structure and sorptive properties of AB facilitate the sorption of the oxidation products and impede their migration to the bulk of electrolyte.

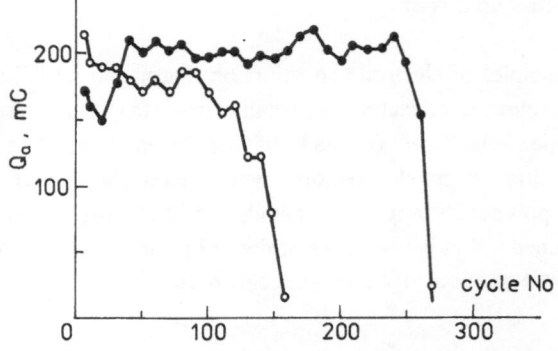

Figure 2. Anodic charge of zinc pellet during cyclic potentiodynamic experiments at 2 mV/s
O   without additive;   ● 5% of AB

Cycle life studies

Investigation of Ag-Zn accumulators with additive AB or without AB was carried out on 5 Ah accumulators as follows: charging at 0,3 A for 19 h and discharging at the C/5 rate i.e. at 1A to the end potential of 1,35V. The average values of the discharge characteristics for the zinc electrode with 5% AB during cycling are shown in Fig. 3. The modified electrodes have maintained their capacity above 200 cycles, whereas electrodes of standard production only performed satisfactorily to about 80 cycles. Self-discharge after one month was only ~ 5%[14].

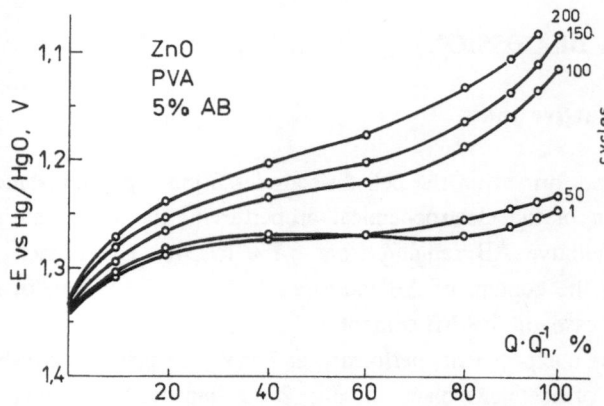

Figure 3. Discharge characteristics of Zn electrode with 5% of AB during cycling
Q_n = 5 Ah;   C/5 rate

The electrical investigation of fibre zinc electrodes (3Ah) with PP/Cd were performed in Ni-Zn cells with sintered nickel electrodes. Low overpotential of the discharge curves confirmed the advantage of this carrier. During cycle life experiments discharge was carried out at the C/5 rate (0,6A) and the electrode operation was controlled by measuring the overpotential at the end of discharge. The experiment was interrupted when for three subsequent cycles the values of overvoltage exceeded 200 mV. The charge process was carried out at 0,52A for 6,5h (coefficient 1,13). Fig. 4 shows the cycling results: for electrodes without the carrier, the number of cycles was ~55 and in the case of the metallized matrix (PP/Cd) ~140 cycles. Investigations showed that a PP matrix of suitable morphology with an evenly metallized Cd coating could be used as a good carrier allowing a high rate of discharge. Additional layers of CdO and CaO placed on the surface of the electrode together with the horizontal segmentation limited the electrode shape change. It seems that the additive of cadmium played a dispersion role for electrode material which has a tendency to densification with cycling. The exterior layer of calcium compound caused a restriction of the solubility of the zinc oxidation products.

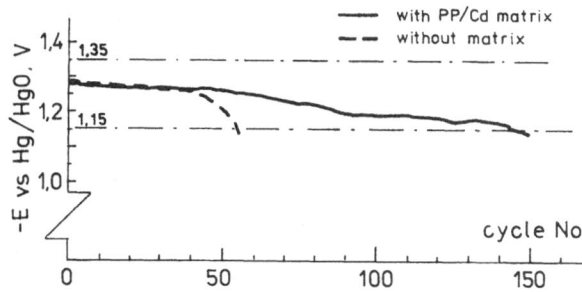

Figure 4. Final discharge potentials of Zn electrodes with PP/Cd matrix during cycling

CONCLUSIONS

5% of AB significantly inhibits active dissolution of zinc, hence, the oxidation products remain in the interior of the electrode owing to the sorptive properties of AB. This assures their close contact with the electrode material. The conductive properties of AB improve the discharge characteristics of pressed zinc electrodes. Zn electrodes with a metallized PP matrix give high discharge rates because of the uniform current distribution in the electrode. Additional oxide layers placed on the surface, as well as horizontal segmentation of the electrode, restrict the electrode shape change. Application of the new PP/Cd matrix causes a twofold increase in durability, whereas modification of the electrode composition gives a threefold improvement. It seems that a combination of the afore-mentioned improvements would be very profitable.

REFERENCES

1. R.E.F. Einerhand, W. Visscher, J.J.M. de Goeij and E. Barendrecht, Zinc electrode shape change, *J. Electrochem. Soc.* 138: 1 (1991).
2. R.E.F. Einerhand, Zinc electrode shape change, PhD Thesis, Eindhoven, (1989).
3. C. Biegler, R.L. Deutscher, S. Fletcher, S. Hua and R. Woods, Accelerated testing of additives in zinc plates of nickel zinc cells, *J. Electrochem. Soc.* 130: 2303 (1983).
4. T.C. Adler, F.R. McLarnon and E.J. Cairns, Low-zinc-solubility electrolytes for use in zinc/nickel oxide cells, *J. Electrochem. Soc.* 140: 289 (1993).
5. J. McBreen and E. Gannon, Bismuth oxide as an additive in pasted zinc electrode, *J. Power Sources* 15: 169 (1985).
6. C. Cachet, B. Saidani and R. Wiart, The behavior of zinc electrode in alkaline electrolytes, *J. Electrochem. Soc.* 139: 644 (1992).
7. G.D. Wilcox and P.J. Mitchell, Electrolyte additives for zinc-anoded secondary cells, *J. Power Sources* 28: 345 (1989).
8. E. Frackowiak and M. Kiciak, The influence of polyethylene glycol on some properties of zinc electrodes, *Electrochim. Acta,* 29: 1359 (1984).
9. J. Jindra, Sealed nickel-zinc cells, *J. Power Sources,* 37: 297 (1992).
10. K. Appelt and K. Jurewicz, Die losliche Zinkelectrode als Anode alkalischer Akkumulatoren, *J. Power Sources,* 5: 235 (1980).
11. G. Bronoel, A. Millot and N. Tassin, Development of Ni-Zn cells, *J. Power Sources,* 34: 243 (1991).
12. A. Duffield, P.J. Mitchell, D.W. Shield and N.Kumar, Evaluation of additives for a secondary zinc electrode in: "Power Sources 11" chap. 17, 253, L.J. Pearce, ed., International Power Sources Symposium Committee, Leatherhead (1987).
13. E. Frackowiak, Modification of secondary zinc electrode, in: "Power Sources 13" chap. 21, 225, T. Keily, B.W. Baxter eds., International Power Sources Symposium Committee, Leatherhead (1991).
14. E. Frackowiak, K. Jurewicz and M. Paszkiewicz, Ag-Zn accumulator of long durability, *Bul. Electrochem.* vol. 9, no.3 (1993) in printing.

HYDROGEN GAS DIFFUSION ELECTRODE IN DILUTE SULPHURIC ACID

SOLUTION

J.J.T.T. Vermeijlen and L.J.J. Janssen

Laboratory of Instrumental Analysis
Department of Chemical Engineering
Eindhoven University of Technology
P.O. Box 513
5600 MB Eindhoven
The Netherlands

ABSTRACT

Experimental data are presented concerning the diffusion limited current density for hydrogen oxidation in a gas diffusion electrode (GDE) under various conditions. These current densities were obtained using mixtures of hydrogen and inert gases.

To elucidate the dependence of the overal mass transport coefficient, a simplified model to describe the transport of hydrogen in a GDE based on literature models was derived. The GDE consists of a hydrophobic and a hydrophilic layer, viz. a porous backing and a reaction layer.

It was found that the transport rate of hydrogen under the experimental conditions is determined by hydrogen gas diffusion in the pores of the porous backing as well as in the macropores of the reaction layer. Diffusion of dissolved hydrogen in the micropores of the reaction layer, through the liquid, is shown to be of little significance.

INTRODUCTION

To describe the potential-current relation for a gas diffusion electrode (GDE) as used in fuel cells, many models for mass transport of electro-active gas have been proposed. Reviews of these models have been presented[1,2].

To elucidate the reactor model for the gas compartment at the backside of the gas diffusion electrode and the rate determining step of the mass transport of electro-active gas in a hydrogen gas diffusion electrode, a mixture of hydrogen gas and inert gas was supplied to the gas diffusion electrode. The limiting current density for the hydrogen oxidation was determined as a function of a large number of parameters, e.g. composition of solution, type of inert gas, flow rate of gas, temperature, gas pressure and liquid pressure.

EXPERIMENTAL

The experimental set-up is shown in Figure 1. Some adaptions to the experimental set-up as described in [3] were made to allow the gas to be saturated with water vapour at the cell operating temperature and to vary the gas and liquid pressure.

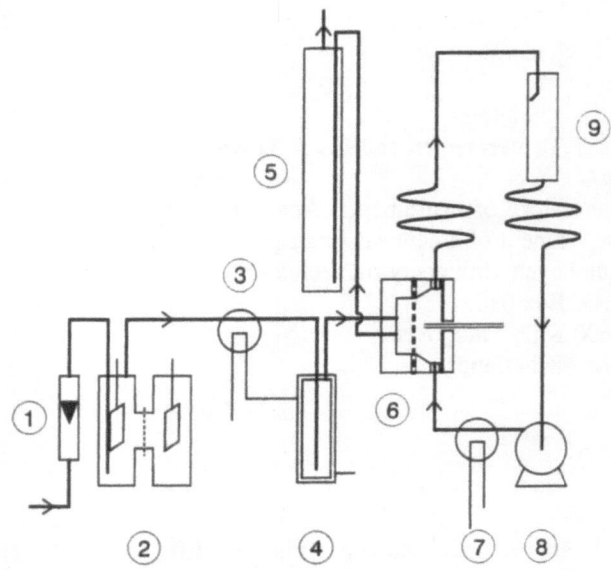

Figure 1. Schematic illustration of the experimental set-up. (1) flowmeter, (2) hydrogen generation cell, (3) heat exchanger for gas, (4) water saturation vessel for gas, (5) water column for gas overpressure, (6) test cell, (7) heat exchanger for solution, (8) solution pump, (9) solution storage vessel.

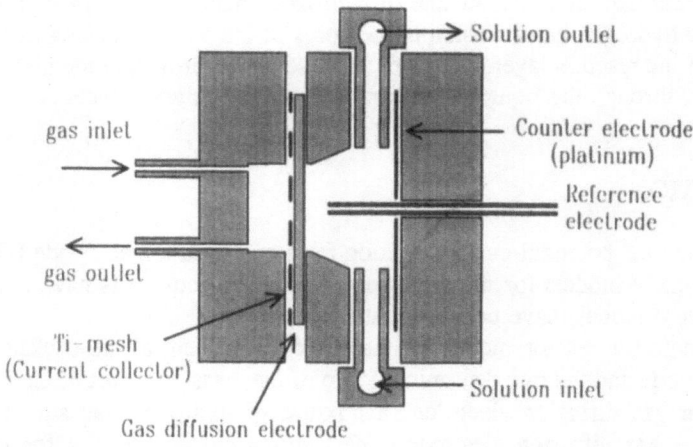

Figure 2. Schematic illustration of the gas diffusion electrode test cell.

The experimental cell is shown schematically in Figure 2. The cell was fitted with Fuel Cell Grade Electrodes on Toray Paper purchased from E-TEK, USA. These gas-diffusion electrodes were loaded with 0.50 mg cm^{-2} platinum. The active (hydrophilic) layer has a thickness of approximately 0.1 mm, whereas the total electrode thickness measures approximately 0.55 mm. A geometric electrode surface area of 20 x 20 mm^2 was exposed to gas and solution.

The solutions used were 0.5 to 9 M H$_2$SO$_4$ prepared from sulphuric acid p.a. (Merck) and deionized water. The solutions were circulated through the solution compartment of the test cell at a flow rate of 5 cm^3 s^{-1}.

The liquid pressure upon the GDE could be varied by means of variation of the height of the solution storage vessel. The temperature of the solution was kept constant by a heat exchanger near the solution inlet of the test cell.

The resulting gas mixture was fed to the gas compartment of the test cell. The inert gases used were nitrogen, helium and argon. Hydrogen generated at constant current from a 4 M KOH solution was added to the inert gas stream.

The gas temperature was controlled by means of a heat exchanger near the gas inlet of the test cell. A thermostatted vessel that could be filled with water was situated in the gas inlet circuit. By means of this cell the inlet gas could be saturated with water vapour at the cell operating temperature.

Cyclic voltammograms were recorded using a Solartron 1286 Electrochemical Interface (ECI) controlled by a microcomputer. The potential range of 1 V between the equilibrium potential of the gas diffusion electrode E_e and the more positive potential E_t $= E_e + 1$ V was scanned at a rate of 5 mV s^{-1} in the ECI's stepped sweep mode at a stepping rate of 1 s^{-1}.

Electrochemical impedance spectra were recorded using the ECI and the Solartron 1250 Frequency Response Analyser (FRA). Ohmic drops between the tip of the Luggin capillary and the gas diffusion electrode were calculated using these impedance spectra. Potentials used in this work are referred to E_e and the overpotential $\eta = E - E_e$ was corrected for Ohmic drop.

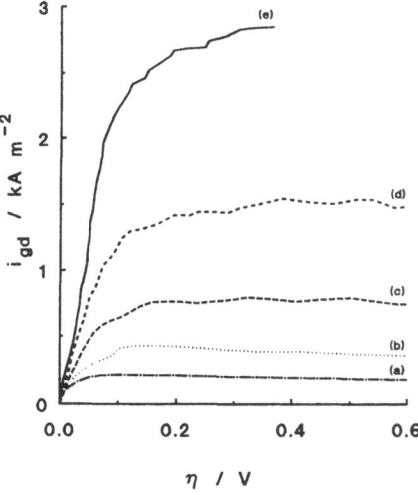

Figure 3. The gas diffusion electrode current density, i_{gd}, as a function of the overpotential, η, at a temperature of 30°C. Volumetric nitrogen flow rate $F_{v,N} = 2.33$ cm^{3}s^{-1}. The hydrogen inlet concentration $c_{in} = $ (a)0.28 mol m^{-3}, (b)0.55 mol m^{-3}, (c)1.08 mol m^{-3}, (d)2.12 mol m^{-3}, (e)4.04 mol m^{-3}.

RESULTS

The hydrogen content of the inlet gas stream was varied by applying currents in the range from 0.005 to 2 A to the hydrogen generation cell. Moreover, the volumetric rate of nitrogen was varied in the range from 0.1 to 15 cm^3 s^{-1}.

Figure 3 shows a typical set of current density-overpotential curves for a gas diffusion electrode with 0.50 mg cm^{-2} Pt loading. The results were obtained for various hydrogen concentrations c_{in} in the gas feed flow. The overpotential η was corrected for the ohmic drop. The current density-overpotential curves clearly show the diffusion limited plateaus. The direction of the potential sweep had practically no effect on the limited current density for high hydrogen concentrations. For low hydrogen concentrations, the mean value of the mean current densities from increasing and decreasing overpotential has been calculated to account for double-layer charging effects.

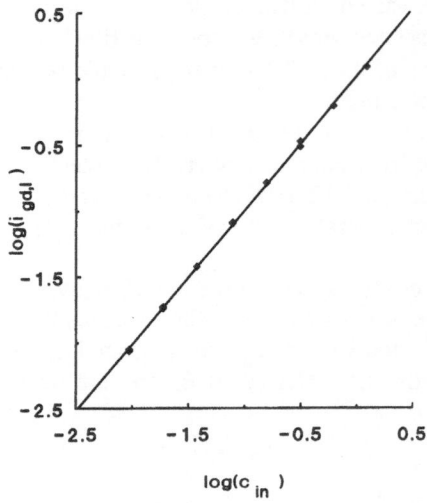

Figure 4. The diffusion limited current density, $i_{gd,l}$, as a function of the hydrogen inlet concentration, c_{in}, on a double logarithmic scale at a temperature of 30°C and at a constant volumetric nitrogen flow rate of 2.33 cm^3 s^{-1}.

Figure 4 shows a curve on a double logarithmic scale representing the diffusion limited current density, $i_{gd,l}$, as a function of the hydrogen concentration in the feed gas at the inlet of the gas compartment of the gas diffusion cell, c_{in}, at a high nitrogen flow rate $F_{v,N}$. From this plot it follows that there is an almost linear relationship between c_{in} and $i_{gd,l}$ at a constant $F_{v,N}$.

Figure 5 shows, on a double logarithmic scale, the diffusion limited current density, $i_{gd,l}$, as a function of the hydrogen concentration in the gas at the inlet of the gas diffusion cell, c_{in}, for constant I_{hp} and various $F_{v,N}$. The almost linear relationship as observed in Figure 4 is absent. The molar flow rate of hydrogen through the gas compartment does not solely determine the diffusion limited current. From the diffusion limited current density the mass transport coefficient for hydrogen to the gas diffusion electrode can be calculated if the flow behaviour of hydrogen gas in the gas compartment at the backside of the gas diffusion electrode is well-known.

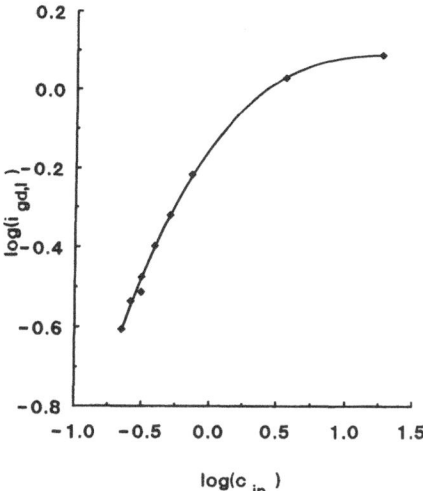

Figure 5. The diffusion limited current density, $i_{gd,l}$, as a function of the hydrogen inlet concentration, c_{in}, on a double logarithmic scale at a temperature of 30°C and at a constant hydrogen production current $I_{hp} = 0.5$ A.

Figure 6 shows the influence of the concentration of sulphuric acid in the solution, c_{SA}, on the diffusion-limited current density, $i_{gd,l}$, at various hydrogen concentrations in the H_2/N_2 mixture, c_{in}, and at a constant volumetric flow rate of nitrogen gas, $F_{v,N}$. The data presented are averaged from multiple measurements.

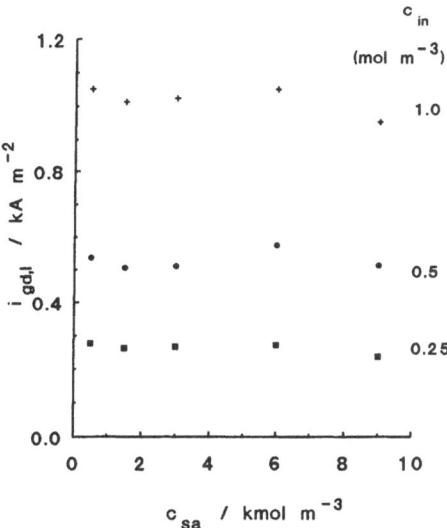

Figure 6. The diffusion limited current density, $i_{gd,l}$, as a function of the sulphuric acid concentration in the solution, c_{SA}, at a temperature of 293 K and various hydrogen gas inlet concentrations. The volumetric nitrogen flow rate = 5.08 cm³ s⁻¹.

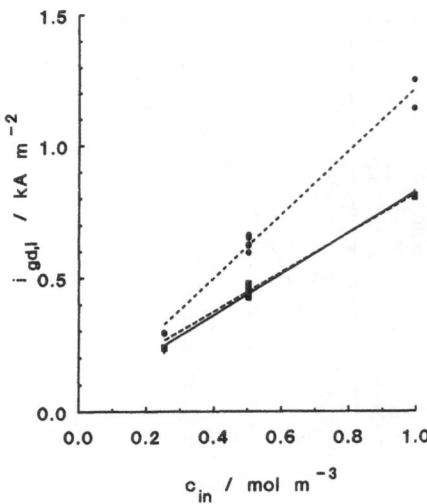

Figure 7. The diffusion limited current density, $i_{gd,l}$, as a function of the gas inlet concentration, c_{in}, at a temperature of 293 K and an inert gas flow of 5.08 cm^3 s^{-1}. Inert gases used are ($\blacksquare$) Ar, ($\bullet$) He, ($+$) N$_2$.

Figure 7 shows the influence of the hydrogen concentration in the inlet gas, c_{in}, on $i_{gd,l}$ at a constant volumetric flow rate of inert gas for N$_2$, Ar and He as inert gases. Figure 8 shows the influence of the temperature on the diffusion-limited current density, $i_{gd,l}$, at a constant hydrogen production current and at a constant N$_2$ gas mass flow rate. As the temperature increases from 293 to 353 K, c_{in} decreases and $F_{v,in}$ increases. Figure 8 shows measurements for gas saturated with water vapour at 293 K, as well as for gas saturated with water vapour at the operating temperature of the cell.

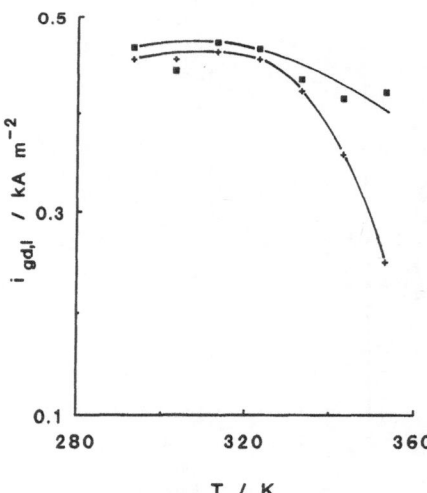

Figure 8. Measured diffusion limited current densities, $i_{gd,l}$, as a function of the cell operating temperature at a nitrogen gas flow rate of 5.08 cm^3 s^{-1} at 293 K and at a hydrogen production current I_{hp} of 0.5 A. The gas fed to the cell was saturated with water at 293 K ($\blacksquare$) or saturated with water at the cell operating temperature ($+$).

DISCUSSION

Both the input and the output of the gas compartment are situated perpendicularly to the electrode surface. This enables good mixing of the gas present in the gas compartment of the gas diffusion electrode test cell. Moreover, the diffusion of hydrogen in the gas is a fast process. Thus, concentration gradients will be levelled rapidly. It is likely that the constant flow stirred tank model (CSTR model) is a reliable model for the behaviour of the gas compartment of the test cell. This means that the hydrogen concentration throughout the gas compartment is equal to the hydrogen concentration in the outlet gas.

It has been found that the mass transport coefficient k_s, calculated from: $I_{gd,l} = n \mathcal{F} A_{gd} k_s^* c_{out}$ (Figs. 4 and 5), is independent of the input hydrogen concentration c_{in} and gas flow rate $F_{v,in}$. Consequently, the gas compartment at the backside of the gas diffusion electrode can be considered as a CSTR.

Various models have been proposed to account for the behaviour of gas diffusion electrodes [1,2,4-10]. A schematic illustration of a gas diffusion electrode is presented in Figure 9. An applicable model to explain the dependencies observed includes a porous backing acting as support, current collector and gas transport layer combined with a reaction layer consisting of a network of drowned and gas-filled pores. To estimate the influence of diffusion coefficients on the overal mass transport coefficient, a simplified model based on e.g. [5-7] using effectiveness factors for the gas filled macropores and the drowned micropores in the reaction layer is derived below. In this model, three stages of hydrogen transport are combined:
1. transport of hydrogen through the gas-filled macropores of the porous backing
2. transport of hydrogen gas in the gas-filled macropores of the reaction layer where simultaneously dissolution of hydrogen gas in the solution present in the micropores of the reaction layer takes place
3. transport of dissolved gas in the solution-filled micropores of the reaction layer where oxidation of dissolved hydrogen on the catalyst sites on the micropore walls occurs simultaneously.

As can be seen from Figure 6., there is very little influence of the sulphuric acid concentration on the diffusion-limited current density for hydrogen oxidation in the GDE. The diffusion coefficient for dissolved hydrogen, D_l, decreases from approximately 0.35×10^{-9} m^2 s^{-1} to 0.20×10^{-9} m^2 s^{-1} when c_{SA} increases from 0.5 to 9 kmol m^{-3} [11]. The solubility of hydrogen in sulphuric acid, as quantified by the Bunsen-coefficient α, also decreases by a factor of approximately 2.5 for the same increase in c_{SA} [12]. This means that H decreases by a factor of 2.5 with increase in c_{SA} from 0.5 to 9 kmol m^{-3}. These changes should result in a large decrease in the diffusion limited current density if mass transport of hydrogen in the liquid layer is of influence on the overall proces. This proces is, therfore, not rate-determining.

Table 1. Interdiffusion coefficients from literature for hydrogen in various inert gases and calculated overall mass transport coefficients for $T = 293$ K.

Inert gas	10^4 x $D_{H2,j}$ (m^2 s^{-1})	k_s (m s^{-1})
Nitrogen (N_2)	0.74	7.33×10^{-3}
Argon (Ar)	0.80	6.94×10^{-3}
Helium (He)	1.40	11.68×10^{-3}

Table 1 shows the influence of the inert gas on the overall mass transport coefficient k_s, calculated using the CSTR model, the results from Fig. 7 , and the interdiffusion coefficients for hydrogen in various inert gases[13]. From Table 1 , it follows that the mass transport coefficient is not proportional to $D_{i,j}$ but almost proportional to $D_{i,j}^{0.8}$.

If the mass transport of hydrogen is completely determined by the macropores in the porous backing, the rate constant k_s is proportional to $D_{i,j}$ [14-16] and if the mass transport of hydrogen through the macropores in the reaction layer, where simultaneously dissolution of hydrogen gas in the solution present in the micropores takes place, is the slowest process it can be derived that k_s is proportional to $D_{i,j}^{0.5}$ [17]. The latter result is consistent with that obtained by Cutlip[5]. Comparing the experimental proportionality between k_s and $D_{i,j}$ with the theoretical relationships, it can be concluded that the diffusion of hydrogen gas in both layers contributes to the overall diffusion resistance. Since only a small influence on hydrogen solubility of the H_2SO_4 concentration has been found (Figure 6), it is concluded that the major resistance for hydrogen transport is located in the macropores of the porous backing of the gas diffusion electrode.

The dependence of the diffusion limited current density, $i_{gd,l}$, on the operating temperature of the cell, T, is shown in Figure 8. for both experiments with gas saturated with water vapour at 293 K, as well as for those with gas saturated with water vapour at the cell operating temperature. The difference in $i_{gd,l}$ is caused by a difference in water vapour content of the inlet gas; in particular at temperatures over approximately 330 K. If the gas inside the pores of the gas diffusion electrode were in equilibrium with the solution inside those pores, the gas inside the pores would have approximately the same composition for both types of measurements resulting in the same diffusion-limited current density. However, a large difference has been found. It is therefore concluded that in this case, with a gas mixture which is not saturated with water vapour, mass transfer of water vapour occurs in the macropores of the gas diffusion electrode. This flow of water vapour is directed from the liquid phase in the hydrophilic layer of the gas diffusion electrode to the gas compartment of the gas diffusion cell. Consequently, the mass transport of hydrogen gas to the reaction layer is hindered by the water vapour flow.

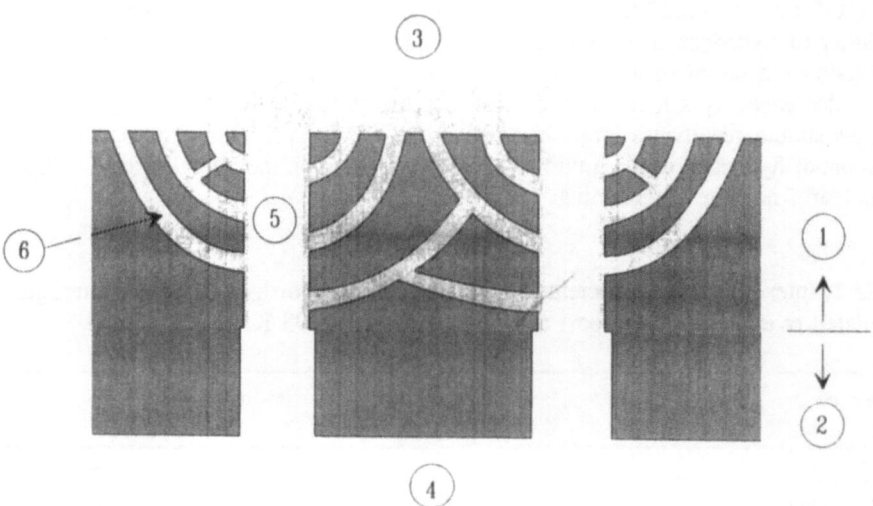

Figure 9. Schematic illustration of a gas diffusion electrode. (1) porous backing, (2) reaction layer, (3) solution, (4) gas, (5) macropore, (6) micropore.

CONCLUSIONS

The method used is well suited to the determination of the mass transfer coefficient for electro-active species from the gas phase to the active sites in a gas diffusion electrode.

Hydrogen gas diffusion in the pores of the gas diffusion electrode is the rate-determining step under current-limiting conditions.

Diffusion of dissolved hydrogen through the liquid phase in the gas diffusion electrode is of little significance for mass transfer limitation.

Acknowledgement

This work was supported by Hoogovens IJmuiden, the Netherlands.

REFERENCES

1. I. Roušar, K. Micka and A. Kimla, "Electrochemical Engineering", Chemical Engineering Monographs 21B, Elsevier, Amsterdam (1986).
2. M.W. Breiter, "Electrochemical Processes in Fuel Cells", Anorganische und allgemeine Chemie in Einzeldarstellungen IX, Springer Verlag, Berlin (1969).
3. J.J.T.T. Vermeijlen and L.J.J. Janssen, Determination of overall mass transport coefficients in gas diffusion electrodes, *J. Appl. Electrochem.* 23:26 (1993).
4. D.M. Bernardi and M.W. Verbrugge, Mathematical model of a gas diffusion electrode bonded to a polymer electrolyte, *AIChE Journal* 37:1151 (1991).
5. M.B. Cutlip, An approximate model for mass transfer with reaction in porous gas diffusion electrodes, *Electrochim. Acta* 20:767 (1975).
6. S.C. Yang, M.B. Cutlip and P. Stonehart, Further development of an approximate model for mass transfer with reactions in porous gas-diffusion electrodes to include substrate effects, *Electrochim. Acta* 34:703 (1989).
7. M.B. Cutlip, S.C. Yang and P. Stonehart, Simulation and optimization of porous gas-diffusion electrodes used in hydrogen oxygen phosphoric acid fuel cells - II, *Electrochim. Acta* 36:547 (1991).
8. L.G. Austin and Satish Almaula, An experimental study of the mode of operation of porous gas-diffusion electrodes with hydrogen fuel, *J. Electrochem. Soc.* 114:927 (1967).
9. Kwong-Yu Chan, G.S. Efthymiou and J.F. Cocchetto, A wedge-meniscus model of gas-diffusion electrodes, *Electrochim. Acta* 32:1227 (1987).
10. W. Jenseit, "Untersuchungen zur morphologischen Charakterisierung von Gasdiffusionselektroden", Thesis, Technischen Hochschule Darmstadt, Darmstadt (1990).
11. E.W. Washburn, ed., "International Critical Tables of Numerical Data, Physics, Chemistry and Technology", Volume 5, McGraw-Hill, New York (1929).
12. C.L. Young, ed., "Solubility Data Series", Volume 5/6, publication of IUPAC, Pergamon Press, Frankfurt (1981).
13. Landolt-Börnstein, "Zahlenwerten und Funktionen aus Physik, Chemie, Astronomie, Geophysik, Technik", 6. Auflage, 2. Band, 5. Teil, Bandteil a, Springer Verlag, Berlin (1969).
14. M.C. Kimble, R.E. White, Yu-Min Tsou and R. Neal Beaver, Estimation of the diffusion coefficient and solubility for a gas diffusing through a membrane, *J. Electrochem. Soc.* 127:2510 (1990).
15. D. Fan, R.E. White and N. Gruberger, Diffusion of a gas through a membrane, *J. Appl. Electrochem.* 22:770 (1992).
16. Yu-Min Tsou, M.C. Kimble and R.E. White, Hydrogen diffusion, solubility and water uptake in Dow's short-side-chain perfluorocarbon membranes, *J. Electrochem. Soc.* 139:1913 (1992).
17. J.J.T.T. Vermeijlen and L.J.J. Janssen, Mass transport in a hydrogen gas diffusion electrode, *J. Appl. Electrochem.* 23:1237 (1993).

Notation

A_{gd} geometric electrode surface area (m^2)

c_{in} concentration of reactive component at the inlet of the gas compartment (mol m^{-3})

c_{out} concentration of reactive component in and at the outlet of the gas compartment (mol m^{-3})

c_{SA} concentration of sulphuric acid in the solution compartment (mol m^{-3})

c concentration of reactive component in a gas diffusion electrode (mol m^{-3})
$D_{i,j}$ interdiffusion coefficient for gas i in gas j (m^2 s^{-1})
D_l diffusion coefficient for electroactive species in solution (m^2 s^{-1})
E electrode potential (V)
E_e equilibrium electrode potential (V)
E_t upper limit electrode potential (V)
$F_{v,in}$ volumetric flow rate at the inlet of the gas compartment (m^3 s^{-1})
$F_{v,N}$ volumetric flow rate of nitrogen at the inlet of the gas compartment (m^3 s^{-1})
$\mathscr{F}$ Faraday constant (A s mol^{-1})
$i_{gd,l}$ diffusion limited current density for gas diffusion electrode (A m^{-2})
$I_{gd,l}$ diffusion limited current for gas diffusion electrode (A)
I_{hp} current for hydrogen production (A)
k_s mass transport coefficient (m s^{-1})
n number of electrons involved in the electrode reaction
T temperature (K)
η overpotential (V)

A NEW CONCEPT OF AN ELECTROCHEMICAL HEAT PUMP SYSTEM: THEORETICAL CONSIDERATION AND EXPERIMENTAL RESULTS

L. Dittmar, K. Jüttner, and G. Kreysa

Karl-Winnacker-Institute of DECHEMA e.V.
Theodor-Heuss-Allee 25, D-60486 Frankfurt am Main

ABSTRACT

A new concept of an electrochemical heat pump system is described which consists of two identical electrochemical cells operating at different temperatures in opposite directions. The principle is based on the exchange of reversible heat ($T\Delta_R S$) of the reaction. The thermodynamics and efficiency of such a system are considered for different cell reactions. Calorimetric measurements, using commercial accumulators and non-commercial redox cells, were carried out to prove the validity of the theoretical predictions with respect to heat exchange and efficiency. The efficiency of a heat pump with the redox system $Cu/CuSO_4//Na_3[Fe(CN)_6], Na_4[Fe(CN)_6]/Pt$ was found to be comparable to that of conventional heat pump systems.

INTRODUCTION

Thermal losses due to diffuse or concentrated heat flux are the main reasons for many industrial processes having exorbitant energy consumption. According to a UN report of 1985, thermal losses from industrial energy consumption in Germany amount to about 40-45% of the primary energy employed[1]. To improve this situation the development of new technologies for the recovery of thermal energy by heat transformation or heat pump systems is of practical importance. In this respect electrochemistry provides interesting, but as yet only tentative, solutions. One well-known example of the direct conversion of heat to electricity is the thermogalvanic cell which takes advantage of the potential difference between two electrochemical half cells at different temperatures[2]. Such cells have been considered as attractive potential energy resources for many years, however to date technical implementation has failed because a thermogalvanic cell requires an electrically conductive, but thermally insulating diaphragm, which is not yet available.

This problem can be overcome by a new concept of an electrochemical system consisting of two identical electrochemical cells operating at different temperatures in opposite directions. The advantage of this system is that there is no electrolytic connection between the two thermal reservoirs. It is based on the principle of the exchange of reversible heat ($T\Delta_R S$) of reactions with the environment. Such a system, operating as an electrochemical heat pump, is compared schematically with a conventional heat pump system in Fig.1.

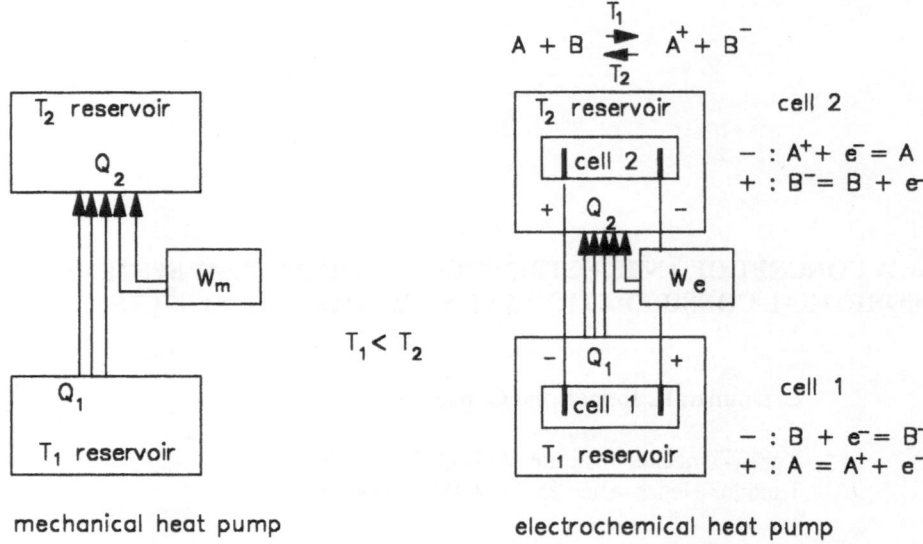

Figure 1. Principles of mechanical and electrochemical heat pump systems

Common to both systems is the principle of heat transfer from the low temperature reservoir T_1 to the high temperature reservoir T_2. For this operation additional energy input, either as mechanical energy W_m in the case of the conventional heat pump, or as electrical energy W_e in the case of the electrochemical heat pump, is needed.

In the electrochemical system both cell reactions take place in opposite directions at corresponding temperatures.

$$A + B \rightleftharpoons A^+ + B^- \tag{1}$$

The Gibbs energy for each cell is given as:

$$\text{cell 1: } \Delta_R G_1 = \Delta_R H_1 - T_1 \Delta_R S_1 \qquad \text{at } T_1 \tag{2}$$

$$\text{cell 2: } \Delta_R G_2 = \Delta_R H_2 - T_2 \Delta_R S_2 \qquad \text{at } T_2 \tag{3}$$

and the sum of both yields the Gibbs energy of the overall reaction of the whole system

$$\Delta_R G = \Delta_R G_1 + \Delta_R G_2 \quad . \tag{4}$$

Provided that $T_1 \Delta_R S_1 > 0$, cell 1 takes up heat from the low temperature reservoir and cell 2 delivers heat $T_2 \Delta_R S_2$ to the high temperature reservoir; the system thus operates as an electrochemical heat pump. A semicontinuous cyclic operation can be achieved by changing the temperatures and directions of the cell reactions as soon as the reactants of each cell have been consumed.

The heat pump efficiency ε_{hp} is defined as:

$$\varepsilon_{hp} = \frac{- T_2 \Delta_R S_2}{\Delta_R G_1 + \Delta_R G_2} . \tag{5}$$

Introducing eqs. (1) and (2) into eq. (5) yields

$$\varepsilon_{hp} = \cfrac{T_2}{T_2 + T_1\left(\cfrac{\Delta_R S_1}{\Delta_R S_2}\right) - \left(\cfrac{\Delta_R H_1 + \Delta_R H_2}{\Delta_R S_2}\right)} \qquad (6)$$

which for a moderate temperature difference ($\Delta_R H_1 + \Delta_R H_2 \approx 0$), leads to

$$\varepsilon_{hp} = \cfrac{T_2}{T_2 + T_1\left(\cfrac{\Delta_R S_1}{\Delta_R S_2}\right)} \qquad . \qquad (7)$$

This equation indicates that the efficiency of the electrochemical heat pump is at its maximum if $|\Delta_R S_2| \gg |\Delta_R S_1|$. For the special case $\Delta_R S_1 + \Delta_R S_2 \approx 0$, eq.(7) results in the reciprocal CARNOT-factor

$$\varepsilon_{hp} = \cfrac{T_2}{T_2 - T_1} \qquad . \qquad (8)$$

The above thermodynamic considerations are valid for the reversible behaviour of the system.

Under real conditions energy losses due to internal inefficiencies must be taken into account. In the energy balance these inefficiencies appear as generated heat (JOULE heat) connected with overvoltages η_{cell1} and η_{cell2} in the different parts of the system. This leads to the real efficiency of an electrochemical heat pump

$$\varepsilon_{hp} = \cfrac{-(T_2 \Delta_R S_2 + v_e F \eta_{cell2})}{\Delta_R H_1 - T_1 \Delta_R S_1 + \Delta_R H_2 - T_2 \Delta_R S_2 + v_e F \eta} \qquad (9)$$

where $\eta = \eta_{cell1} + \eta_{cell2}$.

EXPERIMENTAL

To determine the electrochemical cell heats of reaction calorimetric measurements were carried out in a specially designed differential heat flow calorimeter consisting of a measurement vessel and an identical reference vessel each of 60 cm^3 volume. The experimental set-up is shown schematically in Fig.2.

The heat of reaction was measured by integration of the voltage difference of the thermocouple arrays. The calibration was performed by defined generation of heat with a calibration resistor. The noise level of this device was lower than 0.5mW and the precision better than 3mW.

The experimentally measured heat Q consists of two contributions: the heat of the electrochemical reaction Q_R and JOULE heat Q_J

$$Q = Q_R + Q_J = \frac{I \cdot t}{zF} T \Delta_R S + R I^2 t \qquad (10)$$

By extrapolation of I→0 the reversible heat $T\Delta_R S$ can be obtained from experiments at different currents I, at constant I•t. The internal resistance R of the cell can be considered as invariable due to the high concentration of the supporting electrolyte.

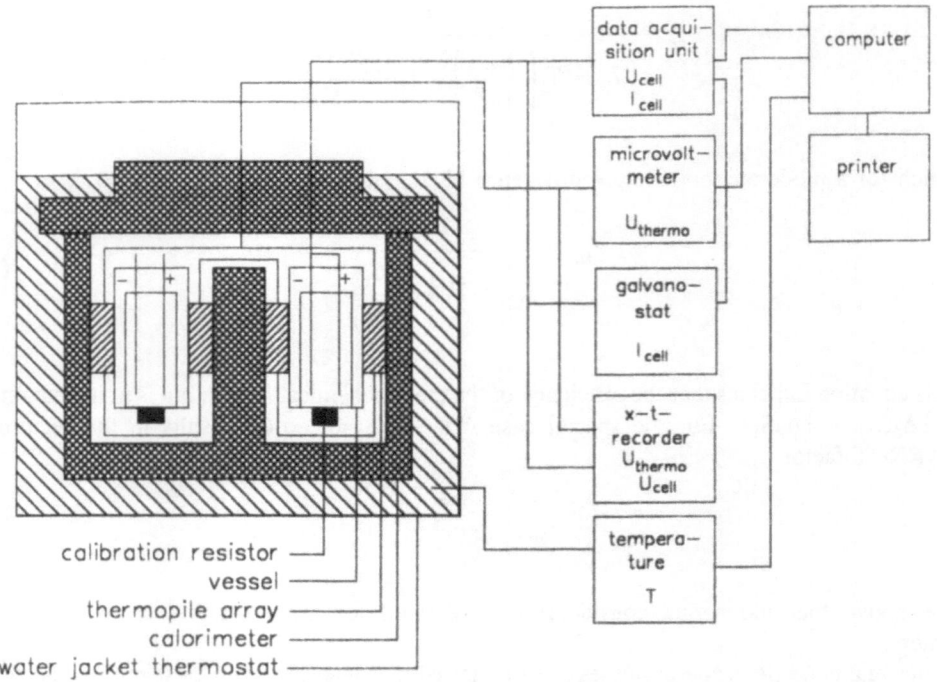

Figure 2. Experimental set-up for calorimetric measurements

Measurements of the different system parameters were carried out by a computer-controlled data acquisition unit (KEITHLEY S500).

Calorimetric investigations were made using commercial D-sized Ni/Cd accumulators (PANASONIC 1.2V/4Ah) and D-sized lead-acid cells (GATES 2V/2.5Ah) as well as non-commercial redox systems. Cells with two electrode compartments divided by an ion exchange membrane were used to measure the redox systems. The geometrical area of flat electrodes used was 12cm² and the electrolyte volume of each cell compartment was 3.6cm³.

RESULTS AND DISCUSSION

Commercial accumulators

At an early stage of the calorimetric measurements with lead-acid cells experimental results gave clear evidence that this type of accumulator is not suitable for electrochemical heat pump systems since the reversible heat exchanged by the system is masked by JOULE heat effects. Moreover, the thermodynamics of the system cannot be satisfactorily described due to the complexity of the reaction mechanism. Therefore, this system will not be further discussed here.

Ni/Cd accumulators appeared to be better candidates for electrochemical heat pumps.

Fig.3 shows thermodynamic data of the Ni/Cd accumulator.

The theoretical heat pump efficiency for the temperature difference of $\Delta T=35K$ is calculated to be $\varepsilon_{hp}=9.68$.

Calorimetric measurements with Ni/Cd accumulators were carried out at different charge and discharge currents I in the range 50mA $\leq$ I $\leq$ 300 mA. The reaction time was varied to keep the overall amount of charge constant (It=0.2Ah). The experimental results obtained at different temperatures $T_1=25°C$ and $T_2=60°C$ are presented in Fig.4.

$$2 \ Ni(OH)_2 \ + \ Cd(OH)_2 \ \underset{\underset{discharge \ T_2}{\rightleftharpoons}}{\overset{charge \ T_1}{}} \ 2 \ \beta-NiOOH \ + \ Cd \ + \ 2 \ H_2O$$

$$\Delta_R G \ = \ \Delta_R H \ - \ T\Delta_R S$$

$T_1 \ = \ 25 \ °C$	$T_2 \ = \ 60 \ °C$
$U_o \ = \ 1.287 \ V$	$U_o \ = \ -1.269 \ V$
$\Delta_R G \ = \ 248.27 \ kJ/mol$	$\Delta_R G \ = \ -244.97 \ kJ/mol$
$T\Delta_R S \ = \ 27.73 \ kJ/mol$	$T\Delta_R S \ = \ -31.95 \ kJ/mol$

electricity consumption : $\Delta_R G_1 \ + \ \Delta_R G_2 \ = \ 3.3 \ kJ/mol$
heat production at T_2 : $Q_2 \ = \ -31.95 \ kJ/mol$

theoretical efficiency : $\varepsilon_{hp} = 9.68$

Figure 3. Thermodynamic data of the Ni/Cd accumulator calculated for two different temperatures

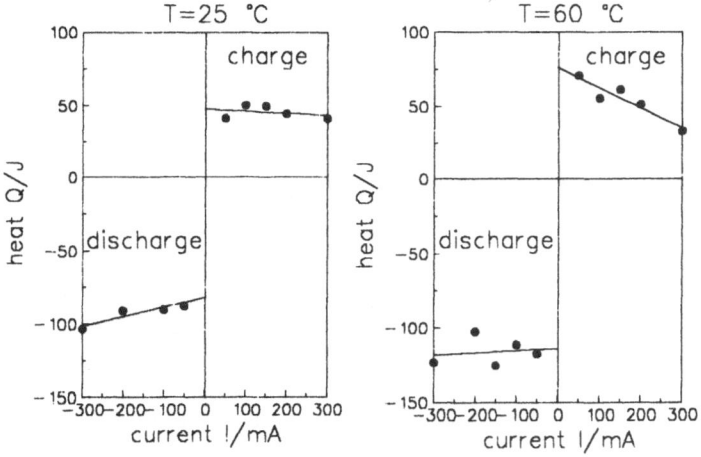

Figure 4. Heat of reaction of the Ni/Cd accumulator at different temperatures as a function of charging and discharging current I, with It=0.2Ah.

The reversible heat obtained from the extrapolation of the Q vs I plot at I→0 in Fig.4 is found to be in good agreement with the calculated values for discharge, whereas significant deviations are observed for charge of the accumulator (c.f. Tab.1).

Table 1. Heat of the reaction for charging and discharging of the Ni/Cd accumulator at different temperatures, theoretical values $T\Delta_R S$, experimental $Q_{(I \to 0)}$ recalculated from the results in Fig.4

Temperature	25°C		60°C	
	charge	discharge	charge	discharge
$T\Delta_R S$/kJ mol^{-1}	27.73	-27.73	31.95	-31.95
$Q_{(I \to 0)}$/kJ mol^{-1}	14.91	-22.49	20.44	-30.45

This deviation can be explained by the fact, that depending on the stage of charging, the oxygen evolution/ reduction cycle is activated as an additional source of heat production. Despite this, the calorimetric data can be used to calculate the efficiency of a heat pump system using Ni/Cd accumulators and taking into account the irreversible heat losses and overvoltages of a half cyclic operation.

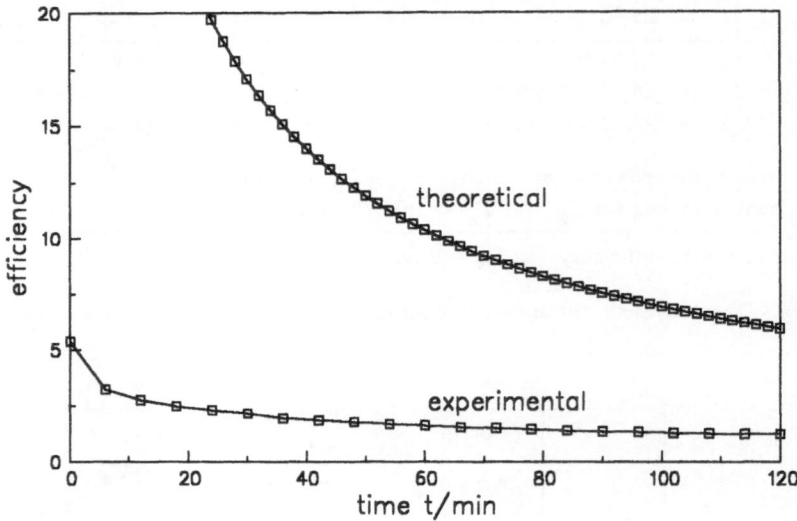

Figure 5. Theoretical and experimental efficiency of a Ni/Cd accumulator heat pump system ($T_1=25°C$, $T_2=60°C$, $I=0.1A$, $It=0.2Ah$)

Fig.5 shows the calculated efficiency as a function of time and, for comparative purposes, the theoretical efficiency calculated from the thermodynamic data taking into account the time dependence of concentration, charge transfer overvoltage and concentration polarisation as described in [3]. The results clearly indicate that the practical efficiency is much lower than the theoretical.

The practical efficiency exhibits acceptable values in the initial period of the experiment and then tends to lower levels of approximately $\varepsilon_{hp} \approx 2$, which is not acceptable for practical applications.

Non-commercial redox systems

The following redox couples Ag/Ag^+, Cu/Cu^{2+} and Fe^{2+}/Fe^{3+} have already been considered as possible candidates for the implementation of electrochemical heat pump systems by Darbyshire et al.[3] Based on theoretical calculations efficiencies ranging between $3.9 \leq \varepsilon_{hp} \leq 5.47$ were predicted.

To implement a practical system an appropriate combination of two redox couples must be found. The main problem is selecting the corresponding anions. Fig.6 shows combinations of different redox couples with corresponding anions.

From a number of combinations three systems appear to be most promising candidates for experimental purposes:

system 1: $Ag/AgNO_3$, HNO_3//membrane//$Fe(NO_3)_2$, $Fe(NO_3)_3$, HNO_3/Pt

system 2: $Cu/CuSO_4$, H_2SO_4//membrane//$FeSO_4$, $Fe_2(SO_4)_3$, H_2SO_4/Pt

system 3: $Cu/CuSO_4$, Na_2SO_4//membrane//$Na_3[Fe(CN)_6]$, $Na_4[Fe(CN)_6]$, Na_2SO_4/Pt

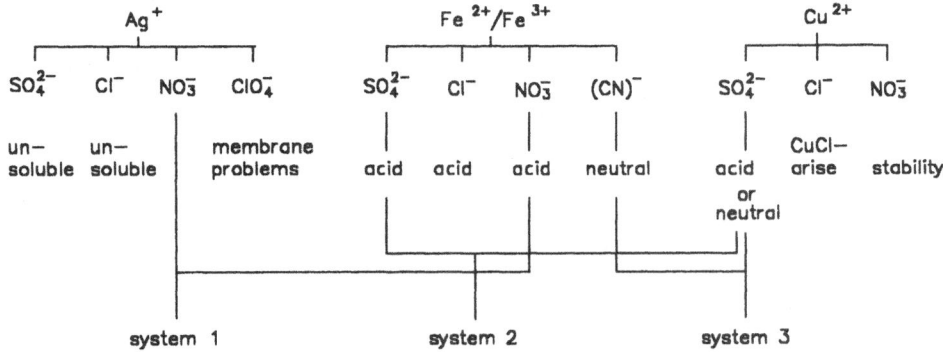

Figure 6. Possible combinations of different redox couples with corresponding anion for the formation of redox cell systems

Another problem requiring closer consideration is the correct choice of an ion exchange membrane, as it is essential to avoid mixing electrolyte solutions. For systems 1 and 2 anion exchange membranes with a very high anion selectivity are required. Unfortunately such membranes are not available at present. Moreover the anion exchange membrane has to sustain the strong oxidizing power of the Ag^+ ion.

System 3 needs a cation exchange membrane with high selectivity. Membranes of this type are in practical use in many industrial processes and are commercially available (eg NAFION®). In system 3, Na^+ ions of the supporting electrolyte are exchanged. System 3 was selected for calorimetric measurements and further investigations. The thermodynamic data of system 3 are shown in Fig.7.

$$Cu^{2+} + 2\ Na_4\ Fe(CN)_6 \underset{T_2}{\overset{T_1}{\rightleftharpoons}} Cu + 2\ Na_3\ Fe(CN)_6 + 2\ Na^+$$

$$\Delta_R G = \Delta_R H - T\Delta_R S$$

T_1 = 25 °C	T_2 = 60 °C
U_o = 0.028 V	U_o = 0.06 V
$\Delta_R G$ = 5.43 kJ/mol	$\Delta_R G$ = 11.61 kJ/mol
$T\Delta_R S$ = 144.72 kJ/mol	$T\Delta_R S$ = −162.62 kJ/mol

electricity consumption : $\Delta_R G_1 + \Delta_R G_2$ = 17.04 kJ/mol

heat production at T_2 : Q_2 = −162.62 kJ/mol

theoretical efficiency : ε_{hp} = 9.54

Figure 7. Thermodynamic data of the the redox cell heat pump (system 3) for two different temperatures

Despite the fact that the absolute value of the reversible heat, $T_2\Delta_R S=162$kJ/mol, is much higher than that of the Ni/Cd accumulator, $T\Delta_R S=31.95$kJ/mol, the value of the heat pump efficiency $\varepsilon_{hp}=9.54$, calculated for a temperature difference of $\Delta T=35$K, is approximately of the same magnitude as that of the Ni/Cd accumulator, $\varepsilon_{hp}=9.68$.

The calorimetric measurements of system 3 were performed at charge and discharge currents in the range $5mA \leq I \leq 20mA$. Analogous to the experiments with the Ni/Cd accumulator, the reaction time was varied to keep the overall amount of charge constant ($It=36As$). The results obtained at different temperatures $T_1=25°C$ and $T_2=60°C$ are plotted in Fig.8.

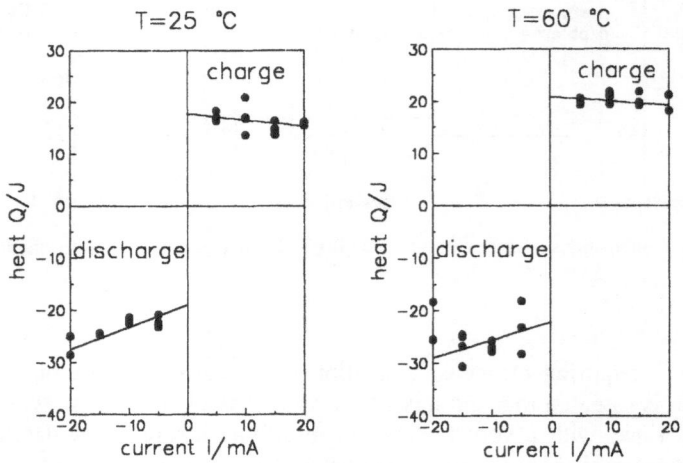

Figure 8. Heat of reaction of the redox cell (system 3) at different temperatures as a function of charging and discharging current I, with $It=36As$.

The scatter of the experimental data in Fig.8 can be attributed to the evaporation of small amounts of water from the non-hermetically closed cells, particularly at higher temperatures. The values of the reversible heat obtained from the extrapolation of the Q-vs I plots at $I \rightarrow 0$ together with theoretical values are summarized in Table 2.

Table 2. Heat of the reaction for charging and discharging the redox cells (system 3) at different temperatures, theoretical values $T\Delta_R S$, experimental $Q_{(I \rightarrow 0)}$ recalculated from the results in Fig.8

Temperature	25°C		60°C	
	charge	discharge	charge	discharge
$T\Delta_R S$/kJ mol^{-1}	144.72	-144.72	162.62	-162.62
$Q_{(I \rightarrow 0)}$/kJ mol^{-1}	95.8	-102.0	112.0	-118.9

The discrepancy between the extrapolated absolute Q values of charge and discharge found for the Ni/Cd accumulator (see Tab.1) obviously does not exist for redox system 3. This indicates reversible behaviour of the cell reaction.

Fig.9 contains the time dependence of the theoretical efficiency, calculated from thermodynamic data, and the experimentally determined efficiency, taking into account the influence of overvoltages.

As expected, the experimentally determined efficiency is lower than the theoretical efficiency. On average, the experimentally determined efficiency attains values of about $\varepsilon_{hp} \approx 4$, which is quite a reasonable value for practical applications.

The overall efficiency of such a system under continuous operation is further decreased due

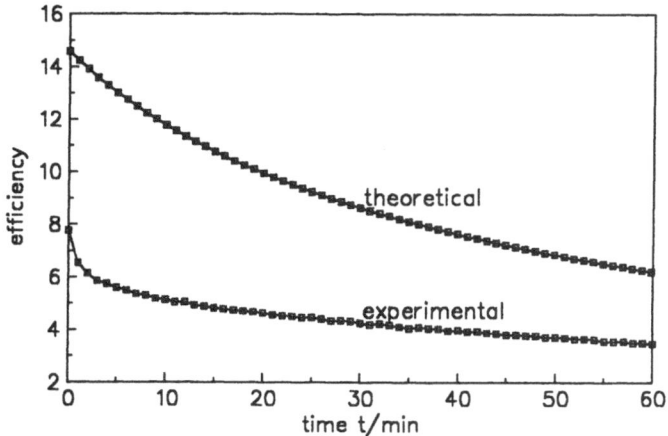

Figure 9. Theoretical and experimental efficiency of a redox cell heat pump (system 3), ($T_1=25$°C, $T_2=60$°C, $I=10$mA, $It=36$As)

to energy losses caused by the differential heat exchange which is necessary after each half cycle operation.

To obtain further information about practical efficiencies, a complete heat pump system, including differential heat exchange, is under construction and will be tested on a laboratory scale.

CONCLUSIONS

Experiments were carried out to test the new concept of an electrochemical heat pump system. The experimental data from calorimetric measurements on commercial Ni/Cd accumulators and non-commercial redox cells were found to be consistent with theoretical calculations. In general it was found that the experimentally determined efficiency is lower than the theoretical efficiency.

The redox system $Cu/CuSO_4$, Na_2SO_4//membrane//$Na_3[Fe(CN)_6]$, $Na_4[Fe(CN)_6]$, Na_2SO_4/Pt proved to be most promising for the implementation of an electrochemical heat pump as its practical efficiency is comparable to that of conventional heat pump systems.

ACKNOWLEDGEMENT

The authors are indebted to the Arbeitsgemeinschaft industrieller Forschungsvereinigungen AiF for financial support of this work under contract number AiF 8465.

REFERENCES

1.   J.N. Agar, "Advances in Electrochemical Engineering", Vol 3, Ch. 2, Intersience, London (1962).
2.   J.-M. Hornut and A. Storck, *J. appl. Electrochem.* (in press).
3.   G. Kreysa, G. F. Darbyshire, Theoretical Consideration of Electrochemical HeatPump Systems, *Electrochim. Acta* 35, 1283 (1990).

COMBINED ELECTROCHEMICAL AND DILATOMETRY
STUDIES ON METAL HYDRIDES

Z. Gavra, M. Abda, and Y. Oren

Chemistry Division
Nuclear Research Center - Negev
Beer-Sheva, PO Box 9001, 84190
Israel

ABSTRACT

The results of combined *in-situ* electrochemical and dilatometric studies made with palladium and a $Pd_{0.85}Ni_{0.15}$ alloy are presented. The metals were charged and discharged electrochemically with hydrogen under constant current conditions. Electrode length changes were measured under normal atmospheric pressure at different charging rates and as a function of hydrogen content. Structural changes were followed by SEM observations. The results are compared to those obtained with hydrogen charging from the gaseous phase. The technique could be used to study other hydrogen absorbing electrodes such as Ti, V, Nb, Zr and intermetallic compounds of the types AB_2 and AB_5 considered for use in metal hydride batteries.

INTRODUCTION

A large variety of hydrogen absorbing metals and intermetallic compounds has been considered for the purpose of storage, purification and gettering of hydrogen. A recent application of metal hydrides is in rechargeable, "environmentally clean", batteries where hydrogen is the charge conveying component.

A major phenomenon which is common to all metal hydrides is the large dimensional changes accompanied by the hydrogen absorption and desorption processes. This is due to the transition of the pure metal to the α phase followed by the formation of the β, hydride phase in which the lattice parameter is changed significantly. For instance, it was found[1] that for palladium, the lattice parameter a changes from a_{pure}=3.89A to a_{α} =3.894A and then to

a_β =4.025A. Accordingly, the relative volume change per unit concentration change of hydrogen in palladium in the range H/Pd 0 to 0.8, was found to be 0.19[1]. With other alloys, as for example from the AB_2 and the AB_5 categories[2], which are recommended for use in hydride batteries, the dimensional changes are much larger and can result in complete disintegration of the metallic structure. Moreover, absorption of hydrogen is also accompanied by dramatic changes in the electrical resistivity of the metals. For palladium, the relative resistivity was found to increase 1.8 times when loaded to H/Pd=0.8 at room temperature[3a]. In metals where dimensional changes do not exceed several percent, as in palladium and its alloys, resistivity changes result mainly from changes in the electronic conductive band. In other metals, where much larger dimensional changes are detected, the destruction of the metallic structure and continuity can result in a complete loss of electrical conductivity.

These changes affect both the mechanical stability and the electrical properties of the hydrides and these phenomena are of great importance from the practical view point when battery construction and hydrogen storage devices are considered.

Simultaneous measurements of electrochemical and dilatometric as well as other properties have the advantage of *in situ* studies of structural and electronic changes within the metal due to loading with hydrogen. Geerken et al.[4] applied these techniques to Pd and $Pd_{80}Si_{20}$. However, only cyclic voltammetry was used, therefore, high degrees of charging and maximum dimensional changes could not be obtained. In this study the results of dimensional changes of $Pd_{0.85}Ni_{0.15}$ alloy due to galvanostatic hydrogen charging and discharging in the entire hydrogen concentration range are presented. This alloy is known for its high chemical and physical stability. It has been studied extensively in the gas phase where the absence of hysteresis in the charging/discharging loop was indicated[5] and the $\alpha+\beta$ region equilibrium pressures were much larger as compared to those for pure palladium[6].

EXPERIMENTAL

A standard Metrhom three electrode electrochemical cell was used for the electrochemical charging/discharging studies of Pd and $Pd_{0.85}Ni_{0.15}$ (prepared in the BNL Hydride Lab.). It contained a platinum auxiliary electrode, SCE as a reference and palladium or $Pd_{0.85}Ni_{0.15}$ wires, 0.005" in diameter and approximately 15 mm long as working electrodes. The cell was thermostated by a Hakke thermostat in the range 5 - 60 °C. The 1M KOH solution was constantly purged with pure (99.999%) argon to prevent any contact with oxygen during the experiments. Prior to each charge, surface oxides were removed from the working electrodes by immersion in 1:1 V/V HNO_3 for 30 minutes.

Charging-discharging cycles under constant current conditions were applied by an EG&G Model 273 Potentiostat/Galvanostat. The current was interrupted for periods of 40 seconds during which the electrode rest potential vs. SCE was monitored. Each cycle took an overnight period. Discharging was terminated when the rest potential became positive with respect to SCE. Rest potentials, potentials during charging and charge were monitored by a digital Nicolet 4049 oscilloscope.

The simultaneous measurements of dimensional, potential and charge changes were conducted in the cell shown schematically in Figure 1.

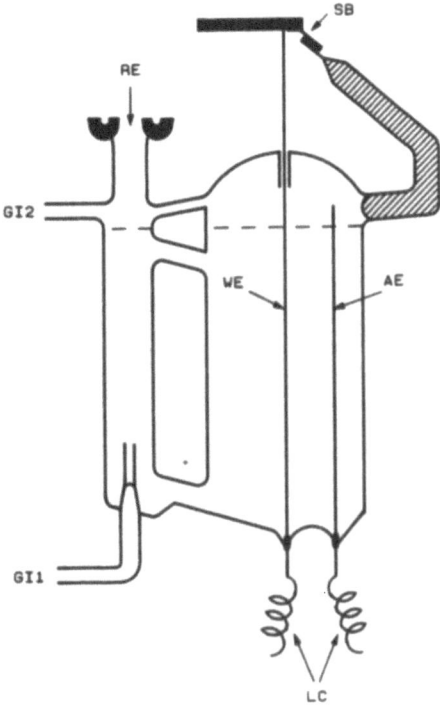

Figure 1. Electrochemical / Dilatometric Cell. WE- Working electrode; AE- Auxiliary electrode; LC-Electrode leads; RE1- Reference electrode; SB- Steel blade; GI1, GI2- Purging gas inlets.

In this cell, SCE was the reference and a porous carbon slab, the auxiliary electrode. The working electrode was made of a 0.6 cm Pd or $Pd_{0.85}Ni_{0.15}$, 0.005" wire, spot welded to two pieces of platinum wire isolated from the solution by shrinkable Teflon tubings. The end of one platinum wire was anchored to the bottom of the cell and the end of the other one to a thin flexible steel blade fixed to a side arm stemming from the top of the cell.

The steel blade kept the working electrode under constant stress thus allowing monitoring any longitudinal change with the tip of a Millitron No. 1306 inductive gauging head placed on it.

RESULTS AND DISCUSSION

The results of long term charging and discharging for both Pd and Pd/Ni alloy at different temperatures are summarized in Table 1. Two points should be emphasized: first, the number of coulombs delivered to the electrodes during the charge period is much larger than that consumed during discharge to the point where the end of the process is indicated by a sharp potential change towards positive values with respect to SCE. This low coulombic charging efficiency results from a larger cathodic H_2 production rate as compared to the penetration rate of H atoms to the metal. The second point is that for both metals, the H/Metal values as calculated from the discharge process, are significantly larger as compared to those obtained from gas phase charging under 1 atm equilibrium pressure and in the same temperature range[3,6]. This phenomenon is most pronounced with the $Pd_{0.85}Ni_{0.15}$ alloy where gas phase charging would allow values which are tenfold lower under similar conditions. This large discrepancy was also found by Luo et al.[7] and shows that electrochemical charging is much more intensive as compared to gas phase charging. The mechanistic explanation for this phenomenon is probably related to the mechanism of the cathodic hydrogen evolution where H atoms are involved. However a more detailed discussion of this issue is beyond the scope of this paper.

Table 1. Results of long term 0.005 A cm^{-2} constant current charging and discharging of Pd and $Pd_{0.85}Ni_{0.15}$.

Temperature °C	Q_{charge} Coulombs	$Q_{discharge}$ Coulombs	H/Metal Discharge	H/Metal Gas Phase
Pd				
6	7.67	1.65	0.72	≈ 0.6
26	11.56	1.63	0.71	≈ 0.6
56	9.51	1.72	0.75	≈ 0.6
$Pd_{0.85}Ni_{0.15}$				
7	6.52	1.64	0.50	<0.05
26	9.79	1.73	0.52	<0.05
56	8.55	1.46	0.44	<0.05

Figure 2 shows typical galvanostatic charging/discharging potential and longitudinal variation patterns of a $Pd_{0.85}Ni_{0.15}$ wire. Charging is accompanied by a 3.5% length increase while the potential remains almost constant. Discharge results in a recovery of the original length while the electrode potential is typified by two regions: a slightly inclining negative potential of the hydrogen loaded metal and a positive potential of the completely discharged metal. Note the coinciding times of the negative-to-positive potential change and the end of the electrode contraction.

The influence of the charging/discharging current on the maximum relative longitudinal change and the charging capacity is shown in Table 2.

Table 2. Relative length changes (RLC) and maximum loading capacities at various charging/discharging currents.

Current Density, A/cm^2	$Pd_{0.85}Ni_{0.15}$ H/Metal	$Pd_{0.85}Ni_{0.15}$ RLC	Pd H/Metal	Pd RLC
0.0042	0.60	0.036	0.9	0.043
0.021	0.58	0.035	0.79	0.040
0.042	0.63	0.037	0.72	0.035

The striking feature shown in Table 2 is the dependency of both RLC and maximum charging capacity on the current for palladium and their constancy for the $Pd_{0.85}Ni_{0.15}$ alloy. At this stage, no clear explanation is available for this behaviour. More work has to be done to show the influence of thermal treatment on dimensional changes of the two metals.

In order to further elucidate the influence of the charging/discharging current on the extent of dimensional changes in the α and $\alpha+\beta$ regions separately, a series of short duration charging experiments was conducted in which the electrode was loaded by applying different current densities to the same charge corresponding to each of the two regions. In Figure 3 , the RLC per charge unit for $Pd_{0.85}Ni_{0.15}$, is shown as a function of the charging current density for the two different phases. The RLC per unit charge for the α region increases with the current below 0.02 mA cm^{-2} and decreases at higher current densities. However, in the

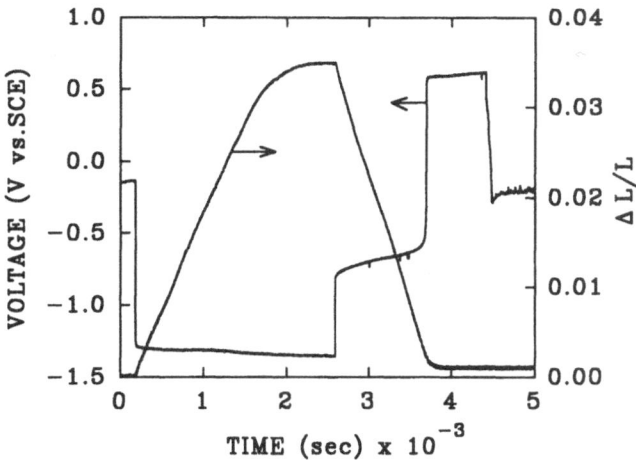

Figure 2. Typical voltage and relative length change as a function of time for $Pd_{0.85}Ni_{0.15}$ at current density 0.021 A cm^{-2}.

$\alpha+\beta$ region, this parameter is essentially constant. These results show that, at least in the range of current densities applied in this study, inefficiencies such as eg. hydrogen atom recombination at the metal surface, are strongly involved in the α phase region. In the $\alpha+\beta$ region, where hydrogen diffusion is faster and its interaction with the metal is stronger[3b], it is likely that less inefficiencies are involved in the charging process and each charge unit contributes the same length change regardless of the rate at which it is supplied to the metal. This is also the reason for the larger contribution of each charge unit to dimensional changes at high current densities (> 0.21 A cm^{-2}) in the $\alpha+\beta$ region as compared to that in the α region.

CONCLUSIONS

A. Electrochemical charging provides an excellent and easy tool for *in situ* studies of the influence of hydrogen loading on the physicochemical properties of metal hydrides.

B. Charge inefficiencies influence dimensional changes of $Pd_{0.85}Ni_{0.15}$ in the α region more than in the $\alpha+\beta$ region. At high current densities each charge unit contributes to the dimensional changes in the $\alpha+\beta$ region more than in the α region.

C. At the same current density, dimensional changes for palladium are larger than those for $Pd_{0.85}Ni_{0.15}$ but the alloy can be charged with hydrogen to a much larger extent compared to gas phase charging.

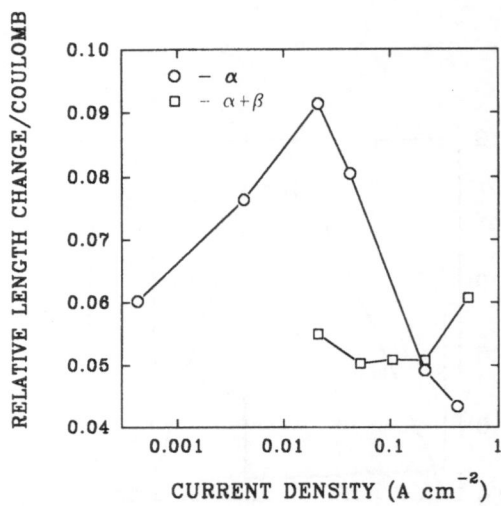

Figure 3. Relative length change per unit charge as a function of current density for $Pd_{0.85}Ni_{0.15}$.

REFERENCES

1. H. Peisl, "Lattice Strains due to Hydrogen in Metals", in "Topics in Applied Physics", Vol. 28, pp. 53-74, G. Alefeld and J. Volkl Eds., Springer-Verlag, Berlin, 1978.

2. J.J.G. Willems, "Metal Hydride Electrodes Stability of $LaNi_5$-Related Compounds", Phillips J. of Research , 39(1), (1984).

3. F.A. Lewis, "The Palladium Hydrogen System", Academic Press, 1967, a). p. 55; b). p. 109.

4. B.M. Geerken, I.A.M. Corbiere and R. Griessen, "A Simultaneous Voltammetric and Dilatometric Study of Hydrogen Absorption in Pd and Amorphous $Pd_{80}Si_{20}$", *J. Phys. Chem. Solids*, 43, 373 (1982).

5. H. Noh, T.B. Flanagan, Z. Gavra, J.R. Johnson, J.J. Reilly, "The Disappearance of Hysteresis for the Hydrid Phase Transition in Palladium-Nickel Alloys", *Scripta Metallurgica et Materialia*, 25, 2177 (1991).

6. Z. Gavra, J.R. Johnson, J.J. Reilly, "Decomposition Kinetics of Palladiun-Nickel Hydrid", J. *Less Common Metals* 172-174, 107 (1991).

7. W. Luo, J.R. Johnson, S.W. Feldberg, J.J. Reilly, "Properties and Discharge Kinetics of an Activated $Pd_{0.85}Ni_{0.15}$ Hydrid Electrode", Paper presented at the Metal Hydrides Gordon Research Conference, July 1993.

GALVANIC DEPOSITION OF PHOTOVOLTAIC

COMPOUND SEMICONDUCTOR COATINGS

D. Becker-Roes, A. Fischer, M. Schimmel and H. Wendt

Technische Hochschule Darmstadt
Institut f. Chem. Technologie
Petersenstrasse 20, D- 64287 Darmstadt, Germany

ABSTRACT

The advantages of galvanic deposition of semiconductors in the production of solar cells are described and possible mechanisms of film growth are discussed. As typical examples fit for the galvanic formation of semiconducting coatings Cu(I) sulfide and Cu(I) selenide have been investigated. The anodic formation of copper (I) sulfide layers composed of relatively large twinned crystals by anodic oxidation of copper in electrolytes containing sulfide anions can be achieved by using high process temperatures and relatively low current densities. The galvanically deposited semiconductor layers are characterised by impedance spectroscopy (copper selenide) and current voltage curves of photovoltaic cells ($Cu_2S/CudS$). Copper (I) selenide layers are p-type doped with a dopant density of $8.43 \ 10^{15} \ cm^{-3}$. The photovoltaic characteristics of Cu_2S/CdS cells are still disappointing as the diode behaviour of the cell is still poor due to a partial short circuit of the diode of unknown origin.

INTRODUCTION

Compound Semiconductor and their Galvanic Deposition

Semiconducting II-VI compounds are of great interest for the production of photovoltaic devices. The high production costs of solar cells based on crystalline silicon are one reason to the detriment of photovoltaics in comparison with alternative energy winning systems like wind and hydropower. The present investigation in thin film technology has been undertaken for two reasons:
(i) to reduce material costs it is necessary to lower the material input. This is done by producing "thin films".
(ii) The electrochemical pathway is a good possibility for lowering production costs and providing good photovoltaic quality of thin film materials on large scale surfaces.

Electrochemical Engineering and Energy, Edited by
F. Lapicque *et al.*, Plenum Press, New York, 1995

In the late 70ies and early 80ies there had been reported a number of investigations on galvanic deposition of compound semiconductors [1]. All these investigations ended with the frustating results, that the semiconductors could either be used as photoelectrodes but were unstable or if converted to photovoltaic cells, gave only poor performance. It is therefore worthwhile to resume the former work with the intention to establish galvanic deposition conditions by which semiconductor layers of higher quality are obtained.

Physical Data of Silicon and Compound Semiconductors

Table 1 shows some physical data of typical direct semiconductor compounds compared with the indirect semiconductor silicon. The high absorption coefficient and the short diffusion length demonstrate that it is unnecessary to prepare thick, semiconducting films. This kind of absorber material needs only several μm to absorb more than 90% of incident light [2]. An absorber of crystallized silicon, in constrast, needs more than 200 μm. The up-to-date technology for thin layer semiconductor preparation is the expensive vacuum sputtering or vacuum evaporation deposition. Galvanic deposition of thin layers of compound semiconductors has the potential of remarkable cost savings.

Table 1. Some physical data on semiconducting materials for photovoltaic devices.

	Thin Film materials (a-Si, Cu_2S, CdS)	c-silicon
Absorption coefficient	10^{-5} cm^{-1}	10^{-3} cm^{-1}
Absorption depth	3 - 4 μm	100 - 200 μm
Diffusion length	2 - 3 μm	50 - 100 μm
Carrier diffusion	0.6 ns	10 μs

Comparison of Silicon and Compound Semiconductors for Photovoltaic Cells

Table 2 compares the different routes for the production of photovoltaic elements. The encapsulated Czochalski process is the first production step in monocrystalline silicon production. In a second expensive step the second crystal must be cut into thin wafers. Much single crystal material is lost in this process. The wafer must be doped and backcontacted to generate an MS-sequence. (MS=metal/semiconductor).

The electrochemical pathway deposits the silicon directly onto the chosen subtrate and an external process followed by deposition of a metal grid results in formation of the MS-sequence. For the electrochemical thin film technology based on semi-conductor compounds like copper (I) sulfide, copper (I) selenide or copper-indium-diselenide the production of an MS-sequence is simpler. In one step the deposited semiconductor film may be doped by a nonstoichiometric codeposition of the dopant elements.

In this manner, it is possible to integrate the doping process into the deposition step by controlling the system parameters: deposition current, working potential and concentration of dopants in the galvanic bath. This process would lead to low material and process costs with the possibility of easy scale up.

Table 2. Minimizing the number of production steps by electrochemical preparation of semiconductor layers.

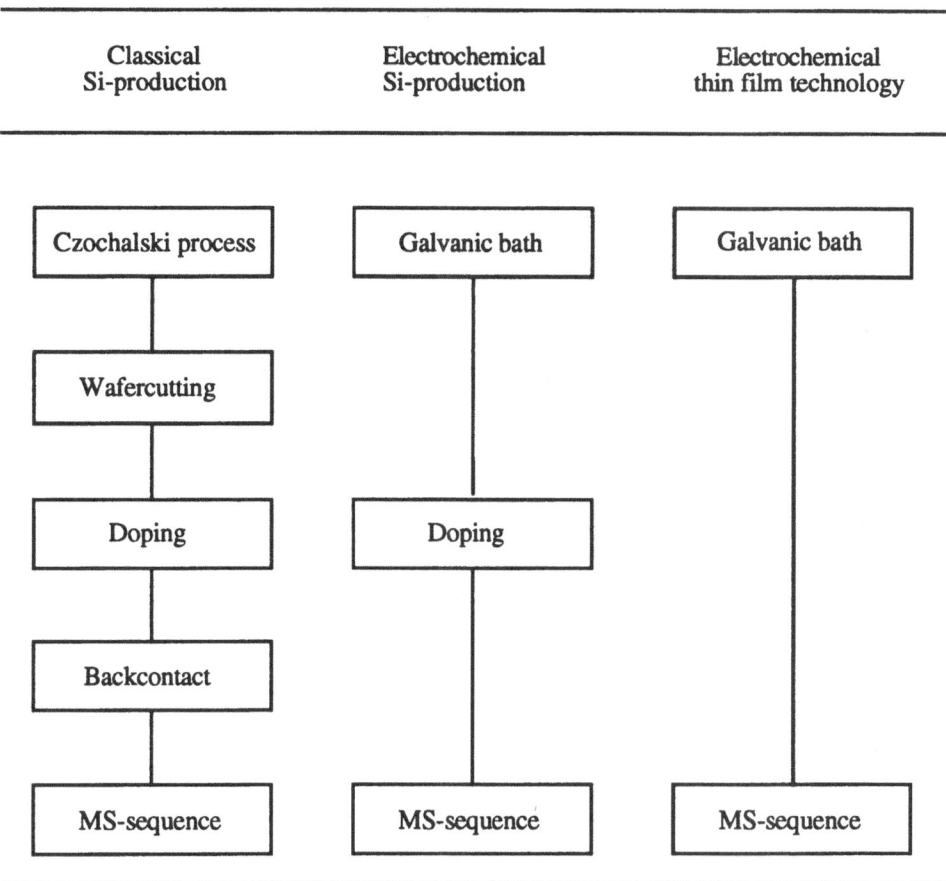

ELECTROCHEMICAL DEPOSITION TECHNIQUES

Cathodic deposition of silicon

$$K_2SiF_6 + 4e^- \longrightarrow 2\,KF + Si^0 + 4\,F^- \qquad (1)$$

The cathodic deposition of silicon from silicon (IV) compounds was carried out in a FLINAK-melt by Rao et al [3] with (LiF, NaF, KF). At 750°C under nitrogen or argon atmosphere, graphite, as a substrate material, is stable under these conditions.

Unfortunately, to assure high coating quality, the current density must be kept below 10 mA cm^{-2} and, consequently, production times for 100 μm coatings exceed a day. Further, due to a mismatch of thermal expansion coefficients of the carbon substrate and the deposited silicon, cracks are formed in the coating by cooling of the coated specimen from 750°C to ambient temperature. These effects render galvanic deposition of Si-coatings practically impossible.

Preparation of Copper-I-Sulfide Coatings by Anodic Oxidation of Copper in the Presence of Sulfide Ions

Compound semiconductors can be deposited galvanically either by anodic oxidation or cathodic reduction. Cathodic reduction and anodic oxidation, however, can only be performed, if the semiconductor has a relatively high intrinsic electronic conductivity. Copper (I) sulfide being a degenerate p-type semiconductor exhibits almost metallic-like conductivity. As shown in equations (2) and (3), copper (I) sulfide (Chalcosite) can be deposited on copper by anodic oxidation of the metal substrate in contact with an electrolyte containing sulfide ions [4,5].

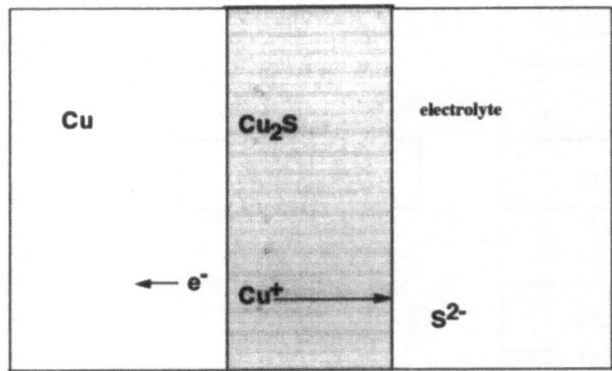

Figure 1. Schematic of the anodic formation of copper (I) sulfide by anodic copper oxidation and precipitation of the sulfide.

As depicted schematically in Figure 1, this reaction consists essentially of three steps:
1. Oxidation of copper to copper (I),
2. Diffusion (not migration) of copper (I) from the metal to the electrolyte through the closed copper (I) sulfide layer,
3. Precipitation and crystallisation of copper (I) sulfide at the chalcogenite/electrolyte interface.

$$2\,Cu^0 - 2e \longrightarrow 2\,Cu^+ \qquad\qquad\qquad (2)$$
$$2\,Cu^+ + S^{2-} \longrightarrow Cu_2S \qquad\qquad\qquad (3)$$

The main problem in electroforming Cu_2S-films consists in obtaining sufficiently pure coatings composed of large crystals as pair annihilation at crystal boundaries and impurity sites must be avoided. Crystal growth rather than formation of new crystals, must be favoured by appropriate choice of process parameters.

The deposition of copper (I) sulfide can be performed at enhanced temperatures at current densities ranging from 0.3 to 0.5 $\mu A\ cm^{-2}$ which allow deposition of a 5 μm in three hours. Figure 2 shows the surface of such a copper (I) sulfide coating with large crystals which are characterized by multiple twinning. High-quality surfaces can only be obtained if the rate of crystal nucleation is by orders of magnitude lower than the rate of crsytal growth, which is accomplished by low current densities and relatively high process temperatures. Crystals with approximately 20-30 μm diameter are obtained. It is even possible to deposit CdS galvanostatically or by electroless reaction of thiourea with Cd-II-cations on these thin coatings, establishing the necessary p-n junction. Although photocurrent yields of the Cu_2S/CdS cell are quite good, the open cell voltages are still very poor.

Figure 2. Surface of copper sulfide coating on copper obtained by anodic oxidation of copper in S^{2-} containing electrolyte; i =3 10^{-4} A cm^{-2}, T=80°C.

Electroless Deposition of Copper (I) Selenide on Copper

In a bath containing seleocyanate anions a copper foil becomes covered by Cu_2Se; the copper oxidation is accompanied by stoichiometric reduction of the selenium donor. Precipitating Cu_2Se layers on copper, according to equation (4) to (6), proceeds without applying external current.

The process is essentially a corrosion process -see Figure 3 - and the coating continues to grow due to the same reaction -diffusion scheme which has been established for copper (I) sulfide.

$$2\,Cu^0 - 2\,e^- \longrightarrow 2\,Cu^+ \qquad\qquad \text{oxidation} \qquad (4)$$

$$SeCN^- + 2\,e^- \longrightarrow Se^{2-} + CN^- \qquad\qquad \text{reduction} \qquad (5)$$

$$2\,Cu^0 + SeCN^- \longrightarrow 2\,Cu_2Se + CN^- \qquad \text{corrosion} \qquad (6)$$

Figure 4 gives an impression of the size of the crystrals of a 5μm thick Cu_2Se layer obtained by electroless depositing the compound semiconductor at 80°C from 0.1 M SeCN$^-$ solutions.

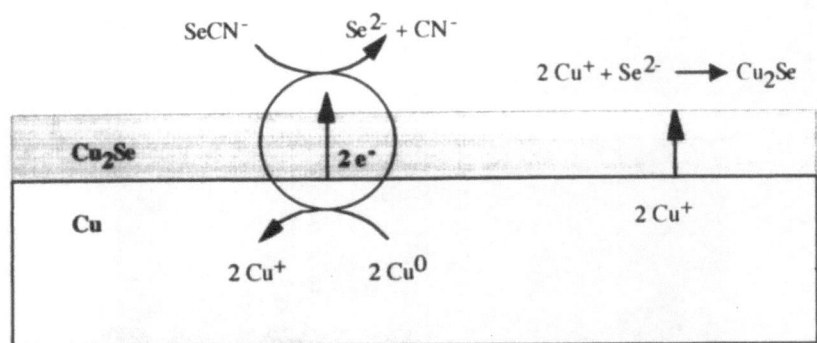

$SeCN^-$ $Se^{2-} + CN^-$

$2\,Cu^+ + Se^{2-} \longrightarrow Cu_2Se$

Cu_2Se

$2e^-$

Cu

$2\,Cu^+$

$2\,Cu^+$ $2\,Cu^0$

Figure 3. Reaction scheme of electroless deposition of copper (I) selenide.

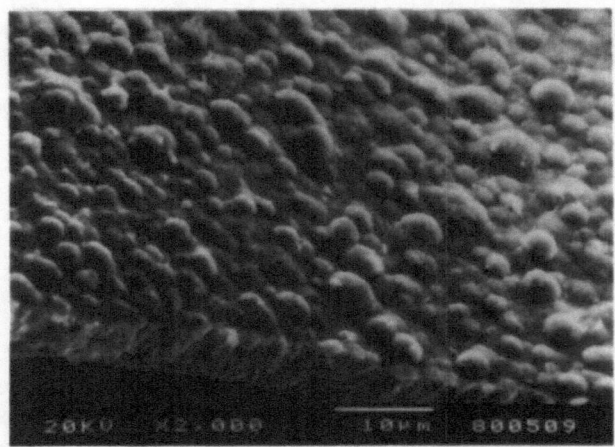

Figure 4. Cu(I) selenide obtained by electroless deposition.

CHARACTERIZATION OF DEPOSITED THIN FILMS

Impedance Spectroscopy

Impedance spectroscopy is a powerful, electrochemical tool for investigating the system "semiconductor/electrolyte". The potential dependent capacitance of the space charge layer in the semiconductor is measured in the potential range from 1V to -1V vs. sat. Ag/AgCl in aqueous electrolyte. The solution contains 2 mM potassium chromate and hydrogen phosphate - hydroxide buffer to keep the pH at a stable value of 9. The Mott-Schottky plot in Figure 5 shows the linear response of the inverse capacitance with the electrode potential. From the slope of the linear correlation one obtains the value of the charge carrier density, which amounts without illumination to 8.43×10^{15} cm^{-3}. The negative slope is characteristic for p-type doping.

For the illuminated semiconductor electrode the majority carrier concentration - in this case the defect electron concentration - is increased in the same space charge region by the photovoltaic effect by more than one order of magnitude as the slope of the Mott-Schottky plot decreases by this factor.

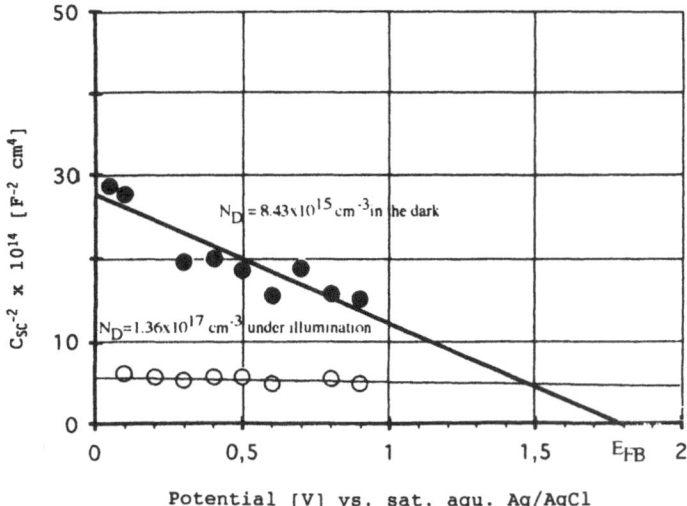

Figure 5. The Mott-Schottky plot of the p-type copper (I) selenide electrode.

Current/Voltage Curves of Photovoltaic Cells

Cu_2S/CdS cells were prepared from Cu_2S-coated copper foils by chemical deposition of CdS. Electroless, i.e. chemical bath deposition of CdS, was achieved by deposition on the coated anodes in solution containing $CdSO_4$ and thiourea. After sputtering of a semitransparent 20 nm gold layer a measurable photoeffect could be observed.

Figure 6 shows as the best example obtained for an ensemble of five cells, the current voltage curve in the dark and under illumination. Under illumination 100 mV open cell voltage - compared to the expected 500 mV - and a limiting current density of 3 mA cm^{-2} were measured. This is a promising result.

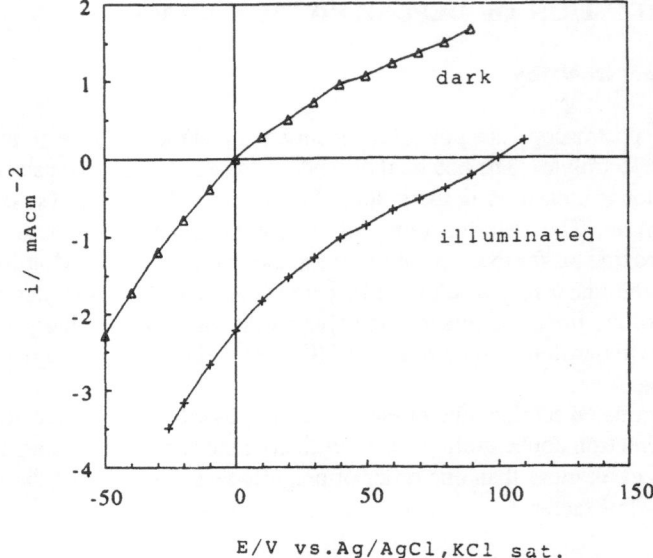

E/V vs.Ag/AgCl,KCl sat.

Figure 6. Current-voltage curve of a Cu_2S/CdS photovoltaic cell composed of anodically generated copper sulfide and a vapour deposited CdS-layer.

There exist numerous measures for improving the cell behaviour. One important issue would be to improve the purity of the copper from which the copper sulfide is formed in order to obtain a higher copper (I)-sulfide purity. Another issue would be to increase the thickness of the CdS-layer and avoid pinholes in it which would short-circuit the depletion layer in the n-CdS. Also the quality of the current collecting semitransparent gold layer must be improved.

Acknowledgements

The authors wish to thank Dipl.-Ing. R. Hoeg (Institute of Chemical Technology Darmstadt) for their helpful collaboration. The authors are indebted to the Ministry of Science and Technology of the state of Hessen - HMWT - for financial support of this work.

REFERENCES

1. G.F. Fulop and R.M. Taylor, Electrodeposition of semiconductors, *Annu. Rev. Mat. Sci.* 15:197 (1985).
2. D. Bonnet, Stromerzeugung durch Solarzellen, in "Technologie -Monitor Solarenergie und Wasserstofftechnik", chap. 1, 3-134, D. Bonnet , W. Fuchs and W. Hoffmann, ed., Batelle Institut, Frankfurt (1990).
3. G.M. Rao, D. Elwell and R.S. Feigelson, Electrowinning of silicon from K_2SiF_6-molten fluoride systems, *J. Electrochem. Soc.* 127:1940 (1980).
4. D. Vasquez Moll, R.G. DeChialvo and A.J. Arvia, Corrosion and passivity of copper in solution containing sodium sulfide, *Electrochim. Acta* 30:1011 (1985).
5. A. Etienne, Electrochemical method to measure the copper ionic diffusivity in a copper sulfide scale, *J. Electrochem. Soc.* 117:870 (1970).

ENERGY CORRELATION OF MASS TRANSFER IN DECAYING ANNULAR SWIRL FLOW

S.YAPICI[1], M.A.PATRICK[2], and A.A.WRAGG[2]

[1]Atatürk Universitesi, Müh. Fak. Kimya Müh. Böl., 25240 Erzerum, TÜRKİYE

[2]School of Engineering, North Park Road, University of Exeter, Exeter, EX4 4QF, U.K.

ABSTRACT

Mass transfer measurements at both the inner and outer walls in decaying swirl flow in an annular duct have been carried out using the electrochemical limiting diffusion current technique. Swirl was generated using axial guide-vanes with vane angles between 15-67° to the duct axis. Pressure measurements at the outer wall were also made to enable estimation of the energy cost of mass transfer enhancement. Overall correlation of mass transfer at the outer wall for a representative length of test section in terms of the energy required to impart swirl to the flow can be expressed as $Sh\,Sc^{-1/3} = 0.0251\,(1 + \tan\,\theta_w)^{-0.18}X^{0.316}$ for the Reynolds number range 3000-50000, where θ_w is the vane angle at the wall and X is an energy dissipation parameter.

INTRODUCTION

Enhancement of transfer rates in convective processes is obtained at the expense of energy dissipated by extra friction caused by non-smooth surfaces and insertions, or by any additional external power. The use of passively-generated, decaying swirl flow is a plausible technique for the enhancement of convective transfer rates in terms of energy cost. Decaying swirl flow is routinely used in a number of engineering applications such as heat exchangers[1] and for improving electrochemical metal removal from dilute solutions[2] . In order to assess the efficient use of the energy required to initiate fluid rotation, researchers have looked for a relationship between mass/heat transfer coefficient and the average mechanical power dissipated per unit mass of the fluid (Ibl[3] and Midoux and Storck[4]). The aim of the present paper is to contribute further to this field by carrying out an energy analysis for decaying annular swirl flow. The detailed approach to the measurements of electrochemical mass transfer rate and pressure drop in decaying annular swirl flow can be found in Yapici[5,6,7].

The mechanical power dissipated in the flow system can be written as

$$P = Q\Delta p \tag{1}$$

and the mechanical power dissipated per unit mass of fluid as

$$\varepsilon = \frac{U \Delta p}{\rho L} \qquad (2)$$

Expressing U in terms of the Reynolds number $Re = d_e U / \nu$ and using the Fanning equation for pressure drop

$$\Delta p = f \frac{L}{d_e} \rho \frac{U^2}{2} \qquad (3)$$

leads to the following

$$\frac{\varepsilon d_e^4}{\nu^3} = \frac{f}{2} Re^3 = X \qquad (4)$$

where X is an energy dissipation parameter. In the literature, the traditional forms of the correlating equations for momentum and mass transfer are respectively

$$f = a \, Re^b \qquad (5)$$

$$Sh \, Sc^{-1/3} = a' Re^{b'} \qquad (6)$$

Using equations (4), (5) and (6), the following general expression can be written to relate transfer rates to the energy required to overcome fluid friction

$$Sh \, Sc^{-1/3} = a' \left\{ \frac{2X}{a} \right\}^{\frac{b'}{(3+b)}} \qquad (7)$$

RESULTS AND DISCUSSION

Energy Correlation

The presure drop across the test section was measured using tappings in the outer wall. For the chosen test section length of approximately 50 equivalent diameters, the total pressure drop, i.e. the sum of that across the test section and that across the swirl generator, for decaying swirl flow is shown in Figure 1 plotted against mean axial velocity. Such data can be correlated in terms of a modified friction factor f_s in the Fanning equation (3) (Ward-Smith[8]), where f_s can be expressed for $15° < \theta_w < 67°$ as

$$f_s = 0.3673 \, Re_s^{-0.28} \, (1 + tan \, \theta_w)^{2.32} \qquad (8)$$

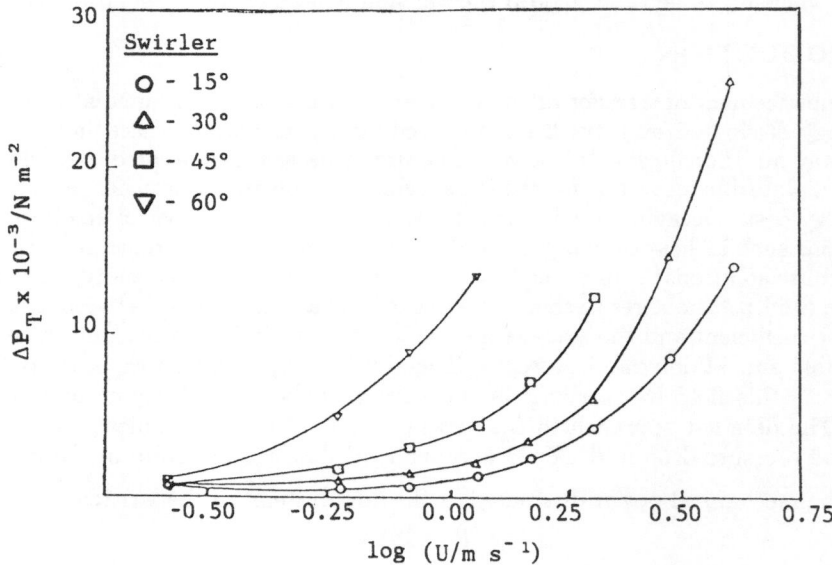

Figure 1 - Total pressure drop in test section including swirler

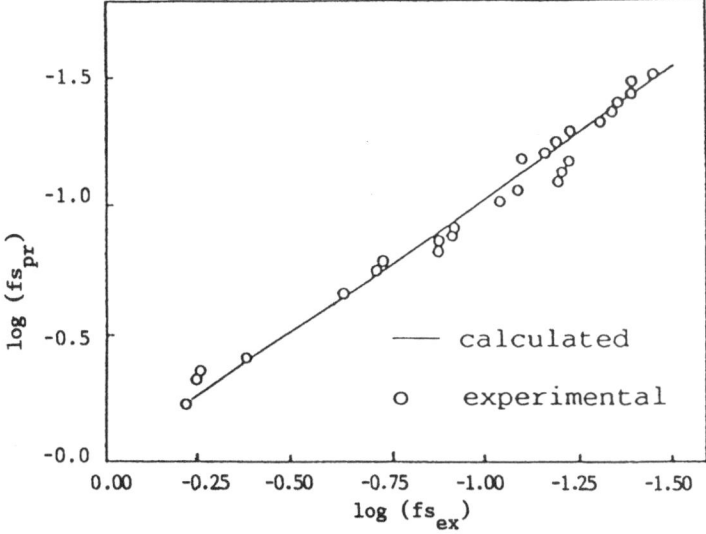

Figure 2 - Comparison of experimental and calculated friction factors

Figure 2 shows that the experimental values and those calculated from the above friction factor correlation equation agree with a relative mean square error of 1.89%.

For the representative length of 50 equivalent diameters of the test section the following mass transfer correlations for decaying annular swirl flow at the outer and the inner wall, respectively, are obtained (Yapici et al ($\underline{7}$))

$$Sh_{si}Sc^{-1/3} = 0.0133\,Re_s^{0.86}(1 + tan\,\theta_w)^{0.53} \tag{9}$$

$$Sh_{so}Sc^{-1/3} = 0.0147\,Re_s^{0.86}(1 + tan\,\theta_w)^{0.55} \tag{10}$$

Noting the general forms of equations (5) to (7), since $a' = 0.0133(1 + tan\,\theta_w)^{0.53}$ for the mass transfer at the inner wall and $a' = 0.0147(1 + tan\,\theta_w)^{0.55}$ at the outer wall,

Table 1 - Energy correlations for duct systems.

System	Correlation Equation	Conditions	Reference
Annulus (inner wall)	$Sh\,Sc^{-1/3} = 0.0227\,X^{0.316}(1 + tan\,\theta_w)^{-0.20}$	$1.8 \times 10^9 < X < 3.7 \times 10^{13}$ $L/d_e < 50$	Present study
Annulus (outer wall)	$Sh\,Sc^{-1/3} = 0.0251\,X^{0.316}(1 + tan\,\theta_w)^{-0.18}$	$1.8 \times 10^9 < X < 3.7 \times 10^{13}$ $L/d_e < 50$	Present study
Parallelpiped channel	$Sh\,Sc^{-1/3} = 0.0204\,X^{0.318}$	$X > 1.2 \times 10^9$	Lin et al (14) + Blasius Eq.
Square duct	$Sh\,Sc^{-1/3} = 0.0277\,X^{0.32}$	$1.2 \times 10^9 < X < 1.5 \times 10^{13}$	Dawson and Trass (15)
Tube	$Sh\,Sc^{-1/3} = 0.0256\,X^{0.321}$		Loughlin et al (16)

and $b' = 0.86$ from equations (9) and (10) at both walls, and $a = 0.3673(1+tan\,\theta_w)^{2.32}$ and $b = -0.28$ from equation (8), the energy correlations for mass transfer in annular decaying swirl flow for the representative length of 50 equivalent diameters can be expressed

$$Sh_{si}Sc^{-1/3} = 0.0227\,(1 + tan\,\theta_w)^{-0.20}\,X^{0.316} \qquad (11)$$

for the inner wall and

$$Sh_{so}Sc^{-1/3} = 0.0251\,(1 + tan\,\theta_w)^{-0.18}\,X^{0.316} \qquad (12)$$

for the outer wall.

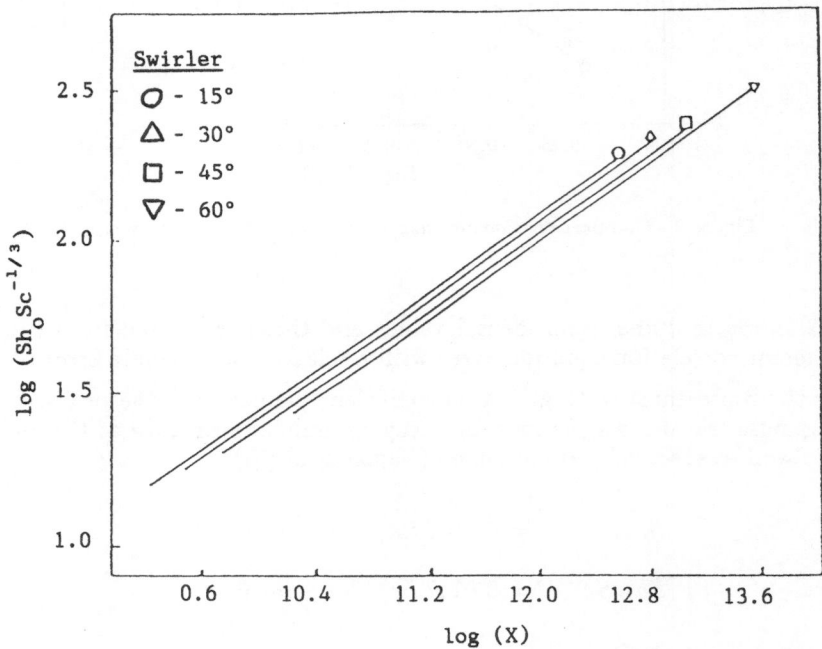

Figure 3 - Comparison of mass transfer energy efficiencies of swirlers.

Some energy correlations for duct systems taken from the literature are given in Table 1. The main coefficients and the exponents on the energy dissipation parameter, X, are closely similar for each system. This indicates that the energy efficiency for each of the arrangements is almost equal.

The comparison of energy efficiencies for different swirlers is shown in Figure 3. The lower vane angle swirlers are more energy effective than those having higher angles and Figure 4 also shows that increase in mass transfer at the outer wall takes place more effectively than at the inner wall, from an energy point of view.

Enhancement Efficiency

For constant pumping power, it is useful to compare the effectiveness of mass transfer enhancement in comparison with the fully developed, no swirl, annular flow condition, such that

$$Q_s \Delta p_s = Q_a \Delta p_a \qquad (13)$$

The enhancement ratio for the process can be expressed as

$$\eta = \frac{k_s}{k_a}\bigg|_{pumping\ power} \qquad (14)$$

For developed annular flow, The Fanning equation (3) can be used with a new friction factor, f_a.

The friction factor for flow without swirl in a simple circular pipe at moderate Reynolds number, is given by the Blasius formula

$$f = 0.316 Re^{-0.25} \tag{15}$$

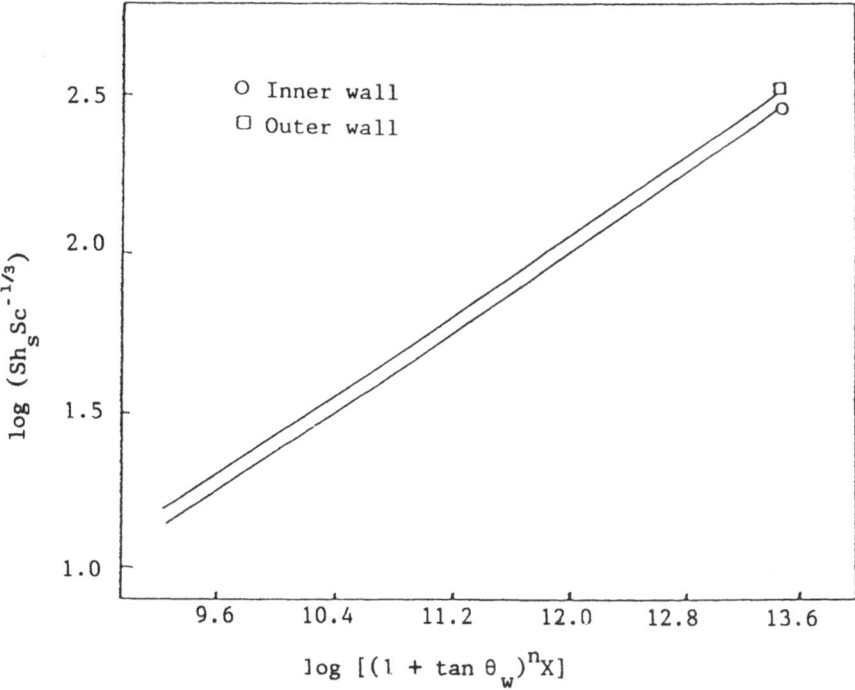

Figure 4 - Comparison of energy efficiencies at the inner and outer walls.

For annular systems, friction factors are approximately 1.05 of those given by the above equation[9]

$$f_a = 1.05 f \tag{16}$$

when based on the equivalent annular pipe diameter $(d_o - d_i)$. Figure 5 shows the good agreement between the measured values of Δp_a obtained using pressure taps on the outer wall and the calculated values of 1.05 times the values from the Blasius formula for a pipe length of $L/d_e \approx 50$ for flows without swirl.

The Reynolds number for flow with no swirl giving rise to the same power consumption as for a flow with swirl can be calculated using equations (3), with friction factors f_s and f_a, (8), (13), (15), and (16). The following expression is obtained

$$Re_a = 1.0378 (1 + tan\,\theta_w)^{0.84} Re_s^{0.99} \tag{17}$$

The mass transfer correlations for fully developed annular flow without swirl, as obtained by Yapici[5], were

$$Sh_{ai} Sc^{-1/3} = 0.0250\, Re_a^{0.80} \tag{18}$$

$$Sh_{ao} Sc^{-1/3} = 0.0126\, Re_a^{0.88} \tag{19}$$

These best fit correlations give plots which lie close to one another and give no firm indication as to whether the mass transfer at the outer wall is higher or lower than that at the inner wall. The character of turbulent annular flow has been discussed by Brighton and Jones[10] who made measurements of velocity profile and shear stress and found that, for an annulus ratio less than 0.56, the radial location of maximum velocity (which is also the location of the point of zero shear stress) always occurs closer to the inner wall, and moves nearer the inner wall as the annulus ratio decreases.

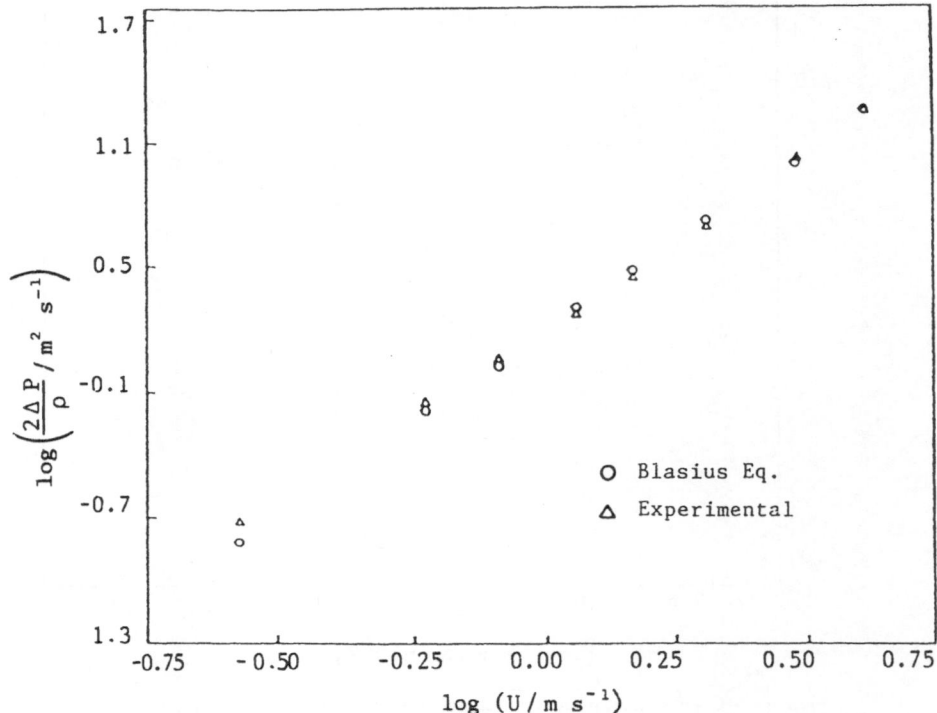

Figure 5 - Pressure drop in test section for fully developed axial flow.

Employing equations (9), (17) and (18), the following expression can be written for the mass transfer enhancement ratio, for the representative length of 50 equivalent diameters, at the inner wall

$$\eta_i = \left. \frac{k_s}{k_a} \right|_P = \left(\frac{Sh_{si}}{Sh_{ai}} \right)_P = 0.5155 \, Re^{0.07} (1 + tan \, \theta_w)^{-0.15} \qquad (20)$$

In the same way, the following expression is obtained for the outer wall from equations (10), (17) and (19)

$$\eta_o = \left. \frac{k_s}{k_a} \right|_P = \left(\frac{Sh_{so}}{Sh_{ao}} \right)_P = 1.1221 \, Re^{-0.01} (1 + tan \, \theta_w)^{-0.19} \qquad (21)$$

As seen from Figures 6 and 7, the mass transfer enhancement ratio increases as the vane angle decreases. Enhancements in mass transfer are between 0.76-1.06 for the inner wall and 0.80-0.98 for the outer pipe when normalised in terms of equivalent power consumption. This suggests that using decaying annular swirl flow generated by axial-vane type swirlers is not an attractive prospect in terms of energy saving. This can be partly explained by the low efficiency of axial-vane type swirlers in producing swirling flow (Beer and Chigier[11]). However, figures 8 and 9, which are plots of the ratio of mass transfer coefficients in swirling annular flow to those in fully developed plain annular flow, show that a substantial increase over the average fully developed mass transfer coefficient can be obtained for the representative length chosen at a given Reynolds number by the use of swirl. This enhancement increases with increase

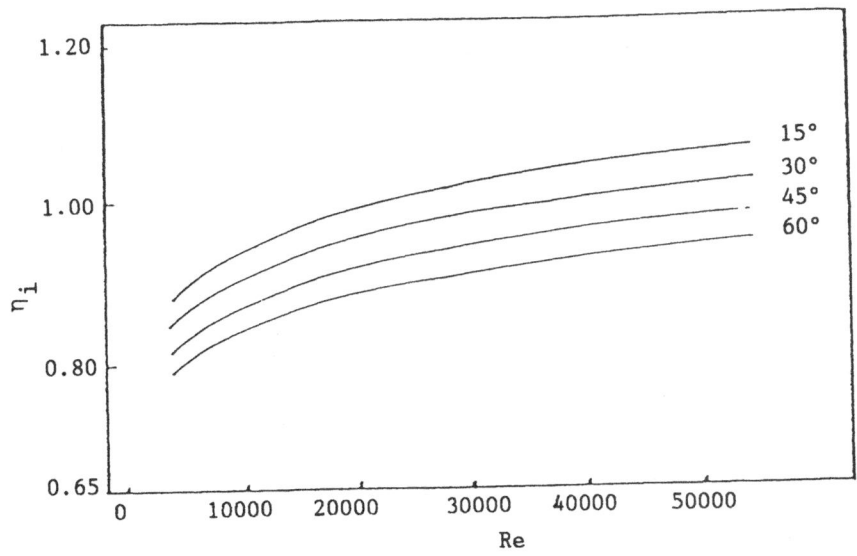

Figure 6 - Inner wall enhancement ratio versus Reynolds number.

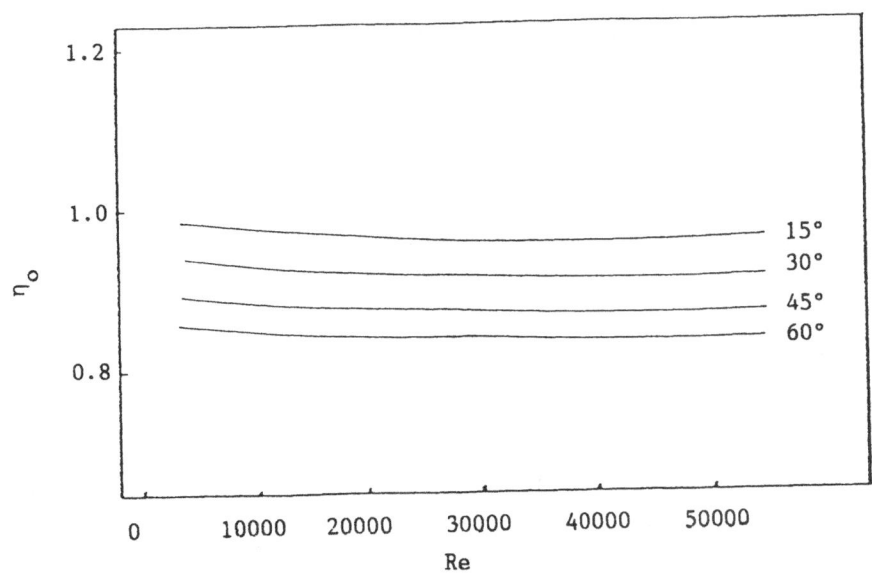

Figure 7 - Outer wall enhancement ratio versus Reynolds number.

in swirler vane angle. While not energy efficient, this may give valuable opportunities in terms of up-rating the performance of existing equipment. Finally, it is interesting to note the way in which the enhancement is affected by Reynolds number; it increases as Reynolds number decreases for the inner wall, and exhibits the opposite but much less pronounced behaviour for the outer wall. In the literature, Ivanova[12] and Koval'nogov and Shchukin[13] have reported enhancement ratios, for heat transfer in swirling pipe flow, on the basis of identical energy consumption for pumping, as 1.03-1.21 for a range of initial swirl angles 20-75° and for $L/d_e \approx 100$ at Re=100000, and as 1.12-1.25 for a range of initial swirl angles 15-75° and for $L/d_e \approx 60$ at Re=10000-90000, respectively.

The reason for these higher ratios compared with the present work is related to the geometry of the swirl generators used for the above studies. While in the work of Ivanova, it is not clear whether the energy required to rotate the swirl generator

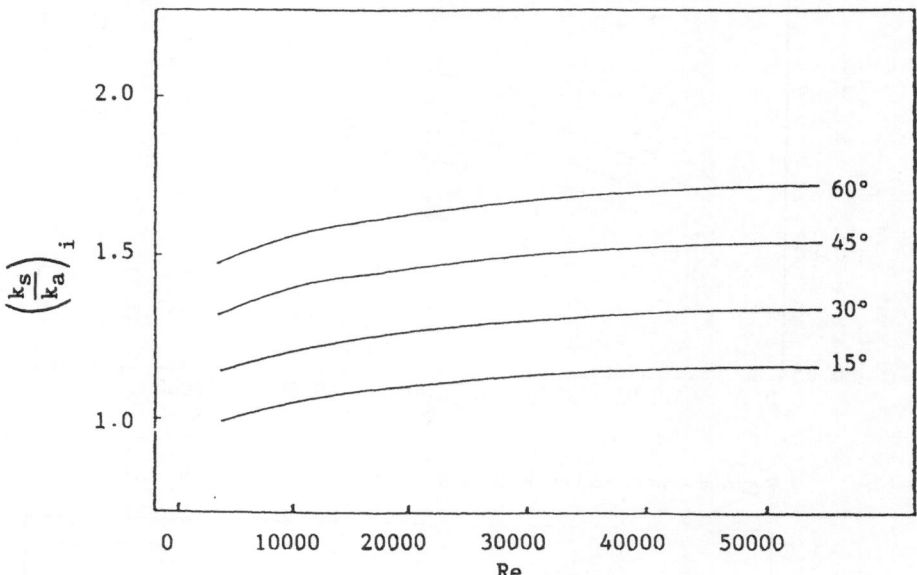

Figure 8 - Inner wall mass transfer enhancement versus Reynolds number.

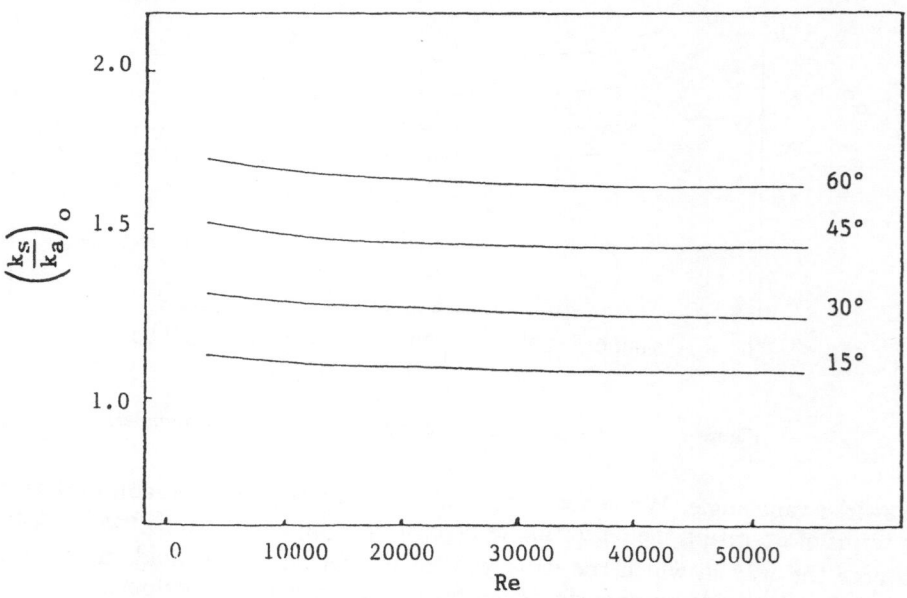

Figure 9 - Outer wall mass transfer enhancement versus Reynolds number.

was included in the calculation, Koval'nogov and Schukin employed swirl generators having vanes with a length of $r/2$ in the radial direction. Since the flow was thus not accelerated from U to $U/\cos\theta$ across the whole cross section, the pressure drop across the swirl generators was reduced. In addition, in the above studies, smooth pipes were used rather than annular channels and therefore the pressure drop over the test section length was inherently lower.

CONCLUSIONS

Using the electrochemical limiting diffusion current technique, mass transfer measurements have been obtained at both the inner and outer walls in decaying annular swirl flow, generated using axial guide-vanes. The effectiveness of mass transfer in relation to overall energy consumption was correlated for a representative test section length of 50 equivalent diameters, using pressure drop measurements at the outer wall and mass transfer measurements at both walls. It was found that from an energy point of view the use of axial-vane swirlers gives disappointing results in terms of enhancement of mass transfer in annular duct flow. However, the work provides a useful framework for the up-rating of capacity of existing equipment by the use of swirl.

NOMENCLATURE

a, a'	Coefficients		P	Mechanical power / $\mathrm{J\,s^{-1}}$
b, b'	Coefficients		p	Pressure / Pa
D_{AB}	Diffusivity		Q	Volumetric flow rate / $\mathrm{m^3\,s^{-1}}$
d	Diameter / m		Δp	Pressure drop / Pa
f	Friction factor		U	Mean velocity / $\mathrm{m\,s^{-1}}$
k	Mass transfer coeff. / $\mathrm{m\,s^{-1}}$		X	Energy efficiency parameter
L	Axial length / m			

Greek Letters

ε	Energy dissipation / $\mathrm{W\,kg^{-1}}$		μ	Dynamic viscosity / Pa s
η	Enhancement ratio		ν	Kinematic viscosity / $\mathrm{m^2\,s^{-1}}$
θ	Vane angle		ρ	Density / $\mathrm{kg\,m^{-3}}$

Subscripts

a	Fully developed annular		o	Outer wall
e	Equivalent		s	Swirl
i	Inner wall		T	Total
L	Axial length		w	Wall

Dimensionless Groups

Re	Reynolds no. (Ud_e/ν)		Sc	Schmidt no. (ν/D_{AB})
Sh	Sherwood no. (kd_e/D_{AB})			

REFERENCES

1 R.F.Lopina and A.E.Bergles, *J.Heat Transfer* 91:434 (1969).

2 F.C.Walsh and G.Wilson, *Trans. IMF* 64:55 (1986).

3 N.Ibl, Convective mass transport., *in:* "Comprehensive Treatise of Electrochemistry, Vol.6," Plenum Press, New-York (1983).

4 N.Midoux and A.Storck, Electrochemical reactors, *in:* "Fluid Transport in Electrochemical Reactor Systems," M.I.Ismail ed., Elsevier, Amsterdam (1989).

5   S.Yapici, "Electrochemical Mass Transfer in Annular Swirl Flow," PhD Thesis, University of Exeter, Exeter, U.K. (1992).

6   S.Yapici, M.A.Patrick and A.A.Wragg, *Int. Comm. Heat Mass Transfer* 21:41 (1994).

7   S.Yapici, M.A.Patrick and A.A.Wragg, Proc. 3rd Int. Workshop on Electrodiffusion Diagnostics of Flows, Dourdan, France, 275 (1993).

8   A.J.Ward-Smith, Internal fluid flow, *in:* "The Fluid Dynamics of Flow in Pipes and Ducts," Clarendon Press, Oxford (1980).

9   K.Rehme, *J. Fluid Mech.* 64:263 (1974).

10 J.A.Brighton and J.B.Jones, *J. Basic Engng.* 86:835 (1964).

11 J.M.Beer and N.A.Chigier, "Combustion Aerodynamics," Applied Science Publishers Ltd., London (1972).

12 A.Ivanova, Procs. Second All-Soviet Union Conf. on Heat and Mass Transfer, Minsk, Vol.1, 243-250 (1964).

13 A.F.Koval'nogov and K.Shchukin, *J. Engineering Physics* 14:239 (1968).

14 C.S.Lin, R.W.Moulton and G.L.Putnam, *Ind. Eng. Chem.* 45:636 (1933).

15 D.A.Dawson and O.Trass, Int. *J. Heat Mass Transfer* 15:1317 (1983).

16 K.F.Loughlin, A.A.Hayamel and L.C.Thomas, *AIChE J.* 31:1614 (1985).

USE OF METALLIC FOAMS IN ELECTROCHEMICAL REACTORS OF FILTER-PRESS TYPE: MASS TRANSFER AND FLOW VISUALIZATION

A. Montillet[1], J. Legrand[1], J. Comiti[1], M.M. Letord[2] and J.M. Jud[2]

[1]Laboratoire de Génie des Procédés
I.U.T. - B.P. 420 - 44606 Saint-Nazaire Cedex - France
[2]E.D.F./D.E.R. - Les Renardières - B.P. n°1
77250 Moret sur Loing - France

ABSTRACT

This article deals with the study of a filter-press type electrolyser (ElectroSynCell of ElectroCell AB) with an undivided configuration obtained by replacing, for the working electrode, the usual inert turbulence promoter by a three-dimensional nickel foam electrode. The mass transfer rates are investigated at the working electrode and at the counter-electrode, which consists of a plane plate covered by a plastic net. The measured mass transfer coefficients are discussed with respect to the available literature data.

The different residence time distribution (R.T.D.) behaviours of the tested configurations (empty channel, channel with plastic nets, channel filled with metallic foam) are explained by use of a visualization technique in the reaction zone.

INTRODUCTION

Optimising filter-press type electrolysers is of major importance in electrochemistry. Electrolytic treatment of dilute solutions or recycling of ionic mediators used in indirect electrosynthesis requires cells characterized by high mass transfer rates, large specific surface areas and small pressure losses[1,2]. Recently, three-dimensional electrodes such as reticulated media, expanded metals or fine mesh grids have been widely tested and used in filter-press type cells. The knowledge of structure of, flow in, and mass transfer to three-dimensional electrodes is insufficient to predict their performances in electrolysers. Previous work has outlined the importance of cell design, i. e. electrode dimensions, inlet and outlet areas. Then, experiment is a necessary step. The main purpose of the present study is to test an undivided configuration, consisting of a three-dimensional electrode coupled with a plane plate electrode, in the ElectroSyncell of Electrocell AB.

Electrochemical Engineering and Energy, Edited by
F. Lapicque *et al.*, Plenum Press, New York, 1995

Membrane-partitioned configurations are generally preferred to undivided ones because they allow a large range of applications including electrodialysis. Working in the undivided mode is, for the instance, limited to simple ionic reactions involved in recycling operations ; its advantage consists in using a single hydraulic circuit. This paper focuses on mass transfer and flow visualization in the reaction zone.

MASS TRANSFER

Scope of the study

This work aims at quantifying mass transfer to the two electrodes of a representative unit cell (R.U.C.) without a membrane. Its principle is based on the imbalance between the surface areas of the electrodes. The working electrode consists of a plane plate with a sheet of foam and the counter-electrode consists of a plane plate with a plastic net turbulence promoter. The ratio of their surface areas is equal to 15.

In the reaction zone, the flow is limited to a single channel. As shown previously[3], the residence time distribution in a R.UC. consisting of plane plates with sheets of foam is radically different from that occuring in a R.U.C. consisting of plane plates with classical turbulence promoters.

In the literature dealing with mass transfer in filter-press type electrochemical cells, two groups of studies can be distinguished :
- those concerning mass transfer to plane plates with turbulence promoters or mass transfer to three-dimensional electrodes made of turbulence promoter type materials (expanded metals, fine mesh grids...),
- those concerning mass transfer to reticulated media : metallic foams, reticulated vitreous carbon (R.V.C.) foams, graphite foams.

Advantages and drawbacks of using turbulence promoter type materials or reticulated media have already been discussed[2,4,5]. The best ratio of specific surface area to generated pressure losses is given by reticulated media whereas turbulence promoters are reputed to enhance mass transfer coefficient and to give a uniform current distribution. To our knowledge, only one study deals with current distribution in metallic foams[6], but the undivided mode was not tested. With the presently studied configuration, a comparison of mass transfer correlations obtained at each electrode with previous data from the literature may indicate a possible interaction between the turbulence promoter and the three-dimensional electrode.

Experimental details

The Electrosyncell is provided with a single R.U.C.. The working electrode consists of a 2.5 mm thickness sheet of foam (grade : 60 pores per pouce, porosity $\varepsilon=0.98$, specific surface area per volume of electrode $A=6400 \text{ m}^{-1}$) bonded to a plane plate. The counter-electrode is a nickel plate covered with a polymeric net from Electrocell AB (diamond mesh, $\varepsilon =0.89$). The length of the reaction zone is 0.28m, the cross sectional area for the electrolyte flow is 5x140 mm.

The mass transfer coefficient between the liquid and the electrode is obtained using the cathodic reduction of ferricyanide ions to ferrocyanide ions. The limiting diffusion current, I, is measured as a function of the flow rate Q_v. The procedure consists in cleaning the electrodes in situ with a 0.1 M solution of sulphuric acid and in cathodically activating the working electrode in a 0.5 M solution of sodium hydroxide (a current of 5 A is applied for one hour). Then, the electrolyte, a mixture of 10^{-3} M $K_3Fe(CN)_6$ and 0.1 M $K_4Fe(Cn)_6$ in 0.5 M NaOH, is introduced to the cell and continuously recirculated at 30°C.

In the reaction zone, the channel electrolyte superficial velocity, U_0, is varied between 0.028 and 0.21 m s^{-1}. Thus the Reynolds number, Re, based on the equivalent hydraulic diameter of the channel, d_h, lies between 300 and 2300.

Calculation of the mass transfer coefficient k_d

The average value of the flow rate in the three-dimensional electrode made of foam and that of the flow rate in the left part of the channel are considered as equal, as shown in a previous pressure drop study[3]. Under limiting diffusion current conditions, the cathodic reduction of ferricyanide ions is a first order reaction.

The mass transfer coefficient at the working electrode was calculated by using the results given by the residence time distribution[7], which has allowed description of the flow in the "channel" of the working electrode as a cascade of 6 continuous stirred tank reactors (C.S.T.R.) of identical volume. For this cascade, the fractional conversion of ferricyanide ions per pass, assuming a faradic yield of one, is :

$$X = 1 - \frac{C_{out}}{C_{in}} = 1 - \left[1 + k_d\, A\, \varepsilon^{-1}\, \tau\right]^{-6} = \frac{I}{n\, F\, Q_v\, C_{in}} \tag{1}$$

The experimental value of X lies between 20 and 40%.

The mass transfer coefficient k_d is deduced from eq. (1) :

$$k_d = \frac{(1 - X)^{-1/6} - 1}{A\, \tau}\, \varepsilon \tag{2}$$

The experimental variation of k_d versus U_0 is given in Table 1.

The mass transfer at the counter-electrode was obtained under conditions of mass transport control. This electrode is operating at the limiting current I, which may be expressed as :

$$I = n\, F\, k_d\, S\, C \tag{3}$$

k_d is calculated using equation (3), assuming a constant value of the bulk concentration C. The correlation k_d versus U_0 is given in Table 1.

Table 1 . experimental variation of k_d versus U_0

plane plate + active sheet of foam ($A = 6800$ m^{-1})	$k_d = 7.3 \times 10^{-5}\, U_0^{0.61}$ (m s^{-1}) (cascade of 6 C.S.T.R.)
plane plate + inert plastic net	$k_d = 1.1 \times 10^{-4}\, U_0^{0.57}$ (m s^{-1})

The values of the exponent of U_0, in the correlations given in Table 1, indicate that the flow regimes in each part of the channel are the same. The flow regime is turbulent. The mass transfer coefficient to the plane plate is greater than the mass transfer to the electrode consisting of the foam and the plate.

In order to take into account the specific surface area of the foam, a comparison of the product k_d A obtained for various foams of same grade (G60) and for stacked nets of fine mesh is given in fig. 1.

Corresponding correlations are given in Table 2. The correlation obtained with nickel foam G60 lies in the range of those previously obtained with RVC or nickel foams in laboratory cells. The exponent of U_0 in the present correlation is greater. This fact may be partly due to the electrolyser entrance effects and to the proximity of the turbulence promoter. The product k_d A obtained with the stacked nets of fine mesh is particularly high for large values of U_0, but the drawback of using this three-dimensional electrode is the high dissipation of mechanical energy.

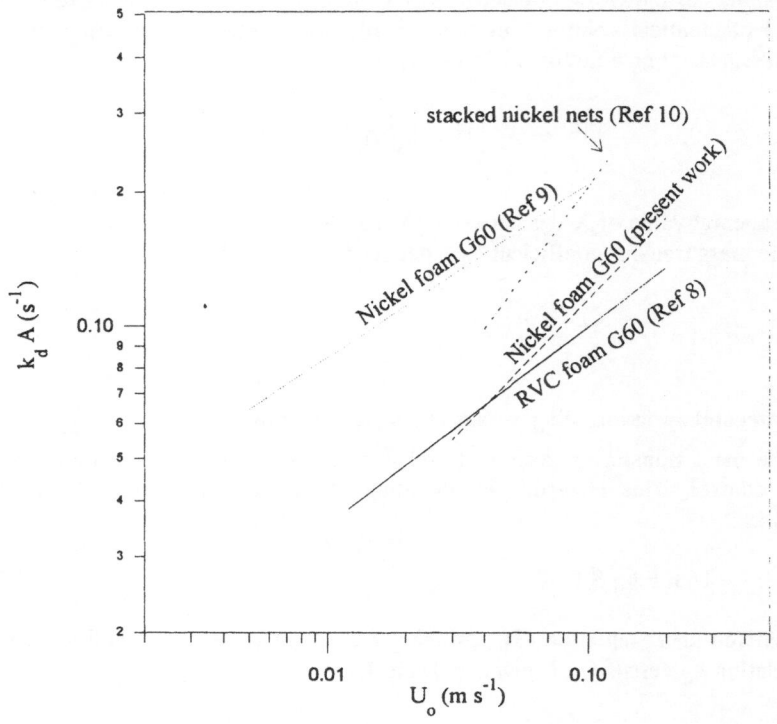

Figure 1 . variation of k_d A versus U_0 for nickel and R.V.C. foams of grade 60 and stacked nickel nets of fine mesh.

Table 2 . comparison of the product k_d A versus U_0 for foams G60 and stacked nets.

Ref. / working electrode	U_0 range (m s^{-1})	k_d A (s^{-1})
(8) RVC foam G60	0.02 -0.2	$0.28\,U_0^{0.45}$
(present work) nickel foam + nickel plane plate	0.03 - 0.2	$0.47\,U_0^{0.61}$
(9) nickel foam G60	0.005 - 0.1	$0.51\,U_0^{0.39}$
(10) 30 stacked nets (fine nickel mesh)	0.04- 0.12	$1.29\,U_0^{0.80}$

In order to emphazise the importance of the cell design and the use or not of a membrane, a number of correlations expressed as Sh versus Re, are gathered in Table 3 and plotted in fig 2.

Table 3 . Comparison of correlations Sh= f(Re) for different R.U.C. and cells.

Ref. / electrode	cell	Re range	$Sh = a\ Sc^{1/3}\ Re^{b}$
([11]) plane plate + net	ElectroSynCell (Electrocell AB)	70 - 800	a=5.57 b=0.40 eq. 6
([12]) plane plate + net (same as in ([11]))	laboratory cell	100 - 1600	a=1.09 b=0.47 eq. 7
([9]) sheet of foam G60	laboratory cell	24 - 480	a=1.81 b=0.39 eq. 8
([4]) plane plate + turbulence promoter	FM01-LC (ICI)	200 - 1000	a=0.74 b=0.62 eq. 9
([4]) plane plate	FM01LC (ICI)	200 - 1000	a=0.22 b=0.71 eq. 10
(present work) plane plate + net (same as in ([11]))	ElectroSynCell (Electrocell AB)	300 - 2200	a=0.77 b=0.57 eq. 11
(present work) plane plate + sheet of foam G 60	ElectroSynCell (Electrocell AB)	300 - 2300	a=0.32 b=0.61 eq. 12
([8]) RVC foam G60	laboratory cell	200 - 3500	a=2.7 b=0.48 eq. 13

Comparing correlations obtained with a plane plate covered with inert turbulence promoter is a way to valid our experimental results. The present correlation lies in the range of data given in the literature. Surprisingly, the correlation obtained in the same electrolyser[11] indicates larger mass transfer .coefficients ; this surprising result was previously emphasized[12]. The difference between this result (eq. 11) and the correlations from Letord-Quemere et al.[12] (eq. 7) and from Brown et al.[4] (eq. 9) is probably due to the difference in the test cells used : design of the inlet zone, dimensions of the channel. The absence of a calming section at the inlet can increase the value of the exponent of Re (eq. 4). The flow regime in the channel of the ElectroSynCell provided with our R.U.C. is turbulent. Considering the three-dimensional electrode, the value of the exponent of Re in eq. 12 is higher than that of Re in eq. 8 and in eq. 13. The same applies to the flow on the channel side of the plane plate : the flow structure is probably different from that observed by Carlsson et al.[11]. Indeed working in the undivided mode and associating a net with a sheet of foam seems to promote a particular flow pattern. The influence of the configuration on the mass transfer to the three-dimensional electrode is difficult to analyse.Through fig 1 and fig 2, data obtained with R.V.C. foam[8] appears respectively lower and higher than the present data for nickel foam sheet. For three-dimensional electrodes, the representation as

Sh Sc$^{-1/3}$ versus Re is not adequate : the real active surface area being unknown, it is evaluated from the dynamic specific area.

Depending on the authors[8,13], the dynamic specific area A lies between 3200-6400 m^{-1}. This fact and the difference in the corresponding equivalent diameters of the cell, leads to the large difference between eq. 13 and eq. 12 in fig 2.

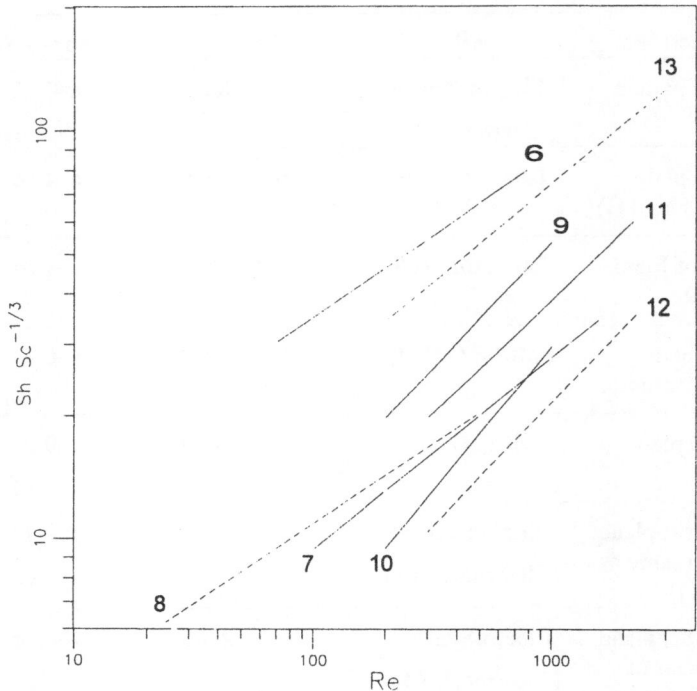

Figure 2 . variation de Sh Sc$^{-1/3}$ versus Re. (___) : plane electrode. (-----) : foam of grade 60. The numbers correspond to equations in Table 3.

Conclusion : As previously emphasized[4,12], the relative value of the mass transfer coefficient depends on the cell design, particularly on the entrance zone. A careful comparison of literature data expressed as Sh Sc$^{-1/3}$ versus Re shows that this way of representation can lead to an arbitrary scatter of data and misleading interpretations. Among other parameters influencing mass transfer rate are the nature of the material and the configuration of the electrode (divided or not). With the presently studied R.U.C., the mass transfer rate to the reticulated medium is rather lower than those previously found in divided R.U.C. but the order of magnitude is unchanged.

FLOW VISUALIZATION

A previous study of residence time distribution R.T.D.[7] has shown a difference between the flow pattern in a channel filled with a net and that occurring in a channel filled with a sheet of foam. The R.T.D. for flow in a channel filled with the foam is identical to that in an empty channel : a cascade of 6 C.S.T.R.s describes the flow in these two cases well. The flow in the channel filled with a net is quasi plug. Though R.T.D. study allows modelling of flow in reactors, this method is too rough and global to give information on the local flow pattern. Therefore, a study of flow visualization has been carried out in order to get a better understanding of mass transport phenomena in electrochemical reactors of filter-press type.

Experimental details

Previous studies were devoted to flow visualization in a narrow channel simulating electrolysers or electrodialysers[14,15]. Studies of flow visualization in commercially available cells are scarce[16]. Among available methods[17], dye injection and the electrode activated pH method are the most convenient for the present study.

Three R.U.C., working in the divided mode, are tested : they are respectively provided with plane plates only, plane plates with acting nets and plane plates with sheets of foam G60. These R.U.C. were described[7] previously. One metallic frame of the ElectroSynCell is replaced by a plexiglass one.

The experimental method can be described as follows :

- dye injection : bromocresol green (1g/l) in 50:50 ethanol-water mixture is the dye solution. It is injected at two injection points located at the inlet of the active area.

- electrode activated pH method : details on the method are given by Quraishi et al.[18]. Thymol blue (2.5 g/l) in 30:70 ethanol-water is the electrolyte. The working electrodes, stainless steel wires, are placed in the injection systems.

For each R.U.C. and technique, four flow rates, respectively 80, 150, 250 and 450 l/h, are tested. The flow in the active area is filmed on a video cassette. A special color printer allows pictures of the flow at any time to be obtained.

Analysis of experiments

The electrode activated pH method was expected to be more interesting as no perbubation of the flow occurs. Indeed, both techniques, dye injection and electrolysis, give similar results. This study allows the flow pattern in the three R.U.C. to be distinguished and the influence of the cell geometry on the flow in the active zone to be demonstrated. Concerning the influence of the cell geometry, experiments show that the behaviour of the entrance area is that of a C.S.T.R.. This result confirms a previous flow modelling[7].

The flow visualization was performed in the channel filled with a turbulence promoter. Using two injectors located in the middle and at the side of the entrance of the active area shows that the polymeric inner frames are not rigid enough to flatten a turbulence promoter against the whole plane plate. Therefore, for some experiments, a plastic sheet of one millimeter thickness was added between the net and the plate in order to ensure a good flattening. With a good flattening, the injected dye follows the rods of the acting net over a distance which depends on the flow rate and then the trace breaks up, as shown in fig 3-a. These observations were previously described by Feron et al.[15]. If the flattening is not perfect, channelling occurs between the acting net and the plate ; its direction is that of the main direction of flow, i. e. vertically. The main interest of these experiments is to show the great lateral dispersion of the dye, whereas the axial dispersion is rather small.

In the R.U.C. constituted by the channel filled with a sheet of foam, the flow is radically different. No lateral dispersion is observed. Whatever the value of the flow rate (80-450 l/h or Re 310-1750), the flow pattern is the same : the streamlines follow the main direction of the flow (fig 3-b). For the injection point located in the left part of the cell, the dye trace is slightly inclined at the bottom of the cell, this is probably due to imperfect distribution of the flow at the outlet of the asymmetrical flow distributor.

In the empty channel, the dye trace shows (fig 3-c) a vortex breakdown pattern. With larger flow rates, the traces become wider. Both lateral and axial dispersion are observed.

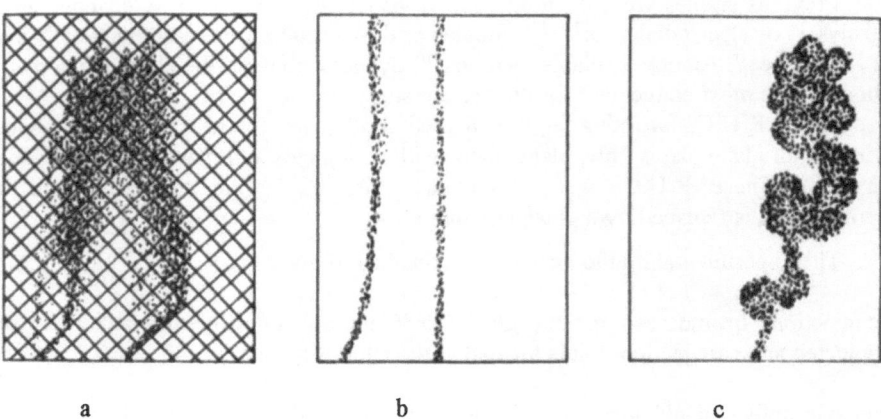

a   b   c

Figure 3 : a : channel filled with a net, dye injection in the middle and in the left (80 l/h). b : channel filled with a sheet of foam , dye injection in the middle and in the left (80 l/h). c : empty channel, dye injection in the middle (80 l/h).
The main direction of flow is vertical, it is oriented from the bottom to the top of the pictures.

Conclusion : The visualization technique used in this work can be regarded as a continuous point source method[19] for determining the transverse dispersion coefficient. The flow pattern observed in a channel filled with a sheet of foam is characterized by straight streamlines with no transverse dispersion of the dye. The flow is channelled in the pores of the porous medium, which can be represented by a bundle of parallel ducts. It seems that the flow interaction between the pores is very weak. This result is confirmed by the value of the tortuosity which is close to one[13] in this type of porous medium. Although the axial dispersion coefficients are similar in a foam filled channel and in an empty channel[7], the flow patterns are completely different. In the empty channel, there is a manifestation of vortex breakdown. This is due to the entrance section of the Electro Syn Cell, which is similar to a multi jet configuration.

The flow visualization in the channel filled with the net allows observation of lateral and axial dispersion phenomena. In previous work[7] the flow was shown to be quasi plug ; this behaviour may result from interactions between lateral and axial dispersion. The plastic net has the same hydrodynamic effect as static mixers. The flow in this channel appears to be the most unsteady : many variations of direction are observed. As the disruption of the mass transfer boundary layer is known to enhance mass transfer rate, the flow visualization confirms that the mass transfer rate should be greater with a net than with a sheet of foam.

NOMENCLATURE

A : specific surface area of the working electrode
C : bulk ferricyanide concentration
C_{in} : inlet ferricyanide concentration
C_{out} : outlet ferricyanide concentration
D : diffusion coefficient

d_h : equivalent hydraulic diameter ($d_h = 2\,l\,h\,(l+h)^{-1}$)
F   : Faraday constant.
h   : channel thickness
I   : limiting diffusion current
k_d : mass transfer coefficient
l   : channel width
n   : number of electrons involved in the reaction
Q_v : flow rate
Re   : Reynolds number (Re = $U_o\,d_h\,\nu^{-1}$)
S   : surface area of the electrode
Sc   : Schmidt number (Sc = $\nu\,D^{-1}$)
Sh   : Sherwood number (Sh = $k_d\,d_h\,D^{-1}$)
U_o : channel electrolyte superficial velocity
X   : conversion
ε : void fraction
ν : kinematic viscosity
τ : mean residence time in a C.S.T.R.

Acknowledgments : The authors wish to thank Yann Provost who performed the visualization experiments.

REFERENCES

1. Storck A. and Wragg A.A. , Electrochemical Engineering - Towards 2000, J. Appl. Electrochem., 21, 463-464 (1991).
2. Storck A. and Coeuret F., in : Eléments de Génie Electrochimique, Lavoisier, Paris (1984)
3. Montillet A., Comiti J. and Legrand J., Axial dispersion in liquid flow through packed reticulated metallic foams and fixed beds of different structures, The Chem. Eng. J. , 52, 63-71 (1993)
4. Brown C.J., Pletcher D., Walsh F.C., Hammond J.K. and Robinson D., Studies of space-average mass transport in the FM01-LC laboratory electrolyser, J. Appl. Electrochem., 23, 38-43 (1993)
5. Walsh F.C., Pletcher D., Whyte I. and Millington J.P., Electrolytic removal of cupric ions from dilute liquors using reticulated vitreous carbon cathodes, J. Chem. Tech. Biotechnol., 55, 147-155 (1992).
6. Langlois S. and Coeuret F., Flow-through and flow-by porous electrodes of nickel foams - Part IV : experimental electrode potential distributions in the flow-through and in the flow-by configurations, J. Appl. Electrochem., 20, 749-755 (1990).
7. Montillet A., Comiti J. and Legrand J., Application of metallic foams in electrochemical reactors of filter-press type : I- Flow characterization, J. Appl. Electrochem., in press (1993).
8. Pletcher D., Whyte I., Walsh F.C. and Millington J.P., Reticulated vitreous carbon cathodes for metal ion removal from process streams - Part I : mass transport studies, J. Appl. Electrochem., 21, 659-666 1991).
9. Langlois S. and Coeuret F., Flow-through and flow-by porous electrodes of nickel foam. II. Diffusion-convective mass transfer between the electrolyte and the foam, J. Appl. Electrochem., 19, 51-60 (1989).
10. Brown C.J., Pletcher D., Walsh F.C., Hammond J.K. and Robinson D., Studies of three dimensional electrodes in the FM01-LC laboratory electrolyser, J. Appl. Electrochem., in press.
11. Carlsson L., Sandegren B., Simonsson D. and Rihovsky M., Design and performance of a modular, multi-purpose electrochemical reactor, J. Electrochem. Soc., 130, 342-346 (1983).
12. Letord-Quemere M.M., Coeuret F. and Legrand J., Mass transfer at the wall of a thin channel containing an expanded turbulence promoting structure, J. Electrochem. Soc., 135, 3063-3067 (1988).
13. Montillet A., Comiti J. and Legrand J., Determination of structural parameters of metallic foams from permeametry measurements, J. Mat. Sci., 27, 4460-4464 (1992).
14. Focke W.W. and Knibbe P.G., Flow visualization in parallel-plate ducts with corrugated walls, J. Fluid Mech., 165, 73-77 (1986).
15. Feron P. and Solt G.S., The influence of separators on hydrodynamics and mass transfer in narrow cells : flow visualization, Desalination, 84, 137-152 (1991).

16. Belfort G. and Guter G.A., An experimental study of electrodialysis hydrodynamics, Desalination, 10, 221-262 (1972).
17. Anonym, Visualized flow ; in summary of flow visualization methods, 1-6. Ed. The Japan Soc. of Mech. Eng., Pergamon press (1988).
18. Quraishi M.S. and Fahidy T.Z., A flow visualization technique using analytical indicators : theory and some applications, Chem. Eng. Sci., 37, 775-780 (1982).
19. Robbins G.A., Methods for determining transverse dispersion coefficients of porous media in laboratory column experiments, Water Res. Res., 25, 1249-1258 (1989).

ELECTROCHEMISTRY IN A PACKED-BED ELECTRODE OF ACTIVATED CARBON PARTICLES: AN INVESTIGATION OF FARADAIC AND CAPACITIVE PROCESSES

Torsten Balduf, Gérard Valentin and François Lapicque

Laboratoire des Sciences du Génie Chimique
CNRS - ENSIC, BP 451, 1 rue Grandville
F- 54001 Nancy, France

ABSTRACT

This paper deals with faradaic and capacitive processes occurring in a packed bed electrode of activated carbon particles. On the basis of the anodic oxidation of sulfite/bisulfite ion in a sulfate medium, activated carbon NC 40 was shown to allow a higher conversion rate than graphite particles of comparable external surface area. However current density has to be kept far below the limiting current density to avoid destruction of the brittle material. Besides, capacitive phenomena at the electrode could be described using an overall model. The accessible area and the specific capacitance of the material could be determined from charge/discharge experiments and voltammetric curves: both variables were shown to be increasing functions of the applied potential. The approach developed for modelling the porous electrode is discussed and compared to published models.

INTRODUCTION

Porous electrodes exhibit high specific area and are therefore used for energy storage[1] and electrolytic treatment of dilute solutions; various designs e.g. stacks of metal grids, packed beds or fluidised beds are encountered. The available electrode area corresponds generally to the external area of the material used, except for carbon materials for which the inner porosity increases the effective electrode area significantly. Activated carbon (AC) has a very high specific surface area, ca. 10^9 m^{-1}, and thus represents an attractive electrode material, even though only a fraction may be available: for instance, an estimation from literature data[2] suggests that the accessible surface area for the adsorption of sulfate ion should represent 20% of the total BET area.

The present investigation deals with the electrochemical behavior of activated carbon particles in a fixed bed for both faradaic and capacitive processes. The anodic oxidation of sulfite and bisulfite ion in a sulfate medium was considered. Experiments carried out on an electrochemical packed bed in batch recycle showed high conversion rates and yields. The

3D electrode was also characterized in terms of accessible area and specific capacitance: the experimental data obtained were treated by a simple model.

EXPERIMENTAL SECTION

The electrochemical device employed is shown schematically in Figure 1. The packed bed measured 100 x 100 x 17 mm. The thickness in the current flow direction was kept low to minimize resistive losses. Activated carbon (NC 40) was in the form of cylindrical pellets, whose diameter and length were 5 mm. The surface area measured with BET instrumentation (Carlo Erba) was near 1160 $m^2.g^{-1}$; the pore volume investigation using N_2 saturation and mercury intrusion revealed the existence of three types of pores: the corresponding ranges of pore size were 10 - 50 nm, 0.5 - 3 μm, and 10 - 25μm respectively. AC particles were thoroughly washed in an ultrasonic bath to remove dust and fines.

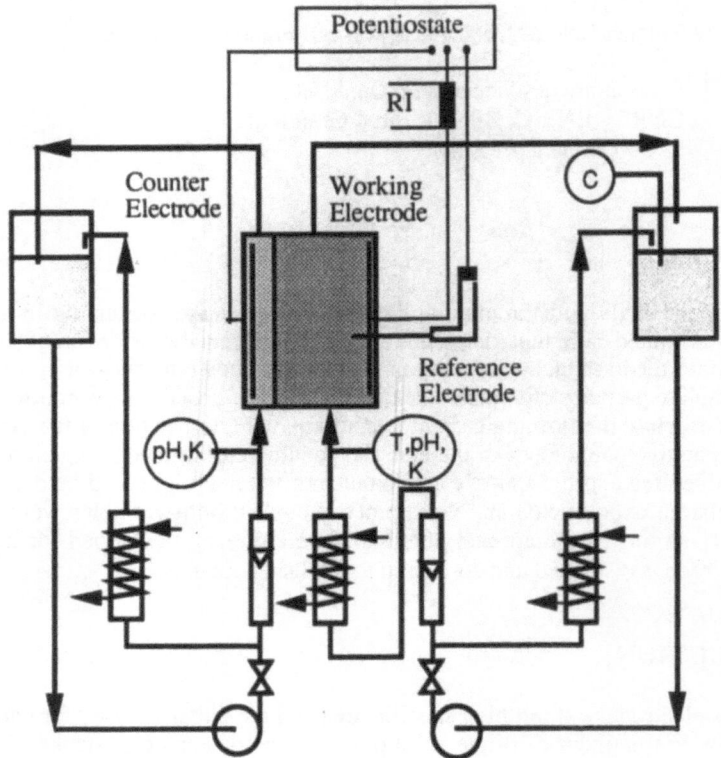

Figure 1. Schematic view of the experimental set-up.

The bed was located inbetween a 417 Nafion® membrane and the current feeder. Both the current feeder and the counterlectrode were platinized titanium expanded metal mesh; a small amount of hydrogen evolution occurs at this counter-electrode. A saturated calomel electrode (SCE), terminated in a flexible silicone tubing and a fine glass Luggin capillary, was inserted into the rear wall of the packed bed compartment. All the experiments were carried out with imposed potential between the packed bed current feeder and the reference electrode. The cell voltage, as well as the current, were continuously recorded and stored by a data acquisition device connected to a personal computer; the charge passed was obtained by numerical integration of the cell current.

The electrolyte solutions were of the same concentration of sodium sulfate (0.1M or 0.5M). Faradaic behaviour of the packed bed anode was studied with addition of 0.1M sodium sulfite. The aqueous solutions were driven continuously in the cell through two hydraulic circuits and the velocity was kept at 0.01 m.s^{-1}. Temperature in the two flow circuits was maintained at 20°C by a thermostated bath.

FARADAIC PROCESSES

When dissolved in aqueous medium, sulfur dioxide can be in the molecular state or in the form of bisulfite or sulfite ion, depending on pH. The solution pH was in the range 4-10 and ionized forms prevailed. Accordingly, the anodic oxidation obeys the following stoichiometry:

$$SO_3^{--} + 2OH^- \rightarrow SO_4^{--} + H_2O + 2e^- \tag{1}$$

$$HSO_3^- + 4H_2O \rightarrow SO_4^{--} + 3H_3O^+ + 2e^- \tag{2}$$

Experimental procedure was carefully investigated: AC particles had to be charged in a sulfate solution at the desired potential for one hour or more prior to the oxidation. Sulfate adsorption had then reached an equilibrium state, depending on the applied potential. This procedure yielded reproducible time variations of conversion and current, as well as the total conversion of sulfite into sulfate. Sulfite solution was introduced after the AC bed pre-charge and the faradaic oxidation occurred. The solution pH was kept to the initial value by automatic addition of 10M NaOH solution to compensate the amount of H$^+$ ions generated by the oxidation. Conversion of tetravalent sulfur was followed by the added volume of NaOH solution and titration of liquid samples; sulfite/bisulfite was shown to be quantitavely oxidized into sulfate by ion chromatography.

Activated carbon was found to allow higher oxidation rate than graphite: conversion of 50% was attained within a time lapse twice as low on AC than on graphite particles of comparable external area.

Cell current was in the range of 1-6 A, depending on the applied potential which was kept below 0.9 V/SCE. For most cases, current yields were over 80% and were not affected by the operating conditions (potential, pH or flow velocity). The current density, calculated on the basis of the external area, was maintained far below the limiting current density to avoid the destruction of the activated carbon by gas evolution in the pores. Accordingly, changes in the flow velocity affected neither the current nor the conversion rate. Current yields higher than one could be observed for the first instants of the batch run: this was attributed to side-oxidation of tetravalent sulfur on sites which had been formed during previous runs or during the charging procedure. Oxidation of tetravalent sulfur is fairly selective below 0.9V/SCE and, as indicated in the relevant literature, side-reactions e.g. esterification, quinone formation or gas evolution, are of moderate significance in this potential range[3,4].

Current and conversion rate are significantly increased by a reduction of the solution pH, as shown in Figure 2. Conversion of bisulfite ion, for pH near 4, is faster than the oxidation of sulfite species, which predominates for pH over 8.

MODEL FOR CAPACITIVE PROCESSES

Numerous investigations have focused on activated carbon packed-bed electrodes and the potential distribution in such a matrix. For most cases, the porous electrode was characterized by a single porosity[5-7]. More recently, AC beds were described using intra-particle microporosity and inter-particle macroporosity[2]. Potential distribution induced by the existence of an electrical field is linked to the transport of charge in the pores.

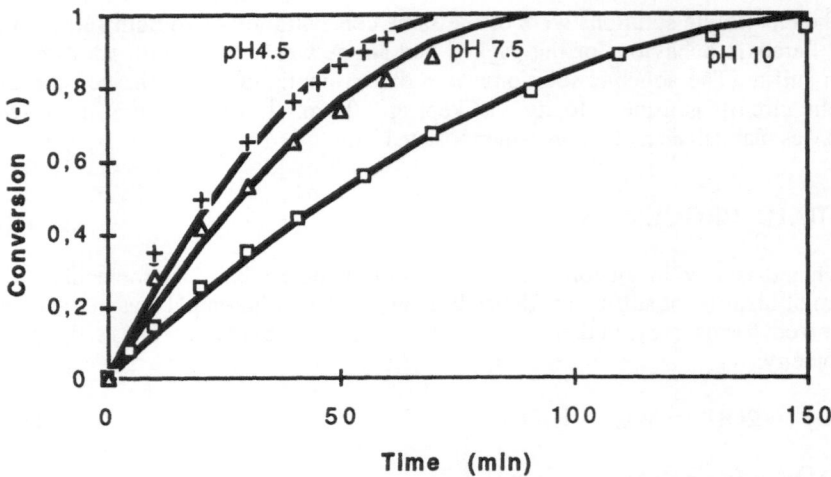

Figure 2. Conversion of sulfite/bisulfite ions at the AC packed bed electrode; E=500 mV/SCE.

It is usually postulated that a large fraction of the pores are too small and tortuous to permit convective transport and diffusion is the main means of mass transport.

In the absence of faradaic processes, the current density in the pores is governed by Ohm's law lumped to the current loss corresponding to the surface charging, which is simply the formation of the double layer. Two parameters are involved:

(i) an overall resistance, R_{el}, taking into account the electrodes, the current feeder and the electrolyte solution;

(ii) an overall capacitance of the electrode, C, corresponding to the charge stored on the accessible surface of the complex electrode.

Parameters R_{el} and C can be estimated from the electrode transient responses to potential variations, as explained below.

Potential step

Consider a system in electrical equilibrium. At t=0, the current feed potential is changed by a potential step ΔV. Assuming a one-dimensional problem for potential distribution and charge transport, Posey and Morozumi[5] established expressions for the local overvoltage and current density in the forms of series expansion of time and the coordinate x. After a certain time, the series can be truncated after the first term and the current is then approximated by:

$$ I = \frac{\Delta V}{R_{el}} \exp\left[-\frac{\pi^2}{8} \frac{t}{R_{el}C} \right] \tag{3} $$

An alternative approach for the complex electrode consists in disregarding the porous nature of the packed bed and the uneven distribution of potential: the electrode is only represented by a resistor in series with a capacitor. The response to a potential step is given by the classic relationship:

$$ I = \frac{\Delta V}{R_{el}} \exp\left[-\frac{t}{R_{el}C} \right] \tag{4} $$

In spite of the different approaches, relations (3) and (4) yield comparable transients of the current density since they only differ by a factor $\pi^2/8$, near 1.2, in the exponent.

Linear voltammetry

When faradaic reaction at an electrode can be neglected, current density for a pure charging process is proportional to the sweep rate, $\partial E/\partial t$:

$$I = C \frac{\partial E}{\partial t} \qquad (5)$$

This relation was successfully used by Gagnon[7] to determine the double layer capacity of AC electrodes. (5) is only valid over a restricted range of potential and the factor C appears to be an overall capacity, as in relation (4).

RESULTS

Potential step

The overall resistance and capacitance have been estimated by use of the potential step technique and the simple relation (4). The operating parameters were:

(i) the concentration of sodium sulfate, 0.1 and 0.5M

(ii) the potential step, i.e. the potential difference ($E_{final}-E_{initial}$), in the range 25-75 mV

(iii) the final potential applied to the current feeder, E_{final}. This potential was kept below 500 mV/SCE to avoid any faradaic process at the carbon surface.

The sodium sulfate solution was introduced in the packed bed reactor one day before the series of experiments at open electrical circuit. The open-circuit potential of the wetted AC bed was near 120 mV/SCE. The potential step was then imposed and the current transient was followed. The observed current corresponded to the establishment of the adsorption equilibrium on the AC surface at E_{final}: steady state was usually attained within one or two hours as the current became negligible after one hour or less. Figure 3 gives an example of current and cell voltage transients. As shown by relation (4), the current measured in the very first instants, gives access to the resistance R_{el}; the overall capacitance is obtained by fitting the current variation to the exponential function in (4).

The overall resistance is weakly affected by the operating conditions: R_{el} is close to 0.09 Ω within 20%. The resistance of the packed bed particles is constituted by three

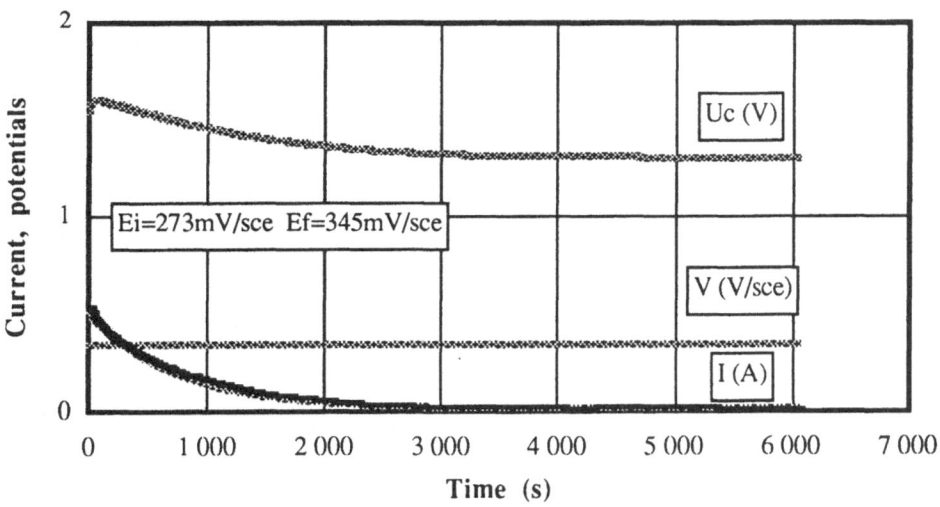

Figure 3. Example of current and cell voltage transients recorded.

resistances in a complex manner, namely the electrolyte solution, the AC particles bed and the current feed. Therefore, the overall structure cannot be easily modelled by an equivalent circuit. However, the R_{el} value can be compared to the resistance of the electrolyte solution in the anode compartment, R_l, regardless of the AC particles. Taking into account the electrical conductivity of the solutions, R_l is estimated at 1.16 and 0.36 Ω for concentrations of sodium sulfate of 0.1M and 0.5M, respectively. This shows that for the prediction of the electrode resistance the complex structure of the electrode has to be taken into account.

The double layer capacitance C is shown to be an increasing function of the initial potential. The specific capacity of the bed, C_a, was deduced from the overall capacitance and the BET area of the AC:

$$C_a = \frac{C}{a_{BET}} \qquad (6)$$

In the most concentrated sulfate solution C_a is found to increase regularly with $E_{initial}$ from 4 $\mu F \cdot cm^{-2}$ at 120 mV/SCE to 16 $\mu F.cm^{-2}$ at 450 mV/SCE. The straight dependance line shown in Figure 4, fits the experimental data well for the three potential steps considered. However, in the case of 0.1M Na_2SO_4 medium, the variation of C_a with potential is different: C_a remains fairly constant near 10 $\mu F.cm^{-2}$ in the range 120-300 mV/SCE; for higher potential values, C_a increases with the potential and is very close to the corresponding capacitance measured in the 0.5M solution.

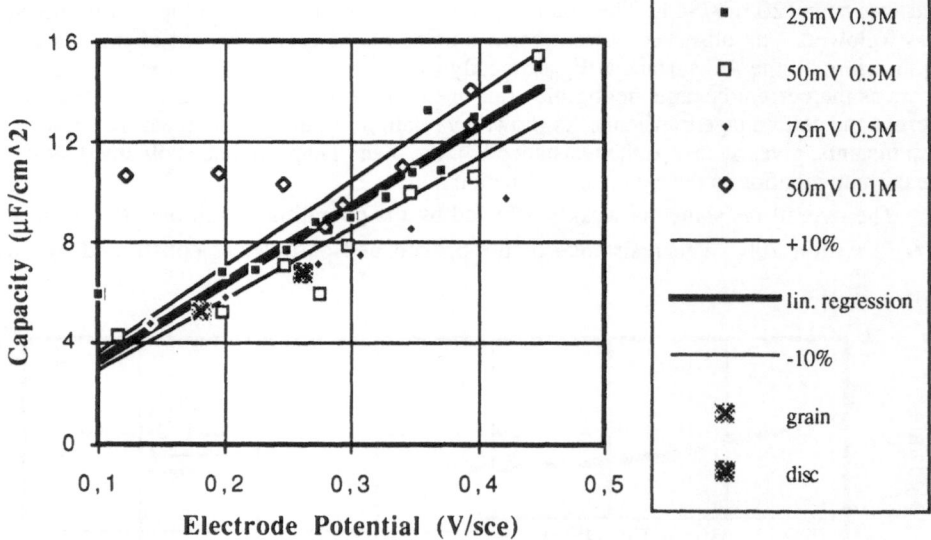

Figure 4. Variation of the specific capacitance of the AC electrode with the initial potential.

Linear voltammetry

Experiments were carried out on two small AC electrodes immersed in a stagnant sulfate solution: a single particle hung by a platinum wire, and a disk AC surface embedded in a Teflon rod. i-E curves were observed to exhibit a significant plateau corresponding to capacitive phenomena for very low sweeping rates: $\partial E/\partial t$ was typically 0.1 $mV.s^{-1}$. The plateau was generally well defined over a range of 200 mV.

The capacitance deduced from (5) corresponds to the potential ranges of the plateaux. The values obtained by this technique are consistent with the potential step data (Figure 4).

DISCUSSION

The accessibility of the activated carbon for capacitive processes can be defined as the ratio of the actual specific capacitance related to the BET area, C_a, to the double-layer capacitance of this material, C_{db}:

$$Acc = \frac{C_a}{C_{db}} \tag{7}$$

This characteristic corresponds to the fraction of surface accessible to adsorption of sulfate ion. Double-layer capacitance of activated carbon has been measured in numerous investigations and the values proposed depend on the nature and the origin of the considered AC, the nature of the adsorbed ion, and the technique used. Reported values for C_{db} vary from 20 $\mu F.cm^{-2}$, acccording to Holze and Vielstich[8], to 42 or even 58 $\mu F.cm^{-2}$ (Card[2]). A good estimate for C_{db} should be near 30 $\mu F.cm^{-2}$, as recommended by Johnson and Newman[9].

The results presented in Figure 4 lead to accessibility values ranging from 0.2 to 0.6, depending on the electrode potential. The lower estimate for Acc is in good agreement with Card's conclusions[2]. This dependence is linked to the charge distribution on the AC surface and to the importance of the various electrical forces in the pores and near the surface, as well to the potential dependence of the capacitance itself, so that the significance of accessibility is not straightforward.

CONCLUSIONS

Packed bed electrodes of AC particles can be used for electrochemical reactions due to the catalytic properties and the large specific area. However, only a small fraction of the real surface is active and the electrochemical conversion probably occurs in the largest pores.

Capacitive phenomena on the AC electrode could be investigated and characterized using a simple R-C model, regardless of the potential distribution in the porous matrix. More realistic models for such an electrode must take into account potential distribution in the pores.

For the example of sulfate ion adsorption, the two techniques employed, showed the fairly large accessibility of the AC surface for charging, depending on the applied potential. Capacitance of AC structures are currently under investigation using other techniques, such as impedance analysis, and the validity of the simple R-C model will then hopefully be confirmed.

REFERENCES

1. K. Fukoshina, "Carbon: Electrochemical and Physicochemical Properties", Wiley and Sons, New York (1988).
2. J.C. Card, G. Valentin and A. Storck, The activated carbon electrode: a new experimentally verified mathematical model for the potential distribution, *J. Electrochem. Soc.* 137:2736 (1990).
3. H. Binder, A. Köhling, M. Richter and G. Sandstede, Über die anodische Oxydation von Aktivkohlen in wässrigen Elektrolyten, *Electrochim. Acta* 9:255 (1964).
4. K.F. Blurton, An electrochemical investigation of graphite surfaces", *Electrochim. Acta* 18:869 (1973).
5. F.A. Posey and T. Morozumi, Theory of potensiostatic and galvanostatic charging of the double layer in porous electrodes, *J. Electrochem. Soc.* 113:176 (1966).

6. J. Newman and W. Tiedemann, Flow-through porous electrodes, in "Advances in Electrochemistry and Electrochemical Engineering", Vol. 11, H. Gerischer and C.W. Tobias ed., Wiley, New York (1978).

7. E.G. Gagnon, The triangular voltage sweep method for determining double-layer capacity of porous electrodes. IV Porous carbon in potassium hydroxide, *J. Electrochem. Soc.* 122:521 (1975).

8. R. Holze and W. Vielstich, Double-layer capacity measurements as a method to characterize porous fuel cell electrodes, *Electrochim. Acta*, 29:607 (1984).

9. A.M. Johnson and J. Newman, Desalting by means of porous carbon electrodes, *J. Electrochem. Soc.* 118:510 (1971).

MODELLING THE PHENOMENA ASSOCIATED WITH THE APPLICATION OF CENTRIFUGAL FIELDS IN FUSED SALT ELECTROLYSIS MANUFACTURE OF LIGHT METALS

Antony Cox, James W. A. Morris and Derek J. Fray

Department of Mining and Mineral Engineering
University of Leeds
England
United Kingdom
LS2 9JT

ABSTRACT

A laboratory scale fused salt electrolysis cell was constructed and successfully used to electrowin zinc.

Centrifugal separation of the products was such that current efficiencies as high as 93 % were attained using electrode spacings of a few millimetres and rotation speeds of 100 rpm.

To assess the feasibility of the technique applied to light metal electrowinning, a low temperature hydrodynamic model was constructed. Almost complete separation of simulated products was achieved by the semi-rigorous physical model using electrode spacings as small as 10 mm at rotation speeds of only 40 rpm.

INTRODUCTION

Whereas fused salt electrowinning techniques are a rapid way of producing high purity light metals such as lithium, sodium and magnesium with a minimum waste of starting materials, they are often criticised for their large energy consumption and low space-time yields relative to pyrometallurgical reduction techniques. Resolving these two problems will result in a major advancement in the light metals industry, with the demand for expensive metals such as lithium rapidly increasing. Both of these problems are now being addressed. This paper is concerned with reducing the energy consumption in fused salt electrowinning techniques to produce light metals.

General aspects influencing the cell energy consumption are the presence of parasitic side reactions, non-uniform current distribution, poor heat insulation, poor space-time yields and high reversible potentials. The factor to be addressed in this paper is the current use of large electrode spacings.

Reducing the inter-electrode spacing from about 100 mm (typical of commercial cells) down to less than 10 mm significantly reduces the resistance between the electrodes. This can easily be done, but separation of the liquid and gaseous products then becomes important since recombination is more likely to occur.

Separation may be achieved by the use of rotating, planar disc electrodes. The technique has already been applied to laboratory-scale fused salt electrowinning of zinc at the University of Cambridge with some success[1], and recently at the University of Leeds with a high degree of success. The technique is currently being applied to the electrowinning of lithium.

In order to optimise the process, the physical interaction of each phase should be studied. This is impractical at high temperature which has led to the development of a low temperature hydrodynamic model to simulate the technique and locate potential zones where recombination is probable. Experimental variables can then be varied to achieve maximum separation and applied to the high temperature cell. Measurement of current efficiency in the latter will indicate the effectiveness of the separation process.

HIGH TEMPERATURE WORK

Electrical contact to the 70 mm graphite electrodes was achieved through mounted graphite brushes via an electrically insulated steel tube and rod (Figure 1). The electrodes were rotated by direct drive through the rod and shaft system from an overhead electric motor. The upper anode plate was perforated to allow the chlorine gas to escape from the gap. The adjustable spacing between the electrodes was made possible by the use of alumina spacer rings.

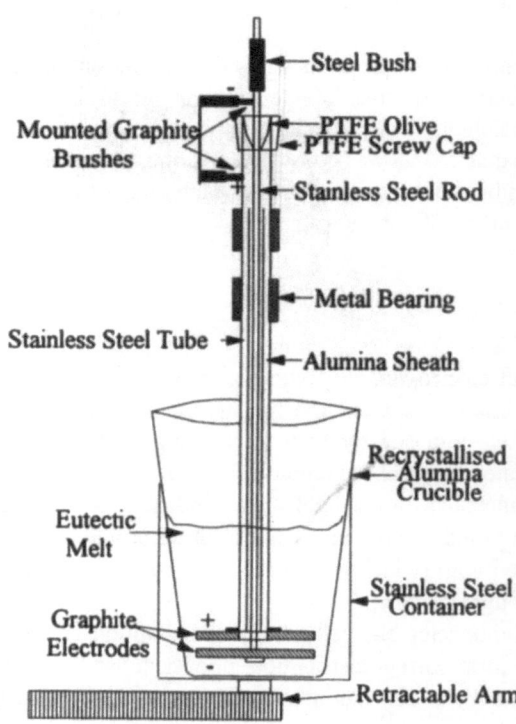

Figure 1. Schematic of high temperature apparatus

The crucible was supported in a steel container on a retractable arm which allowed the cell to be easily placed in and removed from the furnace using an air-hydraulic system.

The chlorine gas produced from the electrolysis was removed from the furnace using an extractor fan which drew the gas into a fume cupboard for subsequent chemical absorption.

Experimental Procedure

A low melting point salt was used consisting of a 50 wt% zinc chloride/potassium chloride eutectic. The density of the eutectic melt was 2020 Kg m^{-3}, the viscosity 0.002 Kg m^{-1} s^{-1} and the melting point was 450 °C.

The cell was operated at 500 °C. The electrodes were immersed in the melt and rotated for 20 minutes for the fluid to reach mechanical equilibrium. A current of 30A (7800 A m^{-2}) was applied for one hour. After electrolysis, the electrode assembly was lifted clear of the cell, the salt allowed to cool and the zinc then extracted into aqueous solution and removed and washed by Buchner filtration. The metal was then dried and weighed.

Results and Discussion

Table 1 indicates that a current efficiency of 93 % results in almost complete separation at a rotation speed of 100 rpm.

Table 1. Effect of rotation speed on separation

Rotation Speed (rpm)	Anode to Cathode Distance (mm)	Immersion Depth of Lower Face of Anode (mm)	Current Efficiency (%)
0	8	25	43
40	8	25	51
70	8	25	60
100	8	25	93

Obviously, the centrifugal force at this speed was optimal in fulfilling the following important separation criteria;

♦ Minimising turbulence effect

♦ Maximising physical separation of phases

♦ Removing metal faster than it was formed

All the above factors determine the extent of the recombination reaction;

$$Zn + Cl_2 = ZnCl_2$$

Using the expression;

$$Re = \frac{\rho r^2 \omega}{\eta}$$

where ρ is electrolyte density (Kg m^{-3}), r is radius of electrodes (m), ω is angular velocity of electrodes (rad s^{-1}) and η is the dynamic viscosity of the electrolyte (Kg m^{-1} s^{-1}), a Rotational Reynold's number of 1.4 x 10^4 was calculated. The threshold value for significant turbulence[2] to occur in parallel concentric disc geometry is given as 1.8 x 10^5. Consequently, turbulence was probably not significant up to rotational velocities of 100 rpm for this system.

Wall jets have been observed[3] inside the inter-electrode gap using a low temperature simulation of the electrode geometry. These wall jets were found to confer separation to the metal and gas phases. The efficiency of the separation is given by the Taylor number;

$$\alpha = \sqrt{\frac{\omega a^2 \rho}{\eta}}$$

where a is the semi-electrode spacing. Table 2 shows how the separation, and therefore the current efficiency, was affected by the rotation speed using 8 mm gaps.

Table 2. Effect of rotation speed on separation

Rotation Speed (rpm)	α
0	0
40	8.23
70	10.88
100	13.00

A maximum Taylor number of 13.0 (at a rotation speed of 100 rpm) was calculated. Jansson[3] states the condition $\alpha > 9$ for adequate separation, and when $\alpha = 16$, separation in the gap would practically, be a maximum with the wall jets occupying only 10% of the gap.

The rate of formation of metal was calculated from an expression derived from Faraday's law;

$$\frac{dm}{dt} = \frac{Mi}{zF}$$

where dm/dt is rate of formation of metal (g s^{-1}), M is relative atomic mass of zinc (65.35g), z is number of electrons transferred and F Faraday's constant. A value of 10.2 mg s^{-1} was calculated.

It is therefore likely that the physical separation of gas and metal phases inside the gap by wall jets, and the rate at which the metal exits the gap (influenced by the centrifugal force) are the limiting factors in determining the current efficiency in the cell. It is these two factors that probably explain the decrease in current efficiency with rotation speed.

With no rotation of the electrodes, a current efficiency of 43 % was still obtained. This may be explained by drops of zinc accumulating on the cathode and "spilling" over the edge of the plate. This is feasible because zinc does not wet graphite or the electrolyte.

Table 3 shows that maintaining the rotation speed at 100 rpm, but decreasing the gap from 8 to 4 mm caused a 20 % reduction in current efficiency. This reflects the increased probability of recombination due to closer proximity of electrolysis products within the gap. This result is almost identical to Copham's[1] when allowance for the size difference in electrodes is made.

Table 3. Effect of electrode spacing on separation

Rotation Speed (rpm)	Anode to Cathode Distance (mm)	Immersion Depth of Lower Face of Anode (mm)	Current Efficiency (%)
100	8	25	93
100	4	25	72

Table 4 shows reducing the immersion depth of the electrolyte resulted in a significant increase in current efficiency. This may reflect a decrease in the down-flow of the fluid - a phenomenon inherent in rotating liquids. Electrolyte down-flow has the effect of "pushing" the vertically rising gas at the anode back into the gap resulting in an accumulation of gas and subsequent recombination. However, the down-flow is proportional to the mass of liquid above the anode which probably explains the improvement in current efficiency when the immersion depth was decreased.

Table 4. Effect of immersion depth on separation

Rotation Speed (rpm)	Anode to Cathode Distance (mm)	Immersion Depth of Lower Face of Anode (mm)	Current Efficiency (%)
100	4	25	72
100	4	10	88

LOW TEMPERATURE PHYSICAL MODEL

It should be emphasised that the high temperature work performed thus far has only been concerned with zinc. This is because the fluid dynamics and chemistry of the system;

$$Zn_{(l)}|ZnCl_{2(l)}|Cl_{2(g)}$$

is much simpler than the system that will eventually be addressed by the project;

$$Li_{(l)}|LiCl_{(l)}|Cl_{2(g)}$$

The choice of the zinc system allowed the novel high temperature rig and the separation technique to be fully tested and compared to the little existing work performed in this field.

To assess the feasibility of the technique applied to fused salt electrowinning manufacture of lithium, a low temperature physical model of the system was constructed.

Hydrodynamic simulation

A semi-rigorous model was used with four states of similarity to be matched; geometric, dynamic, thermal, and kinematic.

Geometric similarity refers to shape conservation. The criterion is that the ratio of any length in one system to the corresponding length in the other system is the same. The model was 2.85 times larger in each dimension than the high temperature cell.

Thermal similarity considers the conservation of temperature gradients in the two systems and was achieved by assuming the real system to be essentially isothermal.

Kinematic similarity embodies motion similarity and is achieved if velocities at the same point in geometrically similar systems are in the same fixed ratio.

Dynamic similarity represents force similarity between the two systems. These forces are inertial and interfacial i.e., centrifugal, gravitational, viscous drag, and surface tension forces. Similarity involves maintaining a constant ratio of these forces at any point in the two systems. However, this is difficult to achieve when trying to match the properties of room temperature liquids with those at high temperature. The electrolyte was represented by 10% sodium nitrate solution which shows a good similarity with the high temperature light metal salt electrolyte in terms of kinematic viscosity, and surface tension to density ratio[5]. Nitrogen was used to model the gas produced at the anode. Paraffin simulated the molten lithium metal by virtue of it being lighter than, and immiscible with, the aqueous "electrolyte". However, due to bonding differences, the surface tension of the metal is several times greater than that of the paraffin. Thus the model goes further than an *ad hoc* simulation, but is not stringent enough to be classified as rigorous. The semi-rigorous model allows qualitative information to be taken from the model as regards phase behaviour, and predictions to be made on the high temperature prototype.

Apparatus

The rotating electrodes were mounted on a stainless steel shaft driven by a 38 W shunt wound motor combined with a motor speed controller (Figure 2).

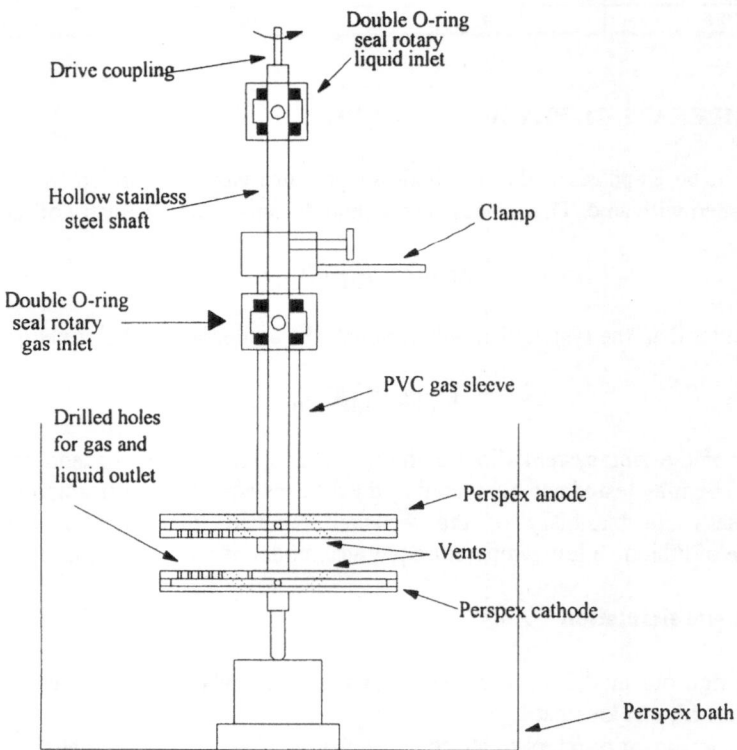

Figure 2. Schematic of low temperature physical model

A 200 mm diameter hollow Perspex^{TM} disc simulated the cathode and was used with either of two different alternate sized hollow Perspex^{TM} discs of 200 and 150 mm diameter, to simulate the anode. The latter were perforated to allow gas to escape, as in the high temperature system.

The model was constructed to simulate the paraffin and gas "generated" at the relevant electrode surfaces. The paraffin was fed to the cell at 0.6 dm^3 min^{-1} via a peristaltic pump and rotary "O"-ring inlet, whilst the gas was introduced via a similar inlet at 5 dm^3 min^{-1}. The paraffin descended down the hollow shaft to an outlet aligned with the interior of the hollow cathode. 0.5 and 1.0 mm diameter holes were drilled in the electrode surface to enable the paraffin to be introduced into the gap from the cathode centre. The gas entered the upper plate in a similar fashion, except that it passed between the outer PVC casing and the shaft before collecting in the hollow anode prior to discharge onto the underside surface of the plate.

Experimental Procedure

The paraffin and gas behaviour were studied for different sizes of stationary and rotating anode with respect to the cathode for varying anode-cathode distances and rotational disc speeds (measured by optical tachometer).

Results and Discussion

The centrifugal fields caused the gas produced at the anode to move inwards, and the paraffin generated at the cathode to move outwards. The gas was seen to concentrate around the disc centre, forming a collar around the central shaft at 40 rpm (Figure 3). This was also observed previously[6].

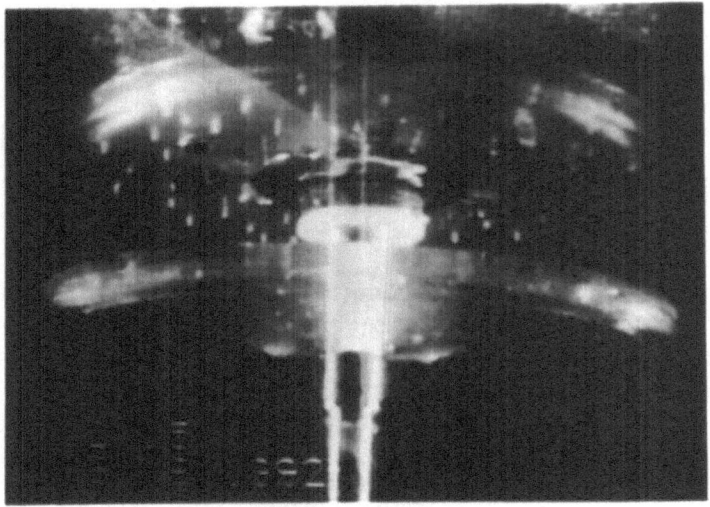

Figure 3. Gas produced at the anode forming a collar around the shaft at 40 rpm

Observations made with regard to the paraffin were primarily concerned with identifying conditions to minimise penetration of the upper anodic zone by the paraffin and, consequently, minimise product interaction. In the high temperature cell, this would result in recombination of products to reform the salt.

Studies of the paraffin movement showed that impingement on the anode was unavoidable under all conditions.

The considerable wetting force known to exist between lithium and a stainless steel cathode prompted an attempt to simulate this effect within the model by coating the lower disc with PTFE (poly-tetra-fluoroethene).

With a small (150 mm diameter) anode rotating in conjunction with a larger (200 mm diameter) cathode complete separation of paraffin was achieved at 28 rpm for a 40 mm gap (Figure 4). Decreasing the gap to 10 mm required the electrodes to be spun at 40 rpm for complete separation.

Figure 4. Paraffin streamers leaving the PTFE coated cathode at 28 rpm for a 40 mm gap

Further work using the smaller stationary anode with the cathode rotating at 35 rpm resulted in complete separation over the range of spacings 40 to 10 mm.

All the above observations may be explained by considering the physical forces operating in the cell at any one instant (Figure 5). Initially, the centrifugal force dominates as the paraffin is constrained to the disc surface until it reaches the periphery zone of the plate whereupon the paraffin film starts to "bulge". The buoyancy and contact forces then become significant as the paraffin forms a streamer pulling away from the cathode. The contact force is the force due to the surface tension acting over the paraffin film/bulge. As the streamer increases in length the contact perimeter decreases and the buoyancy force increases making detachment more likely. Considering the forces present, detachment occurs when;

$$(\text{Centrifugal force} + \text{Buoyancy force}) > \text{Contact force}$$

An increase in the disc rotational speed provides an increased centrifugal force. The resultant paraffin streamer is then modified, making the exit angle from the gap smaller, so diminishing the possibilities of paraffin/anode contact. The wetting effect of the paraffin on the electrode surface greatly assists the separation process by competing with the buoyancy force inside the gap, reducing the probability of impingement into the upper plate zone where the gas formation is occurring.

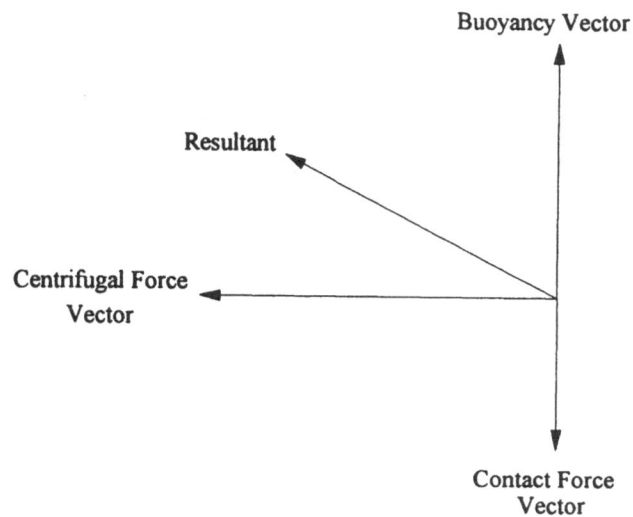

Figure 5. Interaction of component forces within the rotating electrode system

CONCLUSIONS

The centrifugal separation of a heavy metal from its gaseous by-product in a fused salt electrolysis cell incorporating small interelectrode gaps has been successfully accomplished. A low temperature hydrodynamic model has been used to assess the feasibility of applying the technique to light metals (namely, lithium). The physical model suggests that highly satisfactory separation, and therefore current efficiencies, may be obtained in the high temperature cell even when using electrode spacings of only 10 mm.

The next stage will be to apply the technique to a lithium electrowinning cell and continue to use low temperature model observations to guide the high temperature cell experimentation.

The application of the technique in the light metals industry is very promising since the size of the physical dimensions of both high and low temperature cells are almost the same as the proposed industrial light metal electrowinning cell.

The feasibility of extending the technique to a monopolar stack of rotating electrodes is also being considered. This would considerably enhance the space-time yield of the process.

ACKNOWLEDGEMENTS

The authors wish to extend their gratitude to the Science and Engineering Research Council (Swindon, United Kingdom) and Comalco Ltd. (Melbourne, Australia) for their funding of this research work.

REFERENCES

1. P.M. Copham and D.J. Fray, "The use of rotating electrodes in the electrolysis of molten zinc chloride electrolytes", *Met. Trans.* 21B:977 (1990).
2. R.E.W. Jansson and R.J. Marshall, "The rotating electrolyser. II. Transport properties and design equations", *J. App. Elec.* 8:287 (1978).
3. R.E.W. Jansson, R.J. Marshall and J.E. Rizzo, "The rotating electrolyser. I. The velocity field", *J. App. Elec.* 8:281 (1978)
4. J. Szekely, J.W. Evans, and J.K. Brimacombe, "The Mathematical and Physical Modelling of Primary Metals Processing Operations", John Wiley & Sons, New York (1988)
5. E.A. Ukshe, G.V. Polykova, G.A. Medvetskaya, "Dynamics of chlorine and magnesium in electrolysis of fused chlorides", *J. App. Chem. USSR*, 33:2246 (1960)
6. K. Wiemer, "Aspects of molten salt electrowinning", MSc Dissertation, Dept. Materials Science and Metallurgy, Cambridge (1989).

THE ANODIC OXIDATION OF VANADIUM — A POSSIBLE TECHNICAL APPLICATION OF ELECTROCHEMICAL FIXED BED REACTORS

H. Bergmann and K. Hertwig

Institute for Chemical Engineering
Technical University Köthen
D – 06366 Köthen, Germany

Washing solutions for carbon dioxide separation in the nitrogen industry may contain pentavalent vanadium, which, unfortunately, is reduced to the tetravalent form. An electrochemical reoxidation could be an alternative to the conventional oxidation by air. Based upon experimental studies and calculations, with the help of a mathematical fixed bed reactor model, a process quantification was carried out for a given oxidation capacity. Parametric studies allowed process conditions and cell design to be optimized. The results show that the electrical energy costs are relatively low and are not a limiting factor in the processing.

INTRODUCTION

In alkaline aqueous solutions vanadium exists in a number of species and valencies, mainly in the form of vanadate ions. From the electrochemical point of view, these solutions represent interesting systems. It is to be expected that electrochemically directed reactions without phase changes will occur. Unfortunately, no practical application exists and literature data are very rare[1,2]. Nevertheless, a potential possibility of a technical electrolysis results from an analysis of carbonate containing washing solutions for separating carbon dioxide from synthesis gases[3]. These solutions usually contain 20 % potassium carbonate, 5 % potassium bicarbonate and approximately 3 % alcanolamines as well as pentavalent vanadium as a corrosion inhibitor in a concentration of up to 8000 ppm (calculated as V_2O_5). As a side reaction of the washing process pentavalent vanadium ions are reduced to the tetravalent form, making a periodic regeneration necessary. Reoxidation with the help of atmospheric oxygen has a number of disadvantages, above all, the entry and accumulation of undesirable substances. In cooperation with the Nitrogen Plant, Piesteritz, an electrochemical oxidation step was investigated as an alternative to the air oxidation process.

Electrochemical Engineering and Energy, Edited by
F. Lapicque *et al.*, Plenum Press, New York, 1995

EXPERIMENTAL INVESTIGATIONS

The experimental work included the solving of the following particular problems :
- determination of substance properties
- determination of the electrochemical kinetics
- electrolysis in laboratory scale cells.

As will be shown later, the determination of conductivity values of the electrochemical system is of great importance for reactor design calculations. In the present case, the work was focused on the measurement of the conductivity of the washing solution (WS) and of the anode material (graphite), which was proposed for use. The equation

$$\kappa = 13.4 + 0.193(T - 293) \qquad (1)$$

describes the specific WS-conductivity at a V_2O_5 concentration of 8000 ppm.

The graphite conductivity was investigated for several graphite type beds in a special cylindrical apparatus, calculating the resistance from voltage-current measurements across the particle bed, which was located between two polished end plates. The pressure on the bed could be varied. Table 1 shows data for the specific conductivity of one kind of graphite at one temperature and varied outer pressure on the particle bed. The variants 1,2,3,4 and 5 correspond to pressures of 5.4 ; 10.8 ; 16.3 ; 21.7 and 27.1 kN/m^2.

Table 1. Specific conductivity of graphite particles (particle size < 1.6 mm).

current density	pressure				
$A\ m^{-2}$	1	2	3	4	5
$t = 80°C$					
552	52.1	55.4	62.4	67.1	73.8
1105	64.2	67.1	75.1	82.0	93.3
2210	80.2	84.4	93.8	104.9	109.4
3312	92.3	96.7	106.8	115.6	125.5

It can be seen that with decreasing current density and pressure the bed conductivity falls. The better conductivity values at high current densities cannot be explained clearly. Probably, due to local thermal effects, the contact between the particles becomes closer. The measurements showed that, for real working conditions, conductivities between 50 and 100 $S\ m^{-1}$ are realistic.

The electrochemical kinetics were considered with the help of a stirred kinetic cell, a graphite anode (electrode surface 1 cm^2) and a platinum cathode. Most of the experimental measurements were carried out for WS with concentrations of 3750, 5500 and 7000 ppm of tetravalent vanadium at a vanadium concentration of 8000 ppm. Firstly, polarization curves were determined at low temperatures, which resulted in relationships, difficult to quantify. The plots were very similar and suggested the presence of adsorption effects. But, at higher temperatures, which are relevant in practice, unequivocal curves were obtained. Fig. 1 shows the polarization behaviour at $70°C$. The limiting current is linearly dependent on concentration. For the general oxidation process

$$V^{4+} \rightarrow V^{5+} + e \qquad (2)$$

the equation

$$i_E = \frac{k.F.\exp\frac{\alpha F}{RT}\,\eta_A}{1 + \frac{k}{\beta}\exp\frac{\alpha F}{RT}\,\eta_A} \cdot c_{V^{4+}} \qquad (3)$$

was used to describe the anodic kinetics for higher anode overvoltages. α was calculated as 0.4.

The cathodic reaction was calculated using equation (4)

$$U_K = a + b \ln i_K \tag{4}$$

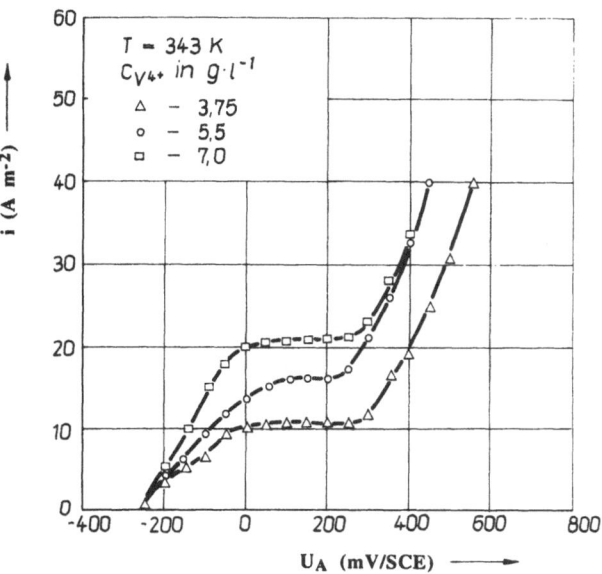

Figure 1. Polarization curves for $T = 343$ K

Preliminary experiments in laboratory scale cells (dimension 0.03 m width by 0.2 m length) were carried out. The size of the graphite particles varied from 0.5 to 5.6 mm. The bed depth of the anodic half cell was changed by replacing frame parts. The cathodic half cell was separated by a PVC-diaphragm. The depth of the cathode space was 1 mm and the cathode was a nickel-plate.

One litre of anolyte was circulated through the cell. There was no catholyte circulation. These preliminary studies had the purpose of obtaining primary information about the possibilities and tendencies of vanadium oxidation. Figure 2 presents cell current cell voltage curves for experiments with different anodic bed depths. This experimental work made it clear that the choice of anode material and the cell current are of primary interest. Especially at higher cell currents desintegration of the anode particles occurred. The calculated values of current efficiency (50 to 70 %) could be correlated with the anodic oxygen evolution as the main side reaction.

MATHEMATICAL MODELLING OF TECHNICAL CELLS

The aim of mathematical modelling was to design technical cells for anodic vanadium oxidation. In particular the following tasks were formulated :
- the calculation of the main dimensions of single cells with anodic graphite beds,
- the calculation of the total number of cells for a given oxidation rate,
- the investigation of the influence of parameters such as concentration and flow rate on the cell unit size,
- the estimation of electrolysis costs.

Initially, all calculations were applied to a basic single cell as shown in Figure 3.

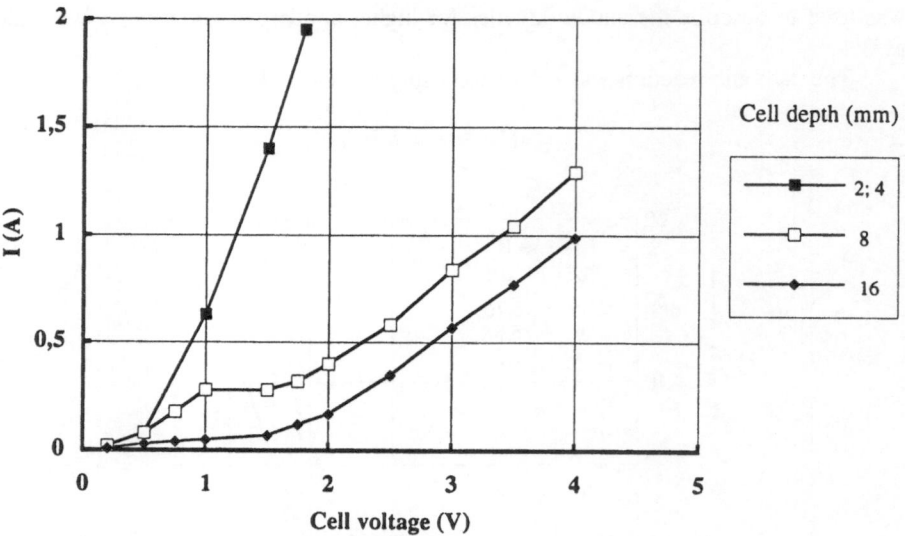

Figure 2. Cell current-cell voltage behaviour of laboratory cells

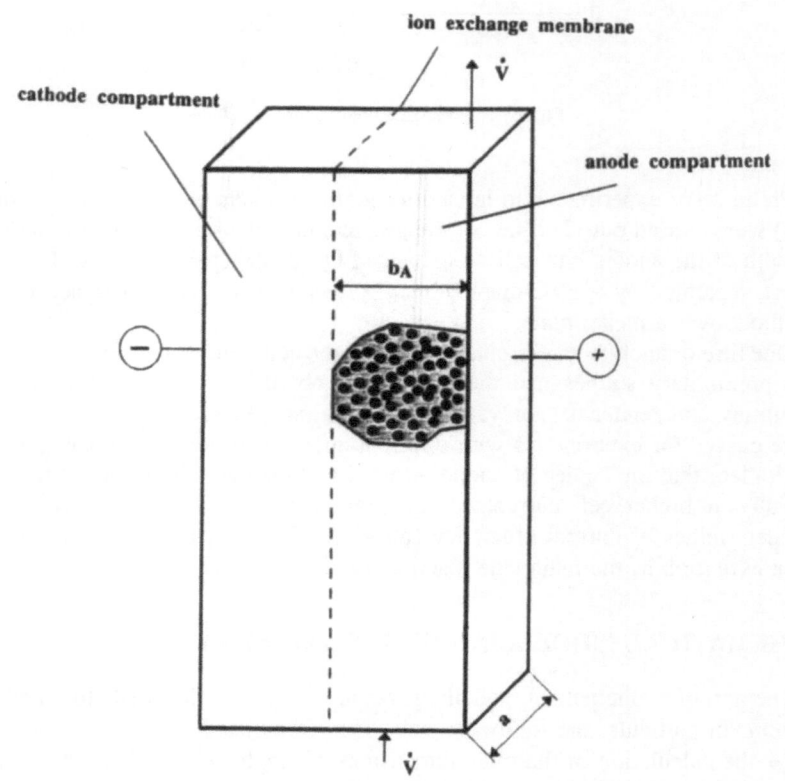

Figure 3. Parameters of a single cell unit

cell height : 0.5 m ; cell width a : 0.10 m ; bed depth b_A : 0.01 − 0.04 m
particle diameter : 2×10^{-3} m ; flow rate $\dot{V}$: $1 \times 10^{-6} − 1 \times 10^{-5}$ m^{3}s^{-1}
particle fraction : 55 %

The cell was considered as an isothermal ideal plug flow reactor at a temperature of $70°C$. For eliminating difficulties in the model simulation for operating conditions with continuously changing inlet conditions, a pseudo-continuous flow regime was chosen. Corresponding to practical demands, it was defined, that 15 m^3 of WS had to be treated in two days, i.e. 0.875 kg V^{4+} oxidized in one hour. The mathematical model of the fixed bed cell has been presented elsewhere[4]. The mass transfer coefficient β, necessary for equation (3), was calculated by equation (5) :

$$Sh = k_1 \ Re^{k_2} \ Sc^{k_3} \qquad (5)$$

A comparison of different correlations for β was done by us in an earlier publication[6]. For determining Sh, Re and Sc the following properties were used : diffusion coefficient $8 \times 10^{-10} \ m^2.s^{-1}$, anolyte density 1270 $kg.m^{-3}$ and viscosity $7.6 \times 10^{-2} \ Pa.s$.

As can be seen from the polarization curves (Fig. 1) oxygen evolution begins at an anode potential of approximately 250 mV. This side reaction must be prevented, because the graphite particles can corrode. Therefore, a maximum anode overvoltage of 150 mV was adopted to avoid model simulation at higher anode potentials.

RESULTS OF MODEL SIMULATION

Within the anode overvoltage limitation mentioned above, the influence of particle conductivity on the behaviour of a single cell (bed depth 40 mm, flow rate $5 \times 10^{-6} \ m^3.s^{-1}$, inlet concentration 5000 ppm V_2O_5) was initially studied. The sample data in Table 2 show the limiting degree of conversion (at an overvoltage of 150 mV) for tetravalent vanadium. It can be seen that, in the expected region, only small differences exist. It is interesting that the four variants differ significantly if the reaction current density across the bed depth (ζ-direction) and the cell length (ξ-direction) are drawn (Fig. 4).

Table 2. Limiting degree of conversion.

κ in $S.m^{-1}$	25	50	100	200
conversion in %	40.5	38.1	36.2	35.5

Even the formation of a minimum is possible. But, in general, all variants are characterized by sufficiently high bed exploitation, which is a consequence of the relatively low bed conductivity, compared to reactors for metal deposition. According to the tetravalent vanadium inlet concentration, two cases were simulated : 3000 ppm and 5000 ppm. Figure 5 presents the variation of outlet concentration for both variants and for varied flow rates and bed depths. The bed length was set at $L = 1 \ m$. Even if the flow rate (and the mass transport condition) is low, deeper beds can give higher degrees of conversion, because the reaction current density does not fall to zero. The increase in outlet concentration becomes more gradual for higher flow rates, which demonstrates the influence of mass transfer enhancement. (It should be remembered, that the simulation was carried out for those cases where an anodic overpotential of 150 mV is only reached at the cell outlet near the separator). The corresponding cell currents for a single reactor are given in Table 3. A tenfold increase in flow rate allows the cell current to increase by approximately 2 to 3 times. For an oxidation capacity of 0.875 kg V^{4+} per hour the number of necessary single cells (n_{ZGE}) results from these data. The graphical illustration of the results (Fig. 6) shows, that for increase in bed depth beyond 40 mm the decrease in the number of single cells is very small.

In the following assessment of cell length influence (or the connection of single cells into cascades), the simulation was implemented only with a bed depth of 40 mm. At larger cell lengths (Table 3) the energy consumption is slightly reduced, but the decrease in the number of single cells is not proportional as the cell length increases. The best proportionality is obtained for doubling of L to 1 m. If, additionally to the cell current and single cell number, the cell voltage is also known, then the annual energy consumption for electrolysis can be calculated. Table 4 shows these values for the given examples. Finally, Table 5 reports values of the annual energy consumption, depending on the flow rate of solution and the bed depth, b_A.

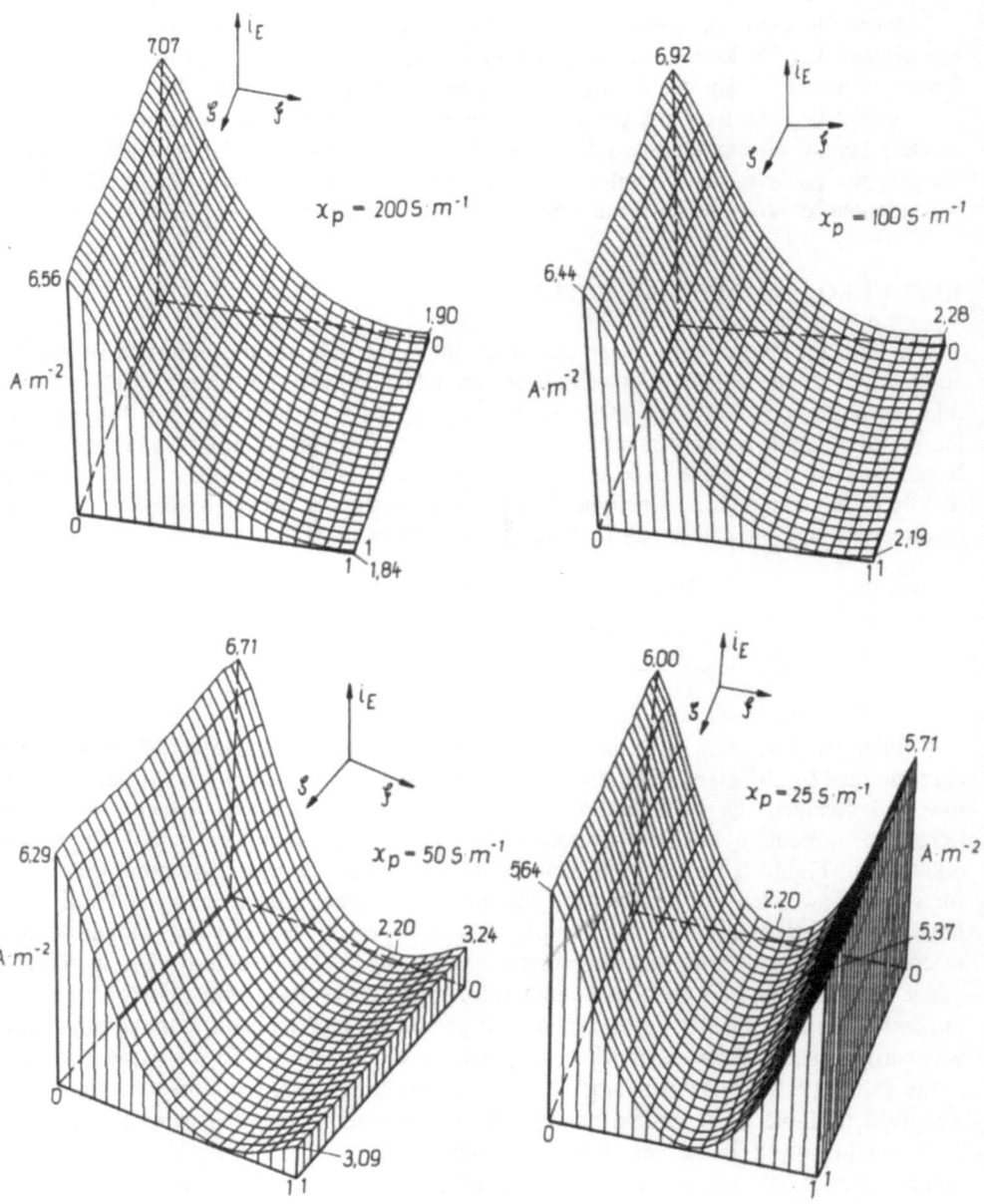

Figure 4. Current density distribution across the anodic bed for varied specific particle conductivity

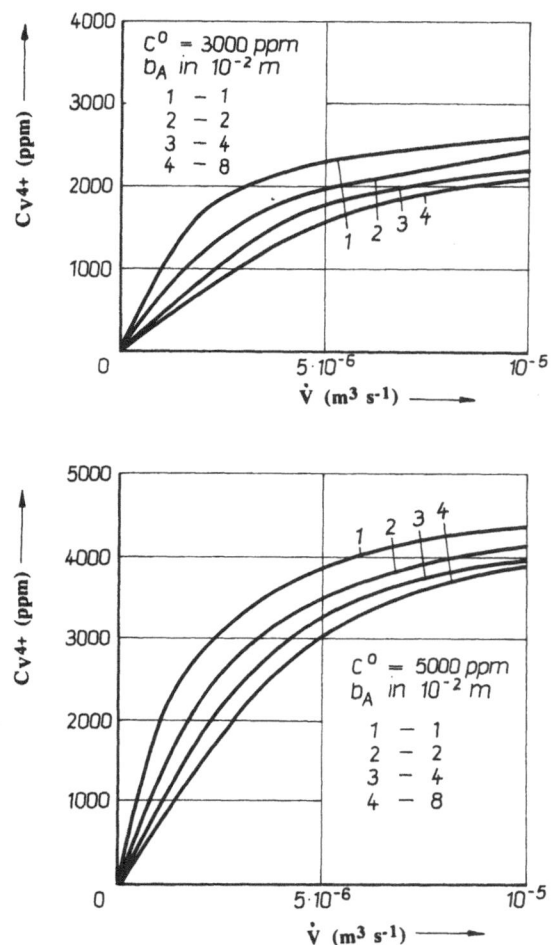

Figure 5. Dependence of V^{4+}-concentration on flow rate and bed depth

Table 3. Cell current per single cell unit (in A)

$\dot{V}$	$C° = 5000\ ppm$				$C° = 3000\ ppm$			
	$b_A\ in\ 10^{-2}\ m$				$b_A\ in\ 10^{-2}\ m$			
$m^3.s^{-1}$	1	2	3	4	1	2	3	4
10^{-6}	3.354	4.024	4.445	4.681	2.019	2.442	2.742	2.900
2×10^{-6}	4.621	5.919	6.850	7.439	2.848	3.780	4.447	4.824
5×10^{-6}	5.838	8.079	9.600	10.638	3.650	5.329	6.583	7.458
10^{-5}	6.389	9.059	10.890	11.054	4.093	6.055	7.874	8.719

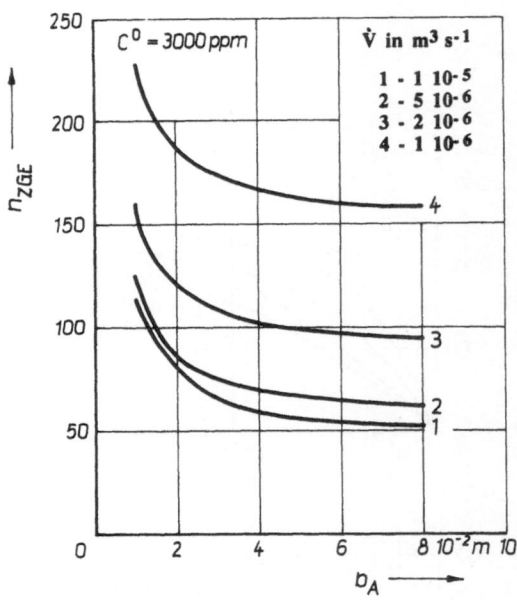

Figure 6. Number of single cell units for several flow rates and bed depths

Table 4. Influence of cell length
$b_A = 4 \times 10^{-2} \ m \quad \dot{V} = 5 \times 10^{-5} \ m \quad C° = 5000 \ ppm$

Cell length L	0.5 m	1 m	2 m
Number of single cell units $(0.05.b_A.L)$	85	48	30
Cell voltage in V	2.229	2.143	2.000
Cell current in A	5.445	9.649	15.309
Electrical energy consumption in kWh (8000 h/a)	8253	7940	7348

Table 5. Annual electroenergy consumption in kWh (8000 h/a)

$\dot{V}$ $m^3.s^{-1}$	$C° = 5000 \ ppm$ $b_A \ in \ 10^{-2} \ m$				$C° = 3000 \ ppm$ $b_A \ in \ 10^{-2} \ m$			
	1	2	3	4	5	6	7	8
10^{-6}	6233	6341	6472	6525	5874	5917	5970	6009
2×10^{-6}	6650	6914	7123	7438	6171	6382	6564	6654
5×10^{-6}	7024	7456	8506	8285	6390	6779	7152	7531
10^{-5}	7058	7724	8156	8317	6556	7555	7559	7893

CONCLUSIONS

The results of practical investigation and model simulation, which was done with a high mathematical safety margin, emphasize a very good possibility for electrochemical vanadium oxidation in washing or similar systems. A relatively small oxidation requirement and specific electrical energy consumption of nearly 1000 kWh per ton of vanadium, results in low energy costs for electrolysis. The calculated size of electrolyzers also allows calculation of the investment costs for commercially available cells.

Last but not least, the simulation results indicate, for different cell working conditions, the maximum anode potential below which the anodic bed outlet must be held to avoid significant graphite corrosion.

NOMENCLATURE

a	bed width	m
a	constant in eq. (4)	V
b	constant in eq. (4)	$V.(ln\ A.m^{-2})^{-1}$
b_A	anode bed depth	m
$c_{V^{4+}}$	V^{4+}-concentration	$ppm\ V_2O_5$
c°	V^{4+}-inlet concentration	$ppm\ V_2O_5$
F	Faraday-constant	$As.mol^{-1}$
i_E	reaction current density	$A.m^{-2}$
i_K	cathodic current density	$A.m^{-2}$
k	constant in eq. (3)	$m.s^{-1}$
k_1	constant in eq. (5)	
L	length of anode bed	m
R	gas constant	$J.mol^{-1}.K^{-1}$
Re	Reynolds number	$Re = \frac{W.d_p}{\nu \varepsilon}$
Sh	Sherwood number	$Sh = \frac{(1-\varepsilon_p)\beta.d_p}{\varepsilon_p.D}$
Sc	Schmidt number	$Sc = \frac{\nu}{D}$
T	temperature	K
t	temperature	$^\circ C$
U_A	anode potential	V
U_K	cathode potential	V
U_Z	cell voltage	V
$\dot{V}$	flow rate	$m^3.s^{-1}$
W	electrolyte velocity	$m.s^{-1}$
ZGE	single cell unit	

Symbols

α	symmetry factor	
β	mass transfer coefficient	$m.s^{-1}$
ε_p	particle fraction	
ζ	dimensionless coordinate	
η_A	anodic overvoltage	V
κ	solution conductivity	$S.m^{-1}$
κ_P	specific graphite conductivity	$S.m^{-1}$
ξ	dimensionless coordinate	
τ	exploitation time per annum	hr

ACKNOWLEDGEMENT

This work was sponsored by the Ministry of Science and Research of Saxony-Anhalt.

REFERENCES

1. K.J. Vetter. "Elektrochemische Kinetik", Springer Verlag, Berlin, 397 (1961).
2. DE OS 2429035.
3. DDR-Patent 148647.
4. J. Wiegand. Großer Beleg, TH Köthen (1990).
5. H. Bergmann. u.A. Dechema Monographien Band 125: 65-82, VCH Verlagsgesellschaft, Weinheim (1982).
6. H. Bergmann and K. Hertwig, *Chemische Technik*. 45, 1: 11 (1993).

ELECTROSYNTHESIS OF AQUEOUS Ti^{3+} SOLUTIONS

G.H.Kelsall, D.J.Robbins* and W.Wang

Department of Mineral Resources Engineering
Imperial College, London SW7 2BP, UK

* Now at: British Nuclear Fuels Ltd
Sellafield Works, Seascale, Cumbria CA20 1PG, UK

ABSTRACT

The reduction of Ti(IV) species in aqueous H_2SO_4 and HCl was studied using an amalgamated lead rotating disc electrode. Whereas a simple one electron Ti(IV) reduction mechanism was operative in HCl, Ti(IV) reduction in H_2SO_4 appeared to involve a Ti(IV)-Ti(III) intermediate, which absorbed visible light strongly, and proceeded via an ECE mechanism. A potentiostatically controlled amalgamated lead plate cathode was used to reduce Ti(IV) species in acidic chloride electrolytes in a membrane-divided reactor operated in batch recycle mode with a continuously stirred tank reservoir. The dependences of the current efficiency for Ti(III) formation on Ti(IV) mass transport rates to the cathode and on its electrode potential, were investigated experimentally and modelled theoretically.

INTRODUCTION

Ti(III) species have been used as reductants for a range of organic and inorganic reactions[1,2], including the reduction of aqueous absorbed SO_2 to elemental sulfur, which might form the basis of a flue gas desulfurisation process[3].

TiO^{2+} ions have been reported as predominating in acidic Ti(IV) solutions in the absence of TiO_2 [4-6]; complexation occurs in chloride electrolytes:

$$TiO^{2+} + nCl^- \Leftrightarrow \left[TiOCl_n \right]^{(2-n)+} \quad ; \quad n = 1 - 4 \qquad [1]$$

$TiOCl^+$ and $TiOCl_2$ species predominate[6,7] at pH < 0 and [Cl$^-$] of 1 - 5 kmol m^{-3}, though there is an equilibrium between TiO^{2+} ions and dihydroxo species, protonated monomers and oligomeric species to some extent[8]. The structure of Ti(IV) species in acidic sulphate electrolytes is less certain, though complexation has been assumed[9] to occur by reactions such as:

$$TiO^{2+} + SO_4^{2-} \Leftrightarrow TiOSO_4 \qquad [2]$$

Electrochemical Engineering and Energy, Edited by
F. Lapicque *et al.*, Plenum Press, New York, 1995

In dilute acids, the main Ti(III) species are $[Ti(H_2O)_6]^{3+}$ ions, whereas complexation occurs in more concentrated HCl [10]:

$$Ti(H_2O)_6^{3+} + nCl^- \Leftrightarrow \left[Ti(H_2O)_{(6-n)} Cl_n \right]^{(3-n)+} + nH_2O \qquad [3]$$

Ti(IV) reduction in acidic solution is usually expressed in terms of the reaction:

$$TiO^{2+} + 2H^+ + e^- \Leftrightarrow Ti^{3+} + H_2O \qquad [4]$$

$$E \quad / V = 0.1013 - 0.1183 pH + 0.0591 \log \left\{ (TiO^{2+}) / (Ti^{3+}) \right\}$$

However, the literature[9] on the electrochemical behaviour of aqueous Ti(IV)/Ti(III) systems is confused and confusing. Casalbore *et al.*[11] investigated aqueous Ti(IV) reduction at a mercury electrode as a function pH; the polarographic wave with $E_{1/2}$ = - 0.75 V vs. S.C.E. indicated an irreversible reduction, though at the higher acidities, the changed behaviour was attributed to an adsorption process prior to Ti(IV) reduction. The kinetics of the Ti(III)/Ti(IV) couple at a glass carbon rotating disk electrode[12] implied charge transfer coefficients of 0.18 and 0.11 for the oxidation and reduction reactions, respectively. Habashy[13] found that the Ti(III)/Ti(IV) couple became reversible at mercury electrodes in concentrated acid sulfate (10 kmol H_2SO_4 m^{-3}). The voltammetric behaviour of Ti(III)/Ti(IV) at glassy carbon electrodes in H_2SO_4 solutions suggested a first order oxidation of Ti(III) to Ti(IV) [14].

Goroshchenko and Godneva[15] reported that the addition of Ti(IV) to Ti(III) in sulphuric acid caused the solution to darken; the absorbance maximum shifted from 527 nm for Ti(III) to 472 nm when the Ti(IV) content was 50 mol %. In addition, in 12 kmol m^{-3} HCl, the formation of "purplish black" mixed oxidation state dimeric species has been reported[16], with an equilibrium constant of approximately 12.

The objective of the presently reported work was to investigate and model the effect, in chloride media, of the Ti(IV) flux to a Pb-Hg cathode on the current efficiency for Ti(III) formation in an electrochemical reactor, operated in batch recycle mode. A model of Ti(IV) reduction by an ECE mechanism in sulfate media will be reported later[17].

EXPERIMENTAL

The electrochemical reactor (Fig.1), made of polymethylmethacrylate (PMMA), was designed with an anolyte compartments (94 x 30 x 200 mm) on either side of a single catholyte compartment (94 x 30 x 200 mm). A 99.99 %+ lead plate cathode (100 x 200 mm, Cookson Ltd.) was amalgamated in the reactor by electrodeposition with mercury, using 0.01 kmol $Hg(NO_3)_2$ + 1 kmol HNO_3 m^{-3} as catholyte. The phase diagram for the Pb-Hg system shows complex and uncertain behaviour, with at least 20 atomic % mercury soluble in the lead lattice[18]. The anode was a platinised expanded titanium mesh (Magneto Chimie B.V., The Netherlands). Cation exchange membranes (Nafion 417, Du Pont Inc.) were sited on either side of the catholyte compartment to prevent re-oxidation of the reductant at the anode. The catholyte disperser consisted of 5 mm diameter holes drilled in the PMMA and covered with a fine (0.25 mm) polypropylene mesh, to enable the use of a fluidised bed electrolyte of glass beads, if required. A reference electrode compartment was bolted onto the side of the catholyte compartment and connected hydraulically to the catholyte by a fine hole drilled into the side of the cell and emerging adjacent to the cathode surface.

Separate hydraulic circuits using uPVC pipe work involved:

a) catholyte: a 3.5 dm^3 glass reservoir connected to a magnetically coupled polypropylene pump (Totton Electrics Ltd.), digital flow meter (George Fisher Ltd.), angle seat valve; quartz fibre optic links between a 10 mm path length quartz optical cell and a diode array UV-visible spectrophotometer (Hewlett-Packard HP8451) enabled direct measurement of the Ti(III) concentration in the 2.5 dm^3 of analytical grade 2 kmol m^{-3} HCl (or 2.5 dm^3 2 kmol m^{-3} H$_2$SO$_4$). Analytical grade Ti(IV) sulfate or chloride (Merck Ltd) was used as reactant in high purity water (Elga Ltd., Elgastat and Prima).

b) anolyte: 3.0 dm^3 glass reservoir connected to a magnetically coupled polypropylene pump (Totton Electrics Ltd.), rotameter (George Fisher Ltd.) and angle seat valve to control the flow rate of 2.5 dm^3 2 kmol m^{-3} H$_2$SO$_4$.

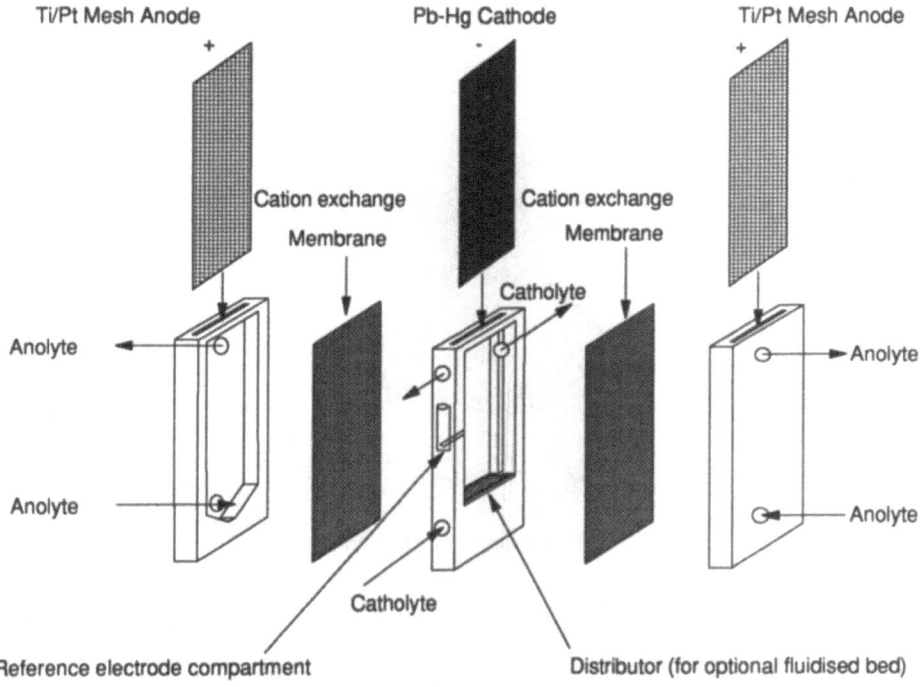

Figure 1. Electrochemical reactor for reductant synthesis at a Pb-Hg cathode.

The cathode potential was controlled using a Sycopel Autostat (6 A / 28 V) potentiostat and a Hi-Tek wave form generator. An IBM-compatible microcomputer, a Strawberry Tree data acquisition card (Adept Scientific Ltd.) and Labtech Notebook software (Adept Scientific Ltd.) was used for data acquisition. An Oxford Electrodes rotating disc electrode system was used in the preliminary voltammetric experiments with a lead disc electrode, amalgamated as described above.

RESULTS AND DISCUSSION

Absorption Spectrophotometry

Fig.2 shows typical UV-visible absorption spectra of Ti(III) and Ti(IV)-Ti(III) solutions. The enhanced absorbance of the latter in sulphuric acid electrolytes probably resulted from intramolecular charge transfer in a mixed oxidation state oxo-bridge intermediate incorporating (hydrogen) sulphate; no such effects were evident in hydrochloric acid electrolytes, as indicated by Fig.3. The existence of such Ti(IV)-Ti(III) species has not been mentioned previously in the literature on the kinetics of the Ti(IV)/Ti(III) couple in sulfate media. Fig.4 shows a possible structure for such Ti(IV)-Ti(III) species, which could be formed by reactions such as:

$$TiOSO_4 + Ti^{3+} + HSO_4^- + H_2O \xrightarrow{k_c} [HSO_4Ti^{IV} {}_O^O Ti^{III} HSO_4]^+ + H^+ \qquad [5]$$

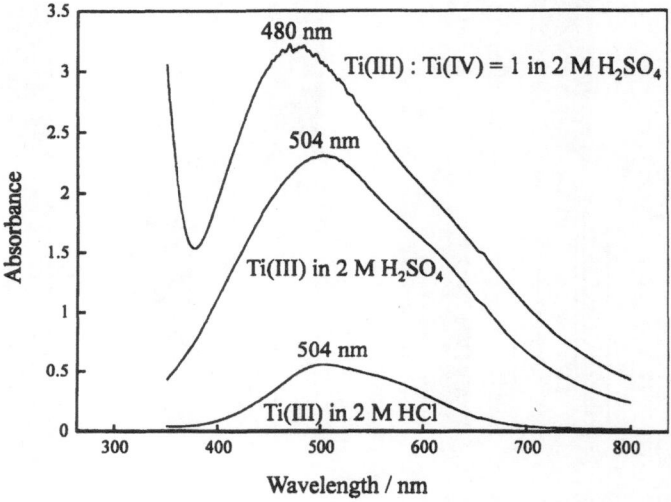

Figure 2. UV-visible absorption spectra of 100 mol m^{-3} dissolved titanium.

However, it is not yet clear whether such reactions involve an equilibrium or are chemically, though not electrochemically, irreversible; this is being investigated presently. If the latter is the case, then the complex could be prepared in its pure state, enabling its absorbance spectrum to be determined. This is a prerequisite for the deconvolution of the absorption spectra measured during Ti(IV) reduction in sulfate media and would enable on-line determination of solution compositions and hence current efficiencies. No such difficulties exist for Ti(IV) reduction in chloride media.

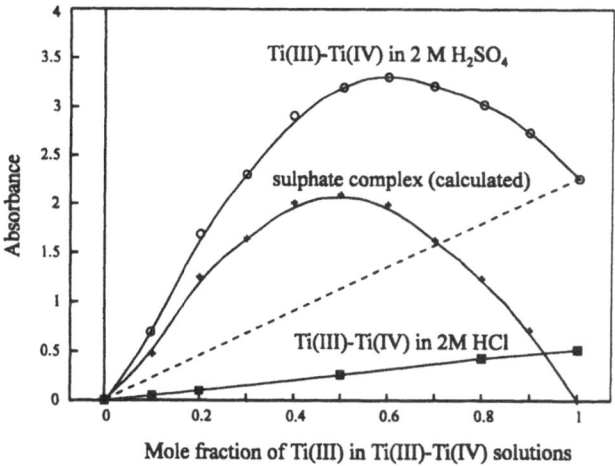

Figure 3. Visible absorbance maxima of 100 mol m^{-3} total dissolved titanium.

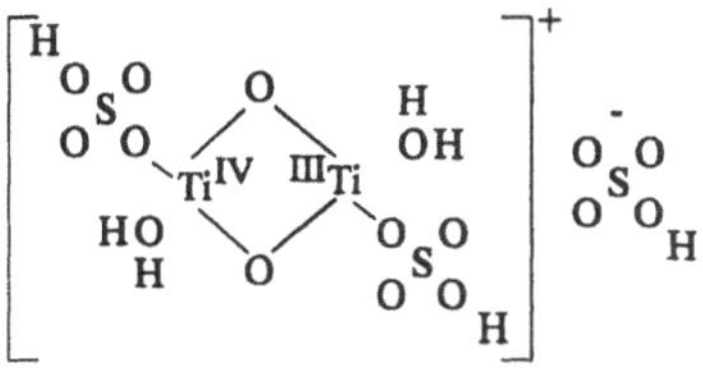

Figure 4. Possible structure of Ti(III)-Ti(IV)-sulphate complex.

Voltammetry and Batch Recycle Reduction of Ti(IV)

The voltammograms shown in Fig.5 for the reduction of Ti(IV) species at an amalgamated lead rotating disc electrode, show that, at potentials < - 0.85 V, the transport limited current density (i_L) in 2 kmol m^{-3} HCl was more than twice that in 2 kmol m^{-3} H$_2$SO$_4$. Data from both types of electrolytes fitted the Levich equation:

$$i_L = 1.554 zFv^{-1/6}D^{2/3}\varpi^{0.5}[Ti(IV)] \qquad [6]$$

but, as implied by Fig.5, the slope of plots of transport-limited current density against square root of rotation rate were radically different. The effective diffusion coefficient for Ti(IV) species in 2 kmol HCl m^{-3} was derived as 1.8x10^{-9} m^2 s^{-1}; according to the Stokes-Einstein equation, this corresponds to an effective hydrodynamic radius of 1.1 nm for the diffusing species, which have been predicted[6] to be an approximately equimolar mixture of TiOCl$_2$ and TiOCl$^+$ ions in 2 kmol HCl m^{-3}. Thus, the Ti(IV) reduction reaction can be represented by:

$$[TiOCl_n]^{(2-n)+} + 2H^+ + e^- \rightarrow Ti^{3+} + nCl^- + H_2O \qquad [7]$$

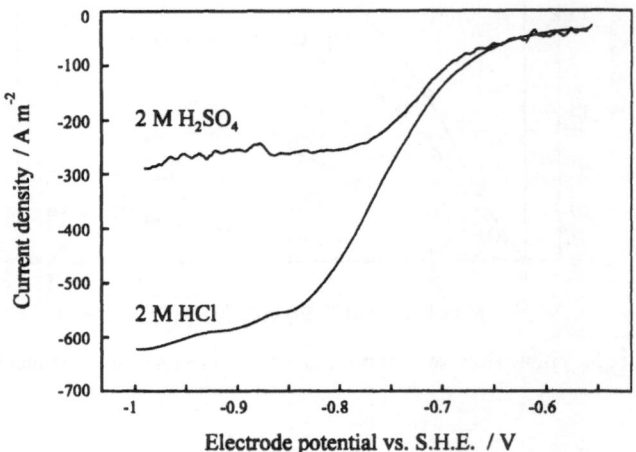

Figure 5. Voltammograms of 100 mol Ti(IV) m^{-3} at a rotating amalgamated lead disc electrode in 2 kmol m^{-3} HCl and 2 kmol m^{-3} H$_2$SO$_4$, 10 mV s^{-1}, 9 Hz.

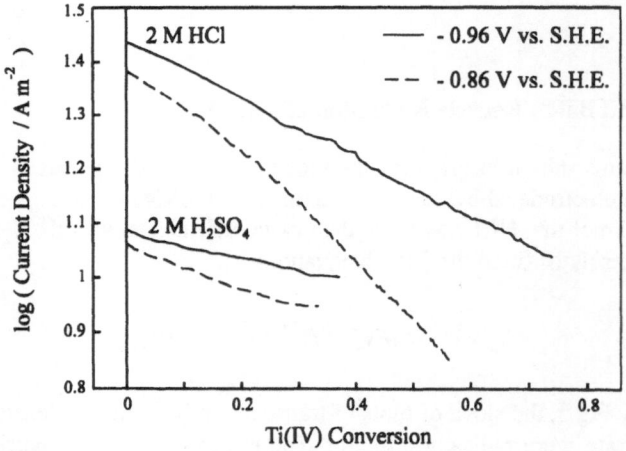

Figure 6. Dependence of partial current density on conversion for batch recycle reduction of Ti(IV) in 2 kmol m^{-3} HCl and 2 kmol m^{-3} H$_2$SO$_4$. Initial [Ti(IV)] = 50 mol m^{-3}, Re = 832.

Differences in electrolyte viscosity accounted for a decrease in the slopes of the Levich plots of only 22 % on replacing HCl with H_2SO_4, whereas the apparent diffusion coefficient of Ti(IV) species in sulfate media was derived as 0.72×10^{-9} m^2 s^{-1}, corresponding to an effective hydrodynamic radius of about 2 nm. Despite the uncertainty about whether $TiOSO_4$ species exist and the exact nature of Ti(IV) species in acidic sulfate solutions, this corresponds to a degree of polymerisation which is unreasonable for a 2 kmol m^{-3} H_2SO_4 electrolyte, particularly if the thermodynamic predictions[6] are essentially correct, though they neglect sulfate complexation, for which no data is currently available.

An alternative hypothesis to explain the Ti(IV) reduction behaviour in sulfate media is that the reaction mechanism is not a simple one electron process, but involves an ECE mechanism, which might be expected if the Ti(IV)-Ti(III) species, discussed above, were formed as intermediates. Unfortunately, the normal voltammetric diagnostic tests for ECE processes were precluded by the electrode materials requirements of a highly irreversible reaction, which needed potentials well into the hydrogen evolution region to achieve transport control; re-oxidation of the Ti(III) product was precluded by the oxidation of Pb-Hg at potentials > -0.4 V vs. S.H.E., though transport controlled oxidation was achieved on a gold electrode only at a potential of 0.9 V vs. S.H.E..

Fig.6 shows comparisons of the decay in current density for Ti(IV) reduction in both 2 kmol m^{-3} HCl and 2 kmol m^{-3} H_2SO_4 electrolytes; the initial current density in the former being more than twice that in the latter is consistent with the data from rotating disc experiments (e.g., Fig.5). Unfortunately, it is not possible to determine the corresponding data for the time evolution of the Ti(IV), Ti(IV)-Ti(III) and Ti(III) concentrations, until the absorbance spectra have been deconvoluted, which requires preparation of solutions of the Ti(IV)-Ti(III) species in pure form. However, initial current efficiencies for Ti(IV) reduction in 2 kmol m^{-3} H_2SO_4 can be estimated as >90 % for 100 mol Ti(IV) m^{-3} by comparison with data obtained in 2 kmol m^{-3} HCl.

Model of Current Efficiency for Ti(IV) Reduction in HCl Media

The time dependence of the quasi-steady state concentrations ($c_{(t)}$) of electro-active species reacted under mass transport control in a plug flow reactor (PFER) operated in batch recycle mode with a continuous stirred tank reactor (CSTR) reservoir of volume V, can be expressed approximately as[19]:

$$C_{(t)} = C_{(0)} \exp\left\{ -\frac{ut}{V}\left[1 - \exp\left(-\frac{k_L 2gwa_e L}{u}\right)\right]\right\} \qquad [8]$$

and hence the mass transport controlled current (I_L) is given by[19]:

$$I_L(t) = zFuC_{(0)} \exp\left\{-\frac{ut}{V}\left[1 - \exp\left(-\frac{k_L 2gwa_e L}{u}\right)\right]\right\}\left[1 - \exp\left(-\frac{k_L 2gwa_e L}{u}\right)\right] \qquad [9]$$

In the presence of hydrogen evolution as a parallel, loss reaction, the rate of which is constant at constant potential during a batch depletion experiment, the current efficiencies (Φ) can be obtained from:

$$\Phi = \frac{I_L}{I_L + I_H} \qquad [10]$$

From Ti(IV) concentration depletion data, $\bar{k}_L$ (6.7×10^{-6} m s^{-1} at Re = 832 and - 0.96 V) was deduced from equation [8] and I_L was then derived from equation [9], which was used to calculate the current efficiencies from equation [10], as a function of the experimentally determined dependence of the hydrogen evolution rate (I_H) on the electrode potential. Some predictions from the model are shown in Fig.7.

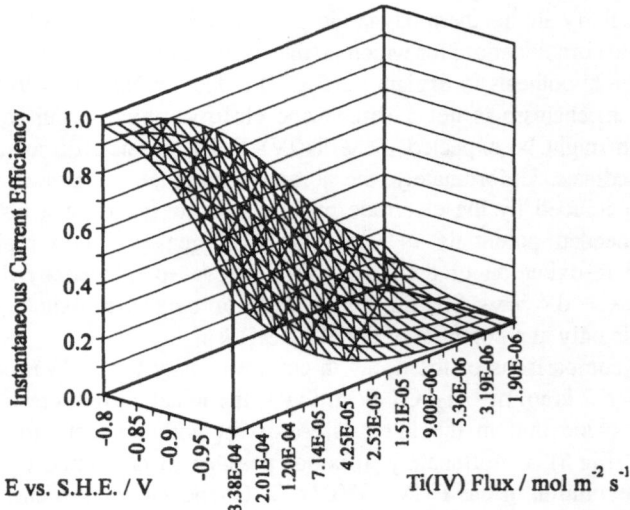

Figure 7. Theoretical instantaneous current efficiencies as functions of electrode potential and flux of Ti(IV), reduced under transport control in parallel with hydrogen evolution on Pb-Hg in 2 kmol m^{-3} HCl.

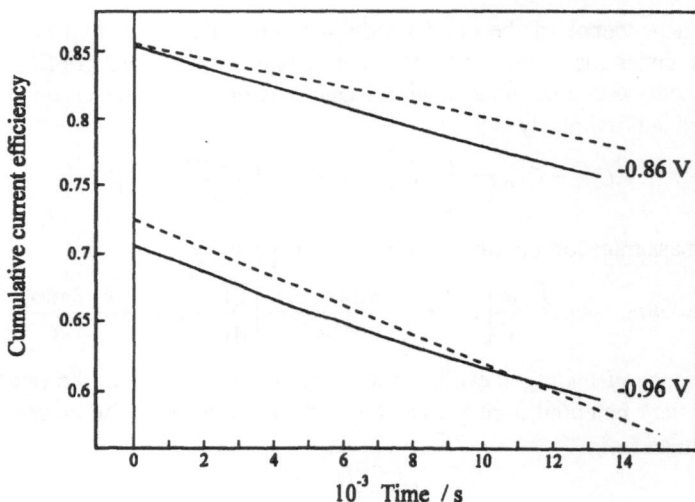

Figure 8. Theoretical (---) and experimental (—) cumulative current efficiencies as a function of potential for reduction of 50 mol m^{-3} Ti(IV). Re = 832.

136

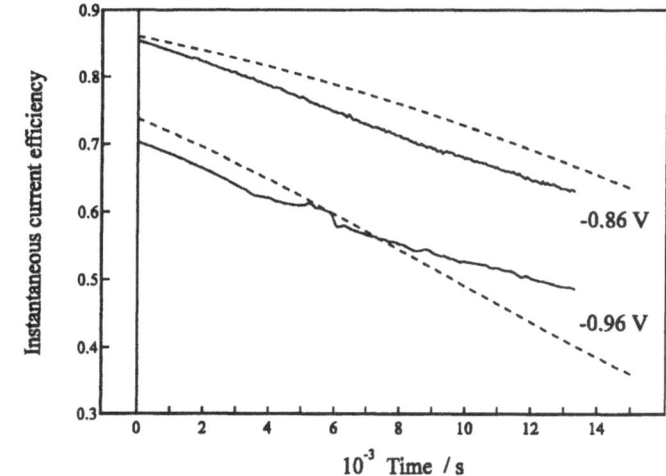

Figure 9. Theoretical (---) and experimental (—) instantaneous current efficiencies as a function of potential for reduction of 50 mol m^{-3} Ti(IV). Re = 832.

Comparison of Model Predictions and Experimental Results

Theoretically derived current efficiencies are compared with their experimental equivalents in Fig.8 and Fig.9, the former showing cumulative and the latter showing incremental current efficiencies, assuming that hydrogen evolution was the only loss reaction and that its rate, determined prior to adding Ti(IV) to the electrolyte, was time invariant. Theoretical predictions of current efficiencies were made as explained earlier. Mean mass transport rate constants (k_L), varying from 5.3 x 10^{-6} m s^{-1} at -0.86 V to 6.7 x 10^{-6} m s^{-1} at -0.96 V, were obtained from the whole partial current data set (0 to 0.7 conversion). Fig.10 shows corresponding data for the decay of current density with increasing conversion of Ti(IV) to Ti(III) as a function of initial Ti(IV) concentration. For the theoretical plots in Fig.10, a value of k_L of 6.5 x 10^{-6} m s^{-1} was obtained over the conversion range 0 to 0.2 for 25 mol Ti(IV) m^{-3}. Calculating the mass transport rate in this way minimised additional factors which caused current enhancement at high conversion and/or at higher initial Ti(IV) concentrations, as discussed below.

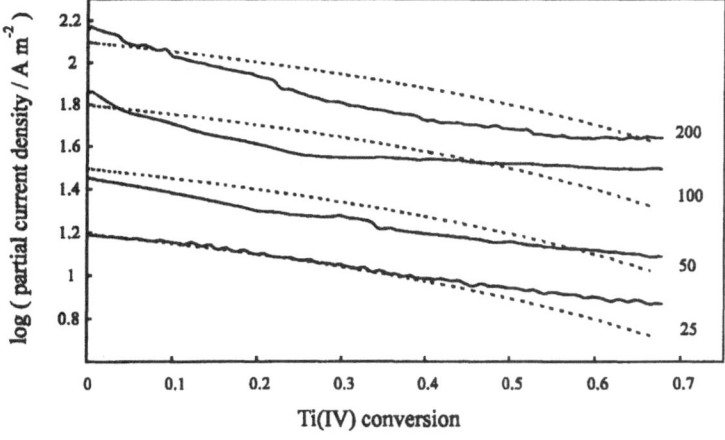

Figure 10. Dependence of theoretical (---) and experimental (—) Ti(IV) partial current densities on conversion and initial concentration (mol m^{-3}). Pb-Hg cathode potential = -0.96 V vs. S.H.E., Re = 832.

At least three processes, not considered in the model, could have been responsible for the deviations between theoretically predicted and experimental behaviour. Firstly, visual observation indicated some time-dependence of the electrode surface coverage of hydrogen bubbles[20], leading to a time-dependent active electrode area. This effect may also be potential dependent, as would the micro-scale turbulence associated with bubble evolution. However, the main reason for higher apparent mass transport rates at -0.96 V than at -0.86 V probably arose from the latter potential resulting in mixed, rather than total transport control, as implied by Figs.5 and 6.

Secondly, Ti(IV) solutions are oxidising to the Pb-Hg cathode under open circuit conditions, which arose as the electrolyte was pumped into the reactor, prior to potentiostatic control being established:

$$2[TiOCl_3]^- + 4H^+ + Pb(Hg) \rightarrow 2Ti^{3+} + PbCl_3^- + 3Cl^- + 2H_2O \qquad [11]$$

Though this chemical reduction of Ti(IV) should not affect the overall current efficiency, the subsequent deposition of the lead from its chloro complex(es), could affect the active surface area and its catalytic behaviour for hydrogen evolution. However, no differences were found in the hydrogen evolution current / potential behaviour of the cathode, prior to, and following its use for Ti(IV) reduction; i.e., impurities from the Ti(IV) electrolyte, if electrodeposited at all, did not remain on the cathode after the reactor was drained following Ti(IV) reduction.

Thirdly, oxygen from the anode reaction, diffusing from the anolyte, through the cation exchange membrane and into the catholyte, would lead to a loss in current efficiency for Ti(IV) reduction. Whilst the diffusion coefficient of oxygen in Nafion materials is significantly decreased with respect to its values in aqueous electrolytes, its solubility is approximately doubled[21], so that its transport rate through such membranes is still significant. This gives rise to a steady state current (predicted as ≈ 2 A m^{-2} for the experimental conditions used) at long times when the dissolved oxygen in the catholyte reaches a plateau value. Rotating disc electrode measurements showed the reaction to involve 4e$^-$ per O_2 molecule at -0.86 V; at the lower potential of -0.96 V there was an increase in the oxygen partial current due to micro-scale convection caused by the enhanced rate of evolution of hydrogen bubbles[20]. Though the rate of homogeneous reactions such as:

$$4Ti^{3+} + 12Cl^- + 2H_2O + O_2 \rightarrow 4[TiOCl_3]^- + 4H^+ \qquad [12]$$

are slow, and indeed was one reason for choosing Ti(III) species for sulfur dioxide reduction[3], the corresponding reactions with H_2O_2 appeared much faster; hence, oxygen reduction at the cathode could be coupled to the homogeneous reaction of Ti(III) and H_2O_2 as an intermediate product of oxygen reduction. A model for current efficiency losses due to oxygen reduction will be reported later[22]; however, its inclusion as another reaction in parallel with hydrogen evolution and Ti(IV) reduction does not fully explain the deviation of experimental data from theoretical expectations.

Despite the lack of detailed knowledge of the species involved, current efficiencies for the transport controlled reduction of Ti(IV) in parallel with hydrogen evolution in acidic chloride media, have been shown to be predicted adequately by a simple model, indicating a cumulative current efficiency of > 90 % for 90 % conversion.

ACKNOWLEDGEMENTS

The authors thank the UK Science and Engineering Research Council for a grant which provided a research associateship for D.J.R., and the British Council and Chinese Government for a research studentship for W.W.

NOMENCLATURE

Symbol	Physical Meaning	Units
a_e	specific surface area of cathode	m^{-1}
c_i	electrolyte concentration of species i	$mol\ m^{-3}$
D_i	diffusion coefficient of species i	$m^2\ s^{-1}$
d_e	Equivalent hydraulic diameter $= 2(g\ w) / (g+w)$	m
E	electrode potential vs. reference electrode	V
F	Faraday constant	$96485\ C\ mol^{-1}$
g	electrode - membrane gap	m
$I_L\ (i_L)$	mass transport limited current (current density)	$A\ (A\ m^{-2})$
k_L	mass transport rate constant	$m\ s^{-1}$
k_c	second order rate constant	$mol^{-1}\ m^3\ s^{-1}$
L	cathode length	m
Re	Reynolds number $= u\ d_e / (2g\ w\ v)$	-
u	electrolyte volumetric flow rate	$m^3\ s^{-1}$
t	time	s
V	catholyte volume	m^3
w	electrode width	m
z	mol electrons per mole of reactant	-
Φ	fractional current efficiency	-
ν	electrolyte kinematic viscosity	$m^2\ s^{-1}$
ω	rotating disc electrode rotation rate	s^{-1}

REFERENCES

1. M.D.Ravi, V.N.S.Pillai, P.N.Anantharaman, Electroreduction of m-nitroaniline to m-phenylene diamine using Ti(III)/Ti(IV) redox system, *Bull.Electrochem.*, 4(3):241 (1988).
2. R.F.Savinell, C.C.Liu, R.T.Galasco, S.H.Chiang, J.F.Coetzee, Discharge characteristics of a soluble iron-titanium battery system, *J.Electrochem.Soc.*, 126: 357 (1979).
3. G.H.Kelsall, D.J.Robbins, SO_2 reduction by electrogenerated reductants. I. Outline of the proposed process, *Trans.I.Chem.E.*, 69B:43 (1991).

4. J.D.Ellis, G.A.K.Thompson, The Cr^{2+} reduction of titanium(IV). Comparisons with the Cr^{2+} reduction of VO^{2+} and evidence for a TiO^{2+} structure in aqueous solutions, pH ≤ 1, *Inorg.Chem.*, 15:3172 (1976).

5. B.S.Brunschwig, N.Sutin, Titanium(III) chemistry: electron-transfer reactions with ground-state poly(pyridine)osmium(III) complexes, quenching reactions with excited-state poly(pyridine)ruthenium(II) complexes, and formal reduction potentials, *Inorg.Chem.*,18:1731 (1979).

6. G.H.Kelsall, D.J.Robbins, Thermodynamics of Ti-(F-)(Fe)-H_2O systems at 298 K, *J.Electroanal.Chem.*, 283:135 (1990).

7. B.I.Nabivanets, L.N.Kudritskaya, Determination of the composition and dissociation constants of titanyl chloro-complexes. *Russian J.Inorg.Chem.*, 12(4):78 (1967).

8. P.Comba, A.Merbach, The titanyl question revisited, *Inorg.Chem.*, 26:1315 (1987).

9. W.J.James, M.E.Straumanis, Titanium; In: "Encyclopedia of Electrochemistry of the Elements", Vol.5, Ch.V-7, 305-395. A.J.Bard (Ed.), Marcel Dekker, New York, (1976).

10. H.J.Gardner, Chloroaquo complexes of titanium(III). The $^2T_{2g} \to {}^2E_g$ absorption band, *Austral.J.Chem.*, 20:2357 (1962).

11. G.Casalbore, P.G.D.Macro, G.Giro, Polarographic study of Ti(IV) in aqueous hydrochloric acid, *J.Electroanal.Chem.*, 111:369 (1980) .

12. M.P.Marani, M.Bartolozzi, M.R.Moncelli, Determination of the kinetic parameters for the Ti(III)/Ti(IV) couple using a rotating disk electrode, *J.Electroanal.Chem.*, 209:275 (1986).

13. G.M.Habashy, Studies of reversibility of Ti(III) and Ti(IV) in acid sulphate medium, *Z.Anorg.Allgem.Chem.*, 306:312 (1960).

14. P.Kiekens, J.Vandenbruwaene, E.Temmerman, Voltammetric behaviour of the Ti(IV)/Ti(III) couple at glassy carbon in H_2SO_4 solutions, *Bull.Soc.Chem.Belg.*, 90:351 (1981).

15. Ya.G.Goroshchenko, M.M.Godneva, Absorption spectra in the visible region of sulphuric acid solutions of titanium sulphates, *Russ.J.Inorg.Chem.*, 6(6):744 (1961).

16. C.K.Jørgensen, Comparative ligand field studies. IV. Vanadium(IV), titanium(III), molybdenum(V) and other systems with one d-electron, *Acta Chem.Scand.*, 11:73 (1957).

17. G.H.Kelsall, W.Wang, I.Cservenyák, Electrosynthesis of titanium(III) in sulfate media, to be published.

18. M.Hansen, "Constitution of Binary Alloys", McGraw-Hill, New York (1958).

19. A.T.S.Walker, A.A.Wragg, The modelling of concentration-time relationships in recirculating electrochemical reactor systems, *Electrochim.Acta*, 22:1129 (1977).

20. H.Vogt, Superposition of microconvective and macroconvective mass transfer at gas-evolving electrode - a theoretical attempt, *Electrochim.Acta,* 32:633 (1987).

21. T.Sakai, H.Takenaka, E.Torikai, Gas diffusion in dried and hydrated Nafions, *J.Electrochem.Soc.*, 133(1):88 (1986).

22. G.H.Kelsall, W.Wang, I.Cservenyák, Simplified model of current efficiency losses due to oxygen during reduction processes, to be published.

A MODEL FOR THE ELECTROCHEMICAL FORMATION OF BROMATE

K. Scott

Department of Chemical and Process Engineering
University of Newcastle upon Tyne
Newcastle upon Tyne, NE1 7RU,England

ABSTRACT

Data for the formation of sodium bromate by the anodic oxidation of sodium bromide solutions is analysed using a mathematical model. The reactor is a parallel plate undivided cell operated in a batch recycle mode. The variation of current efficiency with electrolyte velocity predicted by the model is in reasonable agreement with experimental data.

INTRODUCTION

The chemical reactions caused by the anodic oxidation of bromide are similar to those in the anodic oxidation of chloride, although the kinetic and thermodynamic parameters are different. The subject has been considered by several researchers[1,2] with the work by Cettou et al[3] giving a quite thorough treatment of the subject area. This paper is concerned with the electrochemical production of sodium bromate from sodium bromide. The electrochemical production of sodium bromate is analogous to the industrial production of sodium chlorate. Industrial chlorate cells operate continuously in a recycle mode in which the cell produces the active chlorine which is recirculated through a reaction vessel to complete the transformation to chlorate. Reactors for bromate formation are likely to be similar although the chemical formation of bromate is faster than that of chlorate and thus the required residence time for reaction will be lower. The reactor model considered is for a undivided recirculating batch reactor for which performance data of pilot scale operation has been reported by Millington[4].

REACTION CHEMISTRY

The overall chemical reaction in a bromate cell is given by

$$NaBr + 3\ H_2O \rightarrow NaBrO_3 + 3\ H_2 \qquad (1)$$

Electrochemical Engineering and Energy, Edited by
F. Lapicque *et al.*, Plenum Press, New York, 1995

141

which is a simple way of expressing a series of desirable reactions occurring in the cell, starting with the anodic oxidation of bromide ion

$$2Br^- \rightarrow Br_2 + 2e \tag{2}$$

$$Br_2 + H_2O \rightarrow HOBr + H^+ + Br^- \tag{3}$$

$$HOBr \Leftrightarrow BrO^- + H^+ \tag{4}$$

$$2\ HOBr + BrO^- \rightarrow BrO_3^- + 2\ Br^- + 2H^+ \tag{5}$$

There are several other reactions which can occur as a consequence of anodic bromine formation. These include decomposition reactions of hypobromite ions and hypobromous acid and the formation of tribromide ion

$$Br_2 + Br^- \leftrightarrow Br_3^- \tag{6}$$

There are several possible anodic loss reactions (eg oxygen evolution), with the major one, considered here, being the anodic oxidation of hypobromite ion to bromate.

$$6BrO^- + 3H_2O \rightarrow 2BrO_3^- + 4Br^- + 6H^+ + 3/2O_2 + 6e^- \tag{7}$$

Reaction (7) although leading to the formation of bromate represents a loss in current efficiency and its occurrence should be minimised. Other reactions which can result in a loss of current efficiency for bromate production are the cathodic reductions of the anodically formed bromine species. These can be minimised by the addition of chromate (or bichromate) to the electrolyte, which, through reduction at the cathode, forms a gelatinous membrane coating the cathode. Thus the major cathode reaction is the formation of hydrogen gas

$$2H_2O + 2e^- \rightarrow H_2 + 2OH^- \tag{8}$$

The thermodynamic and kinetic parameters for the reactions are reported by Cettou[2]. There are several general assumptions which are adopted in the reactor model.
1) The kinetics of the anodic reactions (2) and (7) are given by high potential field Tafel approximations.
2) The dissociation of hypobromous acid (reaction (3)) and the formation of tribromide ion (reaction (6)) are fast equilibrium processes.
3) Other electrochemical reactions such as the decomposition of water and back reduction reactions of bromine species are negligible.
4) The chemical formation of bromate (reaction (5)) is relatively slow compared to diffusion mass transport and occurs solely in the bulk electrolyte.

A consequence of assumption (2) (and (4)) is that the chemical reaction rate equation of reaction (5) can be written as

$$r = k\,f\,C_s^3 \tag{9}$$

where C_s is the concentration of active bromine [OBr$^-$ + HOBr] and $f = \dfrac{(K_5/C_{H+})}{(1+K_5/C_{H+})^3}$ with K_5 the equilibrium constant and k the rate constant, which follows an Arrhenius equation. The relative amounts of the different bromine species depends on the pH and the bromide ion concentration. At high bromide ion concentration and at low pH (<8), the formation of tribromide species is favoured. At pH values >8 the formation of OBr⁻ ions predominates. This is the pH range relevant to the formation of bromate.

REACTION MODEL

The reaction model consists of the anodic oxidation of bromide ion to bromine. Bromine then undergoes hydrolysis to hypobromous acid while it diffuses in the mass transport "diffusion" layer of thickness δ. Simultaneously the hypobromous acid dissociates to hyprobromite ions and these species can undergo anodic oxidation to bromate. The chemical formation of bromate occurs in the bulk electrolyte solution. The reactions in the cell are shown schematically below in Figure 1.

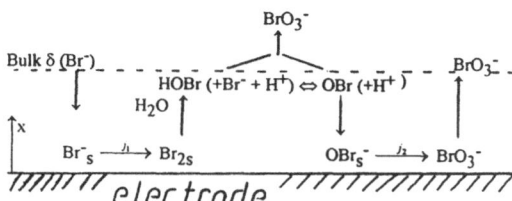

Figure 1 Reaction Model for Bromide Ion Oxidation.

A generalised model of the reaction and diffusion in the electrode boundary layer for species Br⁻, Br₂, HOBr and OBr⁻ and H⁺ is based on the general differential equation

$$D_j \frac{d^2 C_j}{dx^2} = \pm r \tag{10}$$

where r is the appropriate rate of the chemical reaction for the following species

Br⁻ ≡ species A	;	$r = -k\, C_{Br2}$.	Active bromine ≡ species S ; $r = -k\, C_{Br2}$
Br₂ ≡ species 1	;	$r = k\, C_{Br2}$.	Br₃⁻ ; $r = 0$
			H⁺ ; $r = -k\, C_{Br2}$

This model can be solved subject to boundary conditions in the bulk of the cell and at the anode surface. The latter are defined in terms of the partial current densities of the reaction, the sum of which satisfies the current balance $j = j_1 + j_2$. The boundary conditions for the reaction model are generally

at $x = \delta$, $C_j = C_{jb}$ ie bulk concentrations $\quad$ (11) $\quad$ at $x = 0$, $\dfrac{dC_j}{dx} = \pm \dfrac{j_j}{n_j F}$ $\quad$ (12)

where the current densities for reactions (2) and (7) are expressed as

$$j_1 = k_{f1}\, C_A{}^s \tag{13} \qquad\qquad j_2 = k_{f2}\, C_s{}^s \tag{14}$$

where $C_A{}^s$ and $C_s{}^s$ are the concentration of bromide ion and active bromine at the electrode surface and k_{f1} and k_{f2} are electrochemical (high field) rate constants.

143

PRELIMINARY MODEL

A major objective of the model is to determine the effect of various parameters on the current efficiency, t_1, of the "chemical formation" of bromate from reaction (5), ie the fraction of total current used in the formation of bromine and hyprobromons acid, and thus to bromate. The model considered here is a preliminary one and adopts one major assumption; that the electrode potential of reaction (7) is low in comparison to that of reaction (2) and that reaction (6) is mass transport limited. Thus the surface concentration of the "active bromine" approaches zero. The solution of equation (10) for Br_2 (species 1), with the boundary conditions

$$x = \delta, \; C_1 = C_{1b} \; ; x = 0, \; \frac{dC_1}{dx} = \frac{- \, j \, t_1}{2FD} \tag{15}$$

is
$$C_1 = \frac{jt}{nFk_{L1}\varnothing} \frac{\sinh\left[\varnothing\left(1 - \frac{x}{\delta}\right)\right]}{\cosh\varnothing} + \frac{C_{1b} \cosh\left(\varnothing \frac{x}{\delta}\right)}{\cosh\varnothing} \tag{16}$$

where $\varnothing = \delta\sqrt{\dfrac{k_1}{D_1}}$.

The concentration profile for active bromine is obtained from the equation (11)

$$D_s \frac{d^2 C_s}{dx^2} = -k \, C_1 \tag{17}$$

with the boundary condition $x = 0, \; C_s = 0 \; ; x = \delta, \; C_s = C_{sb}$ (18)

$$C_s = \frac{-D_1}{D_s} C_1 + \frac{x}{\delta}\left[C_{sb} + \frac{D_1}{D_s} C_{1b} - B \right] + B \tag{19}$$

where $B = \dfrac{D_1}{D_s}\left[\dfrac{j \, t_1}{nFk_{L1}\varnothing} \dfrac{\tanh\varnothing}{} + \dfrac{C_{1b}}{\cosh\varnothing} \right]$

The flux of active bromine at the electrode represents the electrochemical formation of bromate and is related to the current efficiency according to

$$D_s \frac{dC_s}{dx}\bigg|_{x=0} = \frac{j(1 - t_1)}{F} \tag{20}$$

Combining equation (20) and (21) enables the current efficiency to be obtained as a function of the bulk concentrations of the active bromine and of bromine.

$$t_1 = \frac{1 - \dfrac{k_{LS}C_{Sb}}{\left(\frac{j}{F}\right)} - \dfrac{k_{L1}C_{1b}}{\left(\frac{j}{F}\right)}\left[1 - \dfrac{1}{\cosh\varnothing} \right]}{\frac{3}{2} - \tanh(\varnothing)/2\varnothing} \tag{21}$$

where k_{Lj} represents the mass transfer coefficient, $\frac{D_j}{\delta}$. The expression for the current efficiency of bromate formation is a function of three dimensionless groups, a Hatta modulus, $\varnothing$, and two dimensionless current densities, $j^* = \frac{k_{LS}C_{sb}}{\left(\frac{i}{F}\right)}$ and $j^+ = \frac{k_{L1}C_{1B}}{\left(\frac{i}{F}\right)}$.

The latter are a measure of mass transfer limiting current densities in the absence of chemical reaction, divided by the total applied current density. The general variation of current efficiency with $\varnothing$ and i^* ($i^+ = 0$) is shown in Figure 2 for the case when bromine is consumed completely in the diffusion layer.

The general trend in the behaviour is that with a fixed mass transport rate and chemical rate constant ($\varnothing$ = constant), current efficiency decreases with decreasing current density and increasing active bromine concentration. At a fixed current density, and with increasing active bromine concentration and mass transport rate (j^* = constant) the current efficiency increases with a decreasing chemical rate constant. An increase in the value of mass transfer coefficient increases j^* but decreases $\varnothing$.

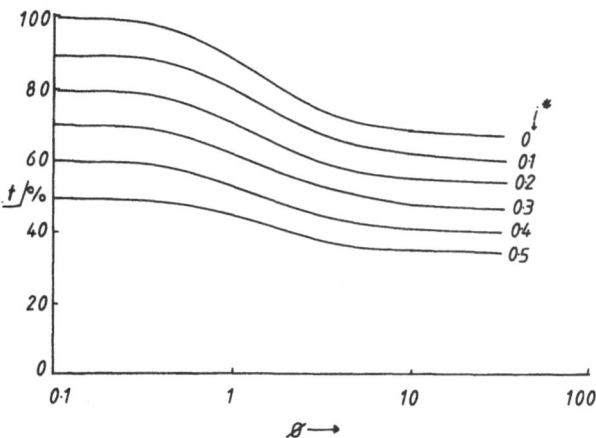

Figure 2   The Variation of Current Efficiencies for Bromide Oxidation ($j^+ = 0$)

REACTOR MODEL

The electrochemical cell reactor is a narrow gap parallel plate undivided flow cell linked to an external electrolyte reservoir. Hydrogen gas is generated in the cell gap and introduces a mixing effect which enhances mass transport to the electrode surface. If a well mixed assumption is adopted in this preliminary model then the reactor material balances on active bromine are

For the reservoir

active bromine: $\qquad C_{S2} = C_{S1} + \tau_m k \, f \, C_{S1}^3 + \tau_m \dfrac{dC_{S1}}{dt}$ $\qquad\qquad$ (22)

bromate (B) $\qquad C_{B2} = C_{B1} - \tau_m k \, f \, C_{s1}^3 + \tau_m \dfrac{dC_{B1}}{dt}$ $\qquad\qquad$ (23)

bromide $\qquad C_{A2} = C_{A1} + \tau_m \dfrac{dC_{A1}}{dt}$ $\qquad\qquad\qquad\qquad$ (24)

For the cell:

active bromide $C_{S1} = C_{S2} + \tau_R k\, f\, C_{S2}^3 + \tau_R \dfrac{dC_{S2}}{dt} - \left(-D_s a\tau_R \dfrac{dC_{S2}}{dx}\Big|_{x=\delta} \right)$ (25)

bromate: $\quad C_{B1} = C_{B2} - \tau_R k\, f\, C_2^3 + \tau_R \dfrac{dC_{B2}}{dt} - \left[-D_B a\tau_R \dfrac{dC_{B2}}{dx}\Big|_{x=\delta} \right]$ (26)

bromide: $\quad C_{A1} = C_{A2} + \tau_R \dfrac{dC_{A2}}{dt} - \left(-D_A a\tau_R \dfrac{dC_{A2}}{dx}\Big|_{x=\delta} \right)$ (27)

Where a is the electrode area per unit volume of cell and τ denotes residence times. It is assumed that the reaction of bromine is complete in the diffusion layer and also that changes in pH are not large. The subscripts 1 and 2 denote the concentrations in the reservoir and reactor respectively. Under conditions where the volume of the reservoir is much greater than that of the cell, the system can be approximated as a batch reactor. Thus we have

$$\frac{dC_s}{dt} = -kfC_s^3 + \left(-D_s a \frac{dC_s}{dx}\Big|_{x=\delta} \right)$$ (28)

$$\frac{dC_B}{dt} = kfC_s^3 + \left(-D_B a \frac{dC_B}{dx}\Big|_{x=\delta} \right)$$ (29)

$$\frac{dC_A}{dt} = -D_A a \frac{dC_A}{dx}\Big|_{x=\delta}$$ (30)

MODEL SOLUTION

Preliminary Model

Earlier a reaction model was developed for the condition of limiting current operation for oxidation of active bromine. The flux of active bromine into the bulk electrolyte can be obtained from this model and combined with the reactor model equations. In the case of the batch reactor equation (28) then becomes

$$\frac{dC_s}{dt} = -kfC_s^3 + \frac{a\,j\,t}{2F}\left(\frac{\tan\varnothing}{\varnothing} - \frac{1}{\cosh\varnothing} \right) + ak_{LS}C_s$$ (31)

Combining equation (31) and (22) with $C_{1b} = 0$ enables the variation in the active bromine concentration to be obtained and, therefore, the current efficiency.

In the steady state, that is with a fixed concentration of active bromine, equation (31) becomes

$$k\,f\,C_s^3 - \frac{a\,j\,t}{2F}\left(\frac{\tanh\varnothing}{\varnothing} - \frac{1}{\cosh\varnothing} \right) - ak_{LS}C_s = 0$$ (32)

This situation is applicable to the pilot scale operation of bromate formation. Equation (22) and (32) now combine to give the following equation for determining current efficiency as

$$\alpha\left(1 - \overline{X}t\right)^3 + \beta\left(1 - \overline{X}t\right) - \overline{Y}\,\overline{X} = 0$$ (33)

where $\alpha = kf\left(\dfrac{i}{k_{LS}F}\right)^3$, $\beta = \dfrac{aj}{F} + \dfrac{\overline{Y}}{\overline{X}}$, $\overline{X} = \dfrac{3}{2} - \dfrac{\tanh(\varnothing)}{\varnothing}$, $\overline{Y} = \dfrac{a'j}{2F}\left(\dfrac{\tanh\varnothing}{\varnothing} - \dfrac{1}{\cosh\varnothing}\right)$

where a' is the electrode area per unit total electrolyte volume. The real solution of equation (33) is

$$\left(1 - \overline{X}t\right) = 3\sqrt{\dfrac{\overline{Y}}{2\overline{X}} + \sqrt{R}} - 3\sqrt{\dfrac{-\overline{Y}}{2\alpha\overline{X}} + \sqrt{R}} \tag{34}$$

where $R = \left(\dfrac{\beta}{3\alpha}\right)^3 + \left(\dfrac{\overline{Y}}{2\overline{X}\alpha}\right)^2$

Figure 3, shows current efficiencies for bromate production predicted using equation (34). The calculations are for active bromine concentrations of 5 mol m^{-3} and less, these being typical values obtained experimentally in the synthesis. The current efficiency increases with an increase in mass transport coefficient and a decrease in the active bromine concentration. This is in general agreement with experimental data[3] although this is far from satisfactory. Thus the application of the more precise model is considered in the following section.

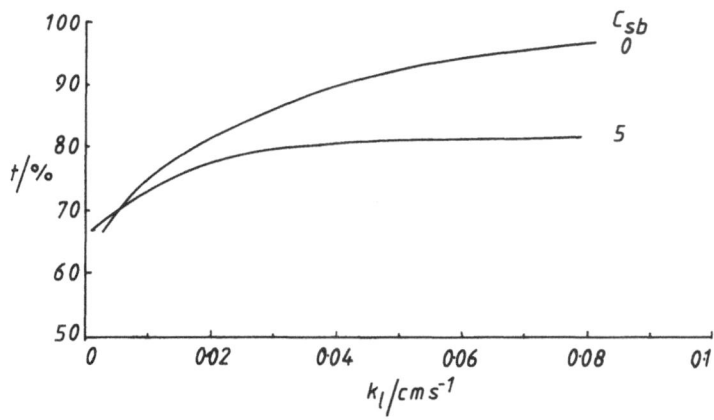

Figure 3 Variation of Steady State Bromate Current Efficiencies with Mass Transport Coefficient. j = 2000 A m^{-2}, k =110 s^{-1}. Concentration of active bromine shown on Fig. (mol m^{-3}).

Diffusion Layer Model

The solution to the diffusion layer model is presented here under conditions of a finite surface concentration of active bromine, i.e. the electrode reaction is under mixed kinetic and mass transport control. It is assumed that bromine hydrolysis is complete within the diffusion layer with a condition of zero flux at the edge of this layer

$$\dfrac{dC_1}{dx}\bigg|_{x=\delta} = 0 \tag{35}$$

The selection of this boundary condition, rather than a zero concentration, did not constrain the concentration profile into giving a finite value of flux[4].

The concentration distribution of bromine in the diffusion layer is found from the solution of equation (11) as

$$C_1 = X C_A^s \cosh\left[\varnothing\left(x/\delta - 1\right)\right] \tag{36}$$

where superscript 's' denotes the surface concentration, where $X = \dfrac{k_{f1}}{k_{L1}\varnothing \sinh(\varnothing)}$

Solving equation (36) with equation (11) gives the concentration distributions of bromide ion:

$$C_A = -\frac{D_1}{D_A}C_1 + \left[C_{Ab} + \frac{D_1}{D_A}XC_A^s\right]x/\delta + \left(C_A^s + \frac{D_1}{D_A}XC_A^s\cosh(\varnothing)\right)\left(1 - x/\delta\right) \tag{37}$$

active bromine:

$$C_s - \frac{D_1}{D_s}C_1 + K_3 x + K_4\left(\frac{k_{f2}C_s^s}{D_S} - \frac{D_1}{D_S}XC_A^s\sinh(\varnothing)\right)(x-\delta)C_{sb} + \frac{D_1}{D_s}XC_A^s \tag{38}$$

An overall balance on bromide ions at $x = \delta$, at the steady state, is

$$D_A\frac{dC_A}{dx}\bigg|_{x=\delta} = -D_s\frac{dC_s}{dx}\bigg|_{x=\delta} - D_B\frac{dC_B}{dx}\bigg|_{x=\delta} \tag{39}$$

The flux of bromate, which is a non-reacting species, is given by

$$-D_B\frac{dC_B}{dx} = k_{f2}C_S^s \tag{40}$$

Equation (39) results in the following expression for the surface concentration of A.

$$C_A^s = C_{Ab}/Q \tag{41}$$

where

$$Q = 1 - \frac{k_{f1}}{k_{L1}\varnothing\sinh(\varnothing)}\left(\frac{D_1}{D_A}(1 - \cosh(\varnothing)) - \frac{D_1}{D_s}\varnothing\sinh(\varnothing)\right)$$

The surface concentration of active bromine is obtained from equation (38) as

$$C_s^s\left[1 + \frac{k_{f2}}{k_{LS}}\right] = C_{Sb} + YC_A^s \tag{42}$$

where $Y = \dfrac{D_1}{D_s}X(1 - \cosh(\varnothing) + \varnothing\sinh\varnothing)$

148

Reactor Model

The approximate batch reactor model for the system requires values for the fluxes of active bromine and bromide. These are obtained from equations (37), (38), (41) and (42) as

$$D_s \frac{dC_s}{dx}\bigg|_{x=\delta} = k_{f2}C_s^s - k_{f1}C_A^s \tag{43}$$

$$D_A \frac{dC_A}{dx}\bigg|_{x=\delta} = k_{LA}C_{Ab}\left[1 + \frac{D_1 X}{Q D_A}\left(1 - \cosh(\varnothing)\right) - \frac{1}{Q}\right] \tag{44}$$

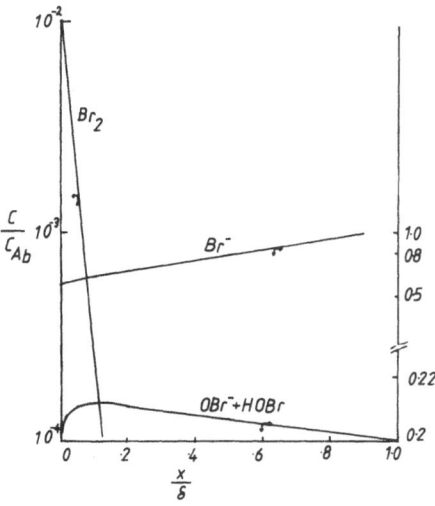

Figure 4 Concentration Profiles of Bromine Species in the Diffusion Layer. Initial Concentration of active bromine 1.0 kmol m^{-3}.

RESULTS

Figure 4 shows the typical concentration profiles of bromide ion, active bromine and bromate in the diffusion layer. Bromine hydrolysis occurs in a region close to the electrode surface and serves to replenish bromide ions in this region and enhance the bromide flux to the electrode in comparison to that which would occur in the absence of the hydrolysis. Active bromide goes through a maximum in concentration due to its electro-activity and mass transport out of the diffusion layer.

The variations in concentrations of bromine species in a batch reactor system is shown in Figure 5. The concentration of the active bromine species rises quite fast initially and approaches a maximum value at a reaction time of approximately one hour under the conditions considered. Bromate formation is initially slow and rises as the concentration of active bromine increases. There is reasonable agreement between the predictions of the model and experimental data. The model also predicts an increase in the current efficiency with current density in line with the predictions of the model.

Figure 6 shows the effect of mass transfer, in terms of the electrolyte velocity, on the current efficiency of bromate formation. The current efficiency increases with velocity and simulation and model are in reasonable agreement. However more reliable data on mass transfer coefficients in this type of cell under the influence of a large rate of oxygen evolution is required to truly compare the model to experiment.

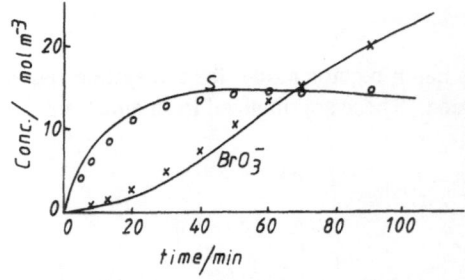

Figure 5 Typical Variation of Active Bromine and Bromate in a Batch Reactor. Initial bromide ion concentration 0.06 kmol m^{-3}, 80 A m^{-2}, 35°C, pH = 9.6. Solidus (-): model and data from ref [2].

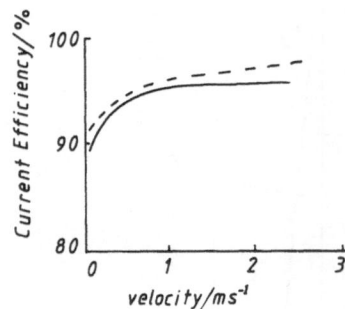

Figure 6 The Effect of Mass Transfer Rate on The Current Efficiency of Bromate. 2000 A m^{-2}, 60°C electrode area 0.05 m^2 (-): data ref [3], (---): Model

REFERENCES

1. N. Ibl and H. Vogt. "Comprehensive Treatise on Electrochemistry".
   Ed. E.Yeager, B. Conway and J. Bockris, Vol 2, Chapter III, Plenum Press, New York (1981).
2. P. Cettou, P.M. Robertson, N. Ibl. The Electrolysis of Aqueous Bromide Solutions to Bromate, Electrochimica Acta 29, 875 (1984).
3. P. Millington, "Electrosynthesis From Laboratory To Pilot, To Production", Chapter 12
   Eds J.D. Genders and D. Pletcher, The Electrosynthesis Co Inc, New York (1990).
4. K. Scott, "Electrochemical Reaction Engineering", Academic Press, London (1991).

FROM VOLTAMMETRY TO INDUSTRIAL PLANT: ELECTROCHEMICAL SYNTHESIS OF DL-HOMOCYSTEINE FROM DL-HOMOCYSTINE — AN EXAMPLE OF SCALE-UP

Gaspar Sánchez-Cano, Vicente Montiel, Vicente García and
Antonio Aldaz

Departamento de Química Física
Universidad de Alicante, Aptdo 99
03080, Alicante, Spain

Eduardo Elías

Prodesfarma SA (D.M.P.)
Ctra Granollers-Girona Km23
08740, Sant Celoni, Barcelona, Spain

ABSTRACT

An industrial process for the electrochemical synthesis of DL-Homocysteine from DL-Homocystine is described. DL-Homocysteine is a key intermediate for the synthesis of Citiolone (a powerful agent used to treat bronchitis and nasal catarrh). The work starts with a fundamental study, the voltammetric behaviour of DL-Homocystine, followed by a laboratory and pilot plant study and finishes with the development of an industrial process (30 tons/year of DL-Homocistine electroreduced). The method avoids the use of hazardous and polluting reductants such as Zn/H^+. The figures of merit are a material yield of 95%, production over $200 kg/m^2 day$ and a energy cost less than 2 kWh/kg.

INTRODUCTION

The cathodic cleavage of disulfides is a very efficient and useful method for obtaining pharmaceuticals such as L-cysteine, S-carboxymethyl-L-cysteine and N-acetyl-L-cysteine[1-8].

Based on this previous work we propose in this paper a new route that is an improved alternative to the synthesis of Citiolone from Methionine (Scheme 2). This route involves an electrochemical key step: the cathodic reduction of Homocystine to yield Homocysteine (Scheme 1).

Basically the route involves three steps: 1)Demethylation, 2) Reduction of Disulfide and 3) Acetylation. Steps 1) and 3) are well-known processes in organic syntheses. The classical route uses the chemical reduction of disulfide by Zn/H^+ treatment as step 2). In fact, up to now, this method had been operated to yield Citiolone in industry. The two main disadvantages of the chemical reduction with Zn/H^+ are the following:

a) <u>High environmental impact</u>

The chemical reduction produces at least 2 tons of zinc salts per ton of Homocystine treated.

b) <u>Hazardous method</u>

The chemical process is exothermic and a notable amount of hydrogen is formed. Strong safety measures have to be taken.

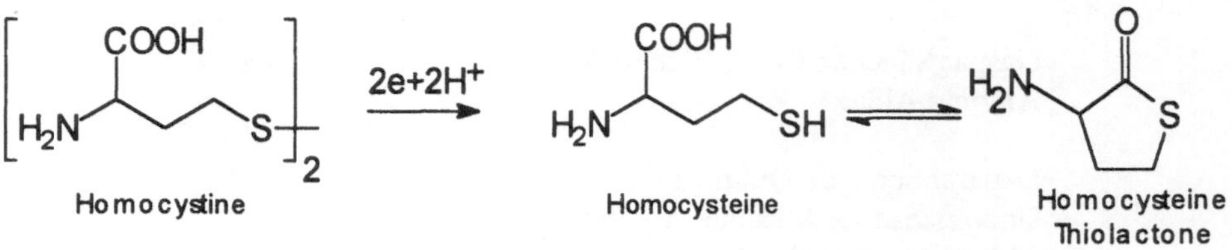

Scheme 1. Electrochemical key step to obtain Homocysteine Thiolactone.

The aim of this paper is to explain the methodology and experimental procedure carried out to eliminate Zn/H^+ and to introduce a new and much more convenient industrial process: the electrochemical reduction of the disulfide bond. This new method avoids the use of pollutants such as Zn metal and dangerous hydrogen evolution. Moreover, the electrochemical process is easy to operate and is without risk. Other important advantages of the electrochemical alternative are that selectivity is higher and the method is cheaper.

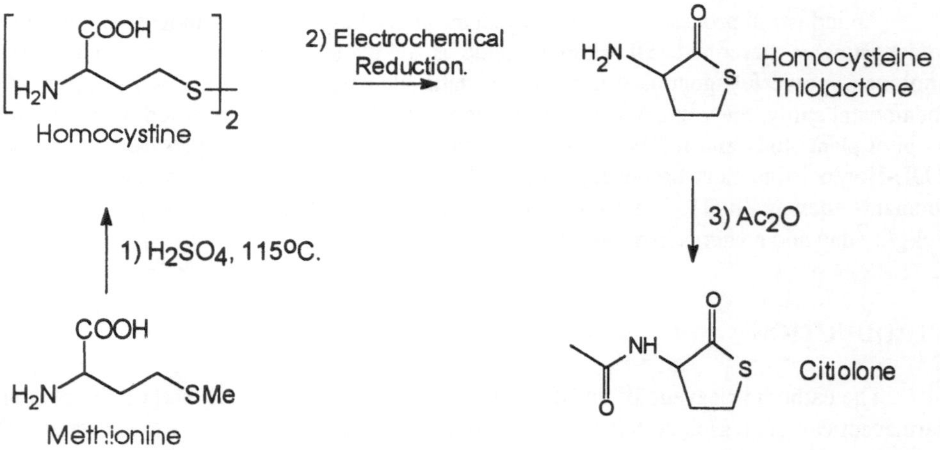

Scheme 2. Electrochemical key step to the synthesis of Citiolone from Methionine.

CYCLIC VOLTAMMETRY STUDY

Cyclic voltammetry is a very powerful electrochemical technique to obtain information about electrode processes[9]. The influence of the cathodic material on the

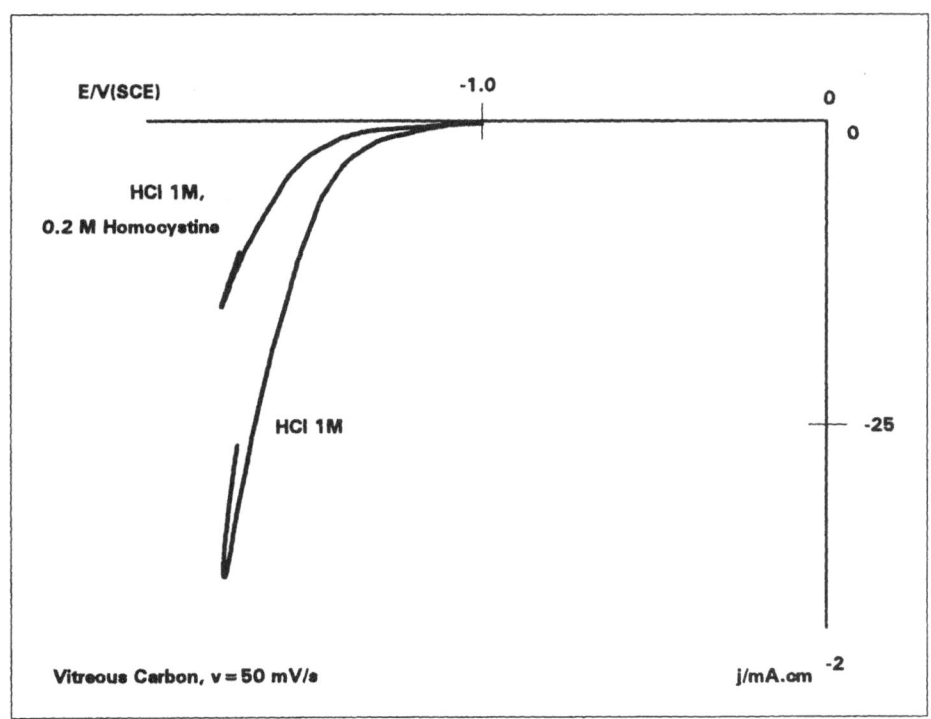

Figure 1. Cyclic Voltammogram for the reduction of Homocystine/HCl on vitreous carbon. 25°C, 50mV/s.

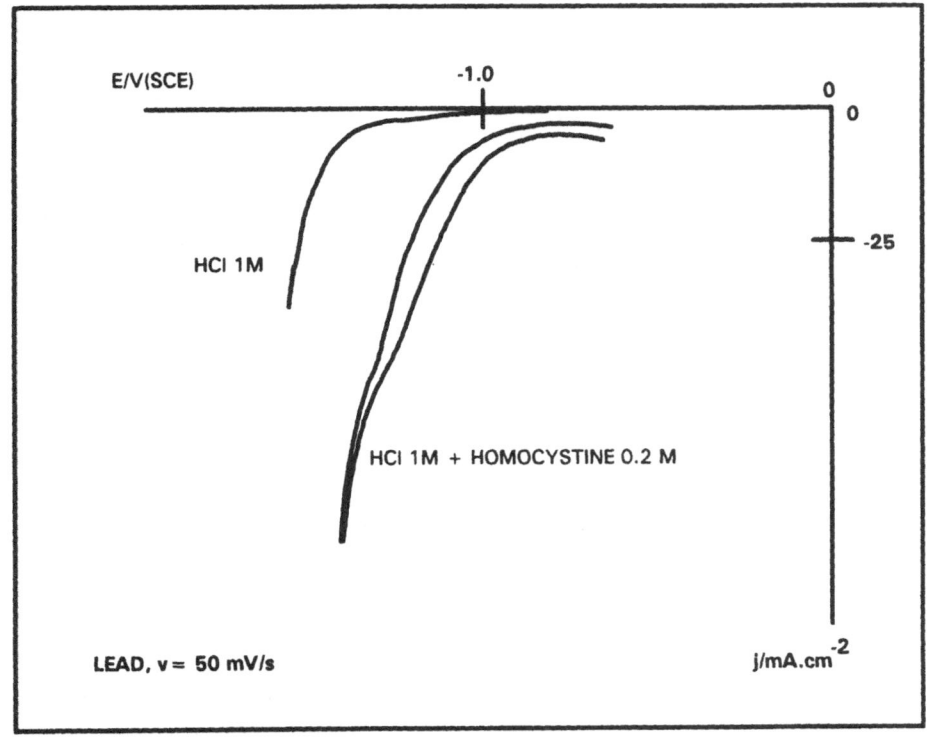

Figure 2. Cyclic Voltammogram for the reduction of Homocystine/HCl on lead. 25°C,50mV/s.

electrochemical reduction of the disulfide bond (Homocystine) in HCl medium has been investigated by cyclic voltammetry. The working electrodes chosen were vitreous carbon and lead that are very appropriated for industrial purposes.

Figure 1 shows two voltammetric curves obtained on vitreous carbon cathode as working electrode in 1M HCl (deoxygenated solution at room temperature). Hydrogen evolution starts at -1.4V vs SCE. The presence of 0.2M Homocystine shifts hydrogen evolution to a more negative potential. This behaviour indicates that Homocystine is adsorbed on the electrode. Reduction of proton and disulfide occur at the same potential; in fact laboratory electrolysis yields homocysteine on a vitreous carbon electrode, although with low current efficiency.

Figure 2 shows the behaviour on a lead cathode. Hydrogen evolution appears practically at the same potential as on a vitreous carbon cathode in 1M HCl. The presence of 0.2M Homocystine produces a new overlapped cathodic peak at -1.28V vs SCE which can be assigned to the reduction of the disulfide bond. This indicates that lead is a good candidate as an industrial electrode for the electrosynthesis of homocystine/homocysteine.

Figure 3 shows some interesting results. If Pb(II) is present at low concentrations, a lead deposit is produced before H^+ reduction. This deposit seems to catalyze the cleavage of the disulfide bond. If a lead coated carbon electrode is used without Pb(II) in solution similar results are obtained.

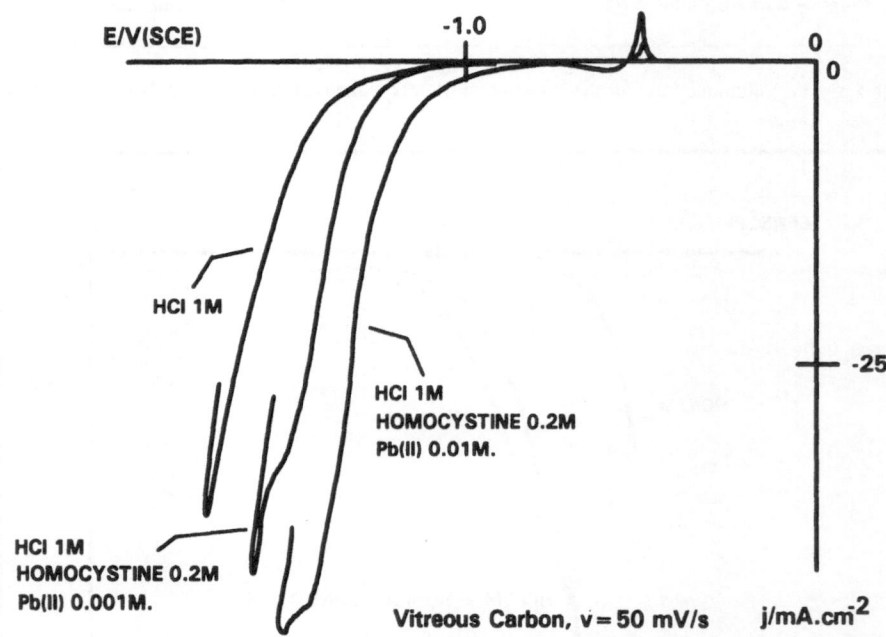

Figure 3. Cyclic voltammogram for the reduction of Homocystine/HCl on modified electrodes. 25°C, 50mV/s.

This basic study concludes that both electrodes, lead and lead coated vitreous carbon, may be appropriate for the electrochemical reduction of homocystine yielding homocysteine. Moreover, the results shown in Figure 3 suggest that carbon felt coated with lead should be a very good three dimensional cathode with big active area and catalytic surface.

LABORATORY CELL STUDY

In order to select the best electrochemical conditions to carry out the cathodic reduction of Homocystine, a series of reactions was programmed. A detailed description of the experimental set-up and the electrochemical cell has been previously reported[10].

The initial experimental parameters selected were: Homocystine 0.79M, HCl 4.5M, Temperature 45°C, Current density 200mA/cm^2. In all cases the electrode materials were carbon felt/carbon for the anode and carbon felt/lead for the cathode (geometric area 20cm^2). The flow rate was 1 liter/minute. Electrolysis were run at constant current in a filter-press cell, cathode and anode being separated by a cation-exchange membrane (Nafion 117) to avoid the electro-oxidation of both homocysteine and homocystine. As anolyte a basic sulfite solution (2M) was employed, which avoids chlorine evolution and hypochlorite formation. The sulfite solution is obtained in the demethylation step of the Citiolone route (Scheme 2) where SO$_2$ is formed and absorbed with a basic solution (scrubber). In this way, the anodic reaction is used to remove a polluting effluent so that the electrochemical reactor has two compatible applications: Electrosynthesis and Waste Water Treatment.

A brief summary of the influence of three typical variables on the synthesis is presented in Table 1: concentration, current density and temperature. It can be deduced that conversion increases when temperature or concentration of homocystine is increased. Conversion is not very dependent on current density because the use of a 3D electrode causes real current densities to be very low.

Table 1. Results of the influence of different conditions on the electrolysis.

Concentration Effect	45°C 200mA/cm^2	Theoretical Charge Passed (%)			
		25	50	75	100
RSSR/HCl					
0.24M/1M		24	47	67	70
0.40M/2M	CV(%)	24	45	64	79
0.79M/4.5M		24	44	74	91

Current Density Effect	45°C RSSR 0.79M/HCl 4.5M	Theoretical Charge Passed (%)			
		25	50	75	100
mA/cm^2					
100		25	49	71	84
200	CV(%)	25	49	70	85
300		24	48	68	83

Temperature Effect	200mA/cm^2 RSSR 0.79M/HCl 4.5M	Theoretical Charge Passed (%)			
		25	50	75	100
(^{o}C)					
10		21	39	55	69
33	CV(%)	20	39	54	67
45		25	49	70	85

The 3D modified electrode is prepared "in-situ" when homocystine reduction is running. The electrolyte dissolves small quantities of lead (from the current feeder) that deposit on the carbon surface. This kind of electrode has other important advantages. Thus, in general, when a three dimensional electrode is used, the cell behaves as a zero gap cell, the ohmic drop decreases and the energy cost (kWh/kg) is minimum. On the other hand the total active area of this kind of electrode yields high current efficiency and the economic parameters (kg/m²day, kWh/kg) are improved. This justifies the choice of 3D electrodes for this synthesis. Presently this modified electrode is being studied for current distribution, mass transport, lead deposition on the surface, different catalysts, etc... and the results will be published later.

Based on the above results (Table I) the following "optimized" conditions were selected: 300mA/cm^2, 45°C, 0.79M Homocystine and 4.5M HCl.

Table 2. Figures of merit. Global Reaction: $aA+ \ldots+ ne \Rightarrow bB+\ldots$

Theoretical Charge Passed(%): (Q / moles A (n/a) 96500) 100.
Conversion, CV(%): (Moles A reacted / Moles A initial) 100.
Current Efficiency, CE(%): (Moles B obtained (n/b) 96500 / Q) 100.
Production, Kg/m²day: $i(\text{mA/cm}^2)$ CE(%) MWB (b/n) (1/11170).
Energy Cost, Kwh/Kg: V (n/b) (1/CE(%)) (1/MWB) 2680.5.
MWB: Molecular Weight B. V: Cell Voltage (Volts). Q: Charge Passed (C).

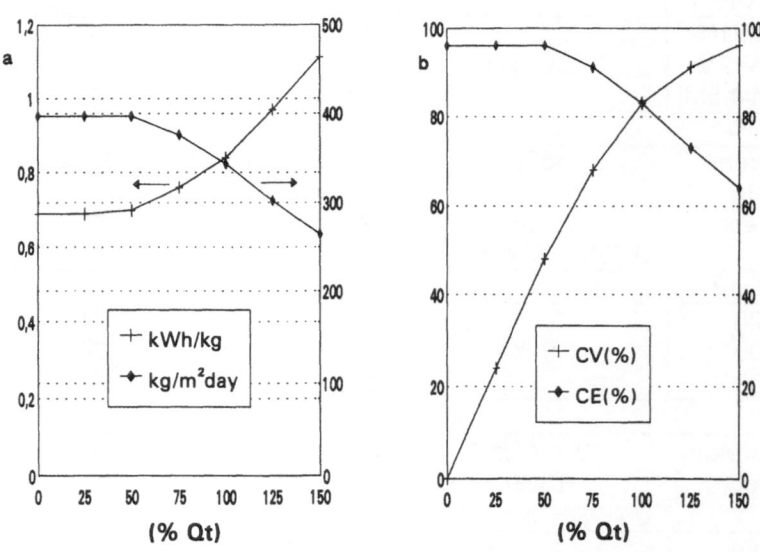

Figure 4. Variation of (a) Energy Cost and Production vs Theoretical Charge Passed(%) and (b) Conversion and Current Efficiency vs Theoretical Charge Passed (%).

Table 3. Optimized conditions for cathodic reduction of Homocystine.

	Theoretical Charge Passed (%)					
	25	50	75	100	125	150
CV(%)	24	48	68	83	91	96
CE(%)	96	96	91	83	73	64
kWh/kg	0.69	0.70	0.76	0.84	0.97	1.11
kg/m^{2}day	397	397	376	343	302	265

kg: Homocysteine Thiolactone Hydrochloride (Molecular weight 154)

Table 3 and figure 4 show the variation of the values of different parameters (Conversion, Current efficiency, Energy Cost and Production, see table 2) versus the theoretical charge passed. The electrolysis is ended when a conversion over 95% is reached. The solution containing Homocysteine is treated to yield Citiolone (Scheme 2).

SCALE UP

The scale-up of the process was carried out using a filter press cell with an electrode of 1.000cm^2 (useful area). The electrolysis conditions were those employed in the laboratory cell and were assumed as "optimized conditions". The aim of the scale-up was to check the performance of the electrosynthesis of Homocysteine in a pilot cell reactor with a total area of 4x1000cm^2. Special attention was taken into account with the figures of merit versus the time of operation. For this purpose 27 reactions were carried out giving a total electrolysis time of 200 hours. Figure 5 shows the evolution of homocystine conversion when the 150% of the theoretical charge (150% Q_t) was circulated. To maintain a high conversion of Homocystine washing of the electrochemical cell, after each electrolysis, was carried out using water at 50°C.

After 200 hours of operation in the batch mode the general condition of the reactor was good. The corrosion of the cathodic current feeder (lead) was low (1 kg/ton Homocystine treated) and the corrosion of the anodic current feeder (carbon) was negligible. Both membrane and felt carbon appeared sound. Four cathodes and four anodes were assembled in bipolar and monopolar mode. A special design of the compartment was developed to avoid shunt current losses[11].

INDUSTRIAL PLANT

Since September 1992 Prodesfarma SA has been operating an Industrial Electrochemical Plant to treat 30 tons/year of Homocystine (electrochemical reactor developed by I.D. Electroquimica SL. Alicante. Spain). After one year of operation the process has resulted in an improved method for obtaining Citiolone (Scheme 2) when compared with the chemical route (reduction with Zn/H$^+$). This new method is safer, cheaper and has avoided the formation of 45 tons of zinc salts.

During this first year (September´92 to October´93), the Electrochemical Plant has treated 22 tons of Homocystine. The total operation time has been 3.000 (three thousands) hours. The average conversion is 95% and the energy cost less than 1.5kWh/kg.

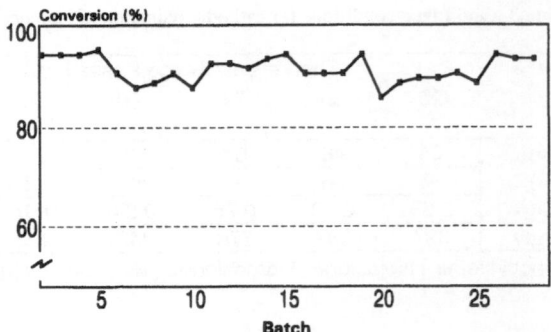

Figure 5. Evolution of the Homocystine conversion for the different electrolyses (27 batches)

REFERENCES

1. Mizuguchi, J; *Bull. Tokyo Inst. Technol.* 4,1, (1965).
2. Wong, C.; *J. Chin. Chem. Soc.* 25, 149, (1978).
3. Asahi Glass Co. Ltd. Jpn Kokai Tokkyo Koho JP 5909, 184 [8409, 184] (Cl. C25B3/04), 18 Jan. 1984, Appl. 82/116, 297, 06 Jul. 1982.
4. Showa Denko K.K. Jpn Kokai Tokkyo Koho JP 8298, 685 (Cl. C25B3/00), 18 Jun 1982, Appl. 80/174, 616, 12 Dec. 1980.
5. Aldaz, A.; Montiel, V.; Sanchez-Cano, G.; Segura, M.; Spanish Patent, No. 9000995. 06 Mar 1990.
6. Kasafirek, E.; Czech. 126,414 (Cl.C07c.), 15 Mar 1968, Appl. 21 Mar 1966.
7. Walsh, F.; A First Course in Electrochemical Engineering. p. 319-327. The Electrochemical Consultancy. England (1993).
8. Sanchez-Cano, G.; Montiel, V.; Aldaz, A.; *Anales de Química.* vol. 85, No. 3, 526 (1989).
9. Heinze, J.; *Angew Chem. Int. Ed. Engl.* 23, 831-847, (1984).
10. Sanchez-Cano, G.; Montiel, V.; Aldaz, A.; *Tetrahedron* Vol. 47, No. 4/5, 877, 886, (1991).
11. Codina, G.; Aldaz, A.; *J. Appl. Electrochem.* 22, 668-674, (1992).

ASPECTS OF THE ELECTROSYNTHESIS OF

2-HYDROXYMETHYLPYRIDINE

A. M. Romulus and A. Savall

Laboratoire de Génie Chimique et Electrochimie, URA 192 CNRS
Université Paul Sabatier
118, Route de Narbonne
31 062 TOULOUSE cedex, France

ABSTRACT

The electrochemical reduction of 2-ethylpicolinate (A) in a sulfuric acid medium was studied on a lead cathode during electrolyses in a laboratory filter press reactor under galvanostatic conditions. The yield of the conversion of (A) to the corresponding 2-formylpyridine (B), 2-hydroxymethylpyridine (C) and picoline (D) was investigated as a function of the medium acidity, current density, concentration of (A) and temperature.

A reaction model including the three electrochemical steps: A → B → C → D is presented. The variations of the charge transfer constants are considered as functions of the reactant conversion and mass transport resistance is integrated. The simulation gives a reasonable description of the behaviour of this complex electrochemical process.

INTRODUCTION

Alcohols derived from aromatic heterocycles are useful synthesis intermediaries for pharmaceutical products. These alcohols can be obtained by electrochemical reduction of the corresponding esters. Several successive and parallel stages make this transformation a very complex process. In the case of pyridine derivatives the reduction of the side chain is privileged with respect to the aromatic nucleus under certain operating conditions: high acidity and use of a lead cathode [1-4]. In this work the electrochemical reduction of 2-ethylpicolinate (A) was studied with a view to defining better operating conditions for obtaining 2-hydroxymethylpyridine (C). The conversion of (A) into 2-formylpyridine (B), then (C) and picoline (D) was achieved on a lead cathode in a sulfuric acid medium during preparative electrolyses under galvanostatic conditions in a two-compartment filter press reactor [5-6]. The main stages in the transformation of the side chain of (A) are represented schematically as follows:

$$\text{ester (A)} + 2e \quad \longrightarrow \quad \text{aldehyde (B)} \qquad (1)$$

$$\text{aldehyde (B)} + 2e \longrightarrow \quad \text{alcohol (C)} \qquad (2)$$

$$\text{alcohol (C)} + 2e \longrightarrow \quad \text{picoline (D)} \qquad (3)$$

The effect on the alcohol yield of the following parameters was examined: sulfuric acid concentration, initial ester concentration, current density, temperature.

Theoretical analysis of such a reaction is extremely interesting with a view to obtaining a calculation tool for predicting the effect of the operating parameters on the results of electrolysis and, in particular, for determining the electrolysis time providing the maximum amount of alcohol. Although galvanostatic electrolysis is generally used in industry, modelling procedures for electrochemical reactions in series and in parallel are mainly based on the potentiostatic mode [7]. Some work does however deal with galvanostatic operation [8-10]. The difficulty with this type of analysis mode is the variation in electrode potential with time. This results in variation in the charge transfer rate constants. The second part of this study proposes simplified bases for a model of galvanostatic operation of a filter press reactor used for the electrochemical reduction of (A).

EXPERIMENTAL TECHNIQUE

Electrolyses of aqueous solutions of 2-ethylpicolinate in the presence of sulfuric acid have been performed on a lead cathode with a view to synthesising 2-hydroxy-methylpyridine [5]. The piston flow filter press type reactor (ELECTROCELL AB, Sweden) had two compartments separated by a cationic membrane (NAFION 423, Dupont de Nemours, USA). Its temperature was regulated to within 1°C. The lead cathode, with a surface area of 14 cm^2, was in contact with a turbulence promoter. It was cleaned with emery cloth and rinsed with distilled water before each electrolysis operation. The anode was made of platinum-plated titanium (IMI Titanium, UK).

Catholyte and anolyte were recycled by pumping through two closed loops respectively located between the reactor cathode and anode compartments, and between two thermostatically controlled 200 cm^3 storage tanks; these two tanks make it possible to add or sample the liquid, or to separate dihydrogen or dioxygen. The anolyte was an aqueous solution of sulfuric acid at the same concentration as the catholyte. The initial volumes of catholyte and of anolyte were each 150 cm^3. For electrolyses occuring from 3 and 12 hours, no notable variation in these volumes was observed. The electrolyte flow in each loop was 12.8 cm^3 s^{-1} (circulation speed: 0.14 m s^{-1}) for all the experiments. The current densities used in galvanostatic mode electrolyses were in the range 500 - 3000 A m^{-2}, the initial ester concentrations between 0.24 and 1 mol L^{-1}, the sulfuric acid concentrations between 0.4 and 9 mol L^{-1} and the temperature between 20 and 80°C.

The variation in the amounts of reagent and of electrolysis products was monitored by taking ten 0.5 to 1 cm^3 samples from the cathode storage tank at regular intervals. Each sample was cold neutralized (0°C) by an aqueous solution of sodium hydroxyde up to pH 7, centrifuged and then filtered at 0.45 µm before being analyzed. The pyridine derivatives were assayed by HPLC (HEWLETT-PACKARD 1090 apparatus with diode strip) on a BROWNLEE ODS column with a length of 220 mm and a diameter of 2.1 mm, protected by an identical pre-column 30 mm long. Detection was accomplished at 230 nm and the pyridine was used as standard. The eluent was a mixture of acetonitrile (26% volume) and of an aqueous solution buffered to pH 7 (6.80 g of KH_2PO_4 and 1.16 g of NaOH per liter), circulating at a rate of 0.2 cm^3 s^{-1}. Products of the pyridine nucleus reduction were assayed using a gas phase chromatograph (HP 5890) coupled with a mass spectrometry detector (HP 5971 A).

RESULTS OF PREPARATIVE ELECTROLYSES

Figure 1 shows an exemple of the variation in the number of moles of pyridine derivatives with time, for a given set of operating conditions [5,6]. The time was replaced by the parameter Q, which is the ratio of the electrical charge that had passed through the cell from the beginning of electrolysis to the initial number of moles of (A). This parameter allows easy comparison of the effects of variations in the current density and in the initial ester concentration; the theoretical value of 4 F mol^{-1} corresponds to the number of moles of electrons required to reduce one mole of (A) into one mole of (C). Figure 1 is representative of the main scheme consisting of three successive reductions of the ester function (reactions (1), (2), (3)). In addition the presence of picolinic acid (E), resulting from hydrolysis of the ester in the acid medium, was also observed. Modifications in the operating conditions (current density, initial ester concentration, sulfuric acid concentration, temperature) did not induce any notable change in the general shape of the curves plotted for each electrolysis. Formation of aldehyde (B) and of alcohol (C) started early in the run and the concentrations of (B) and of (C) passed through a maximum, that of (C) taking place after that of (B). In an acid medium ester (A) is therefore reduced into aldehyde (B), into alcohol (C) and then into picoline (D), before the occurrence of the aromatic nucleus reduction. The ester conversion is nearly complete (>90%) for charge Q higher than 3 F mol^{-1}. The picolinic acid (E), on the other hand, which is formed in small amounts is only slightly reducible and its conversion remains very incomplete.

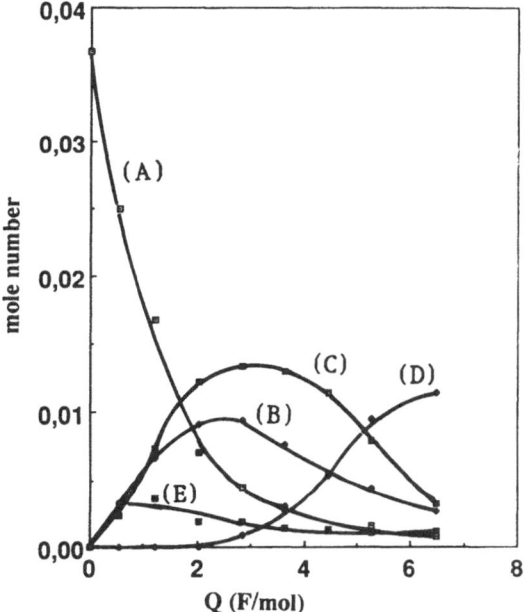

Figure 1. Variation of the mole numbers of reactants (A), (E) and products (B), (C), (D) as functions of the charge. $C_A° = 0{,}24$ mol L^{-1}; $C_{H_2SO_4} = 3$ mol L^{-1}; $j = 571$ A m^{-2}; $\Theta = 50°C$.

Reduction of the aromatic nucleus is significant after the appearance of the picoline; depending on the operating conditions, it leads to hydrogenation monomers (dihydropicoline, tetrahydropicoline, etc.) detected by CPV qualitative analysis coupled with mass spectrometry, or to polymers whose presence is revealed by the formation of a thin viscous layer on the cathode. The mass balance ratio is defined as the ratio of the sum of the moles of (A), (B), (C), (D) and (E) measured by HPLC, and the initial number of moles of (A). This ratio is higher than 0.9 before the formation of picoline, and then decreases significantly. The maximum alcohol (C) yield value, which is a key variable, was studied systematically as a function of the operating conditions.

Effect of the medium acidity

The first reason for the presence of sulfuric acid is to solubilize the pyridine derivatives by protonation of the nitrogen situated on the pyridinic cycle; it also favours the reduction of the side chain with respect to that of the aromatic nucleus.

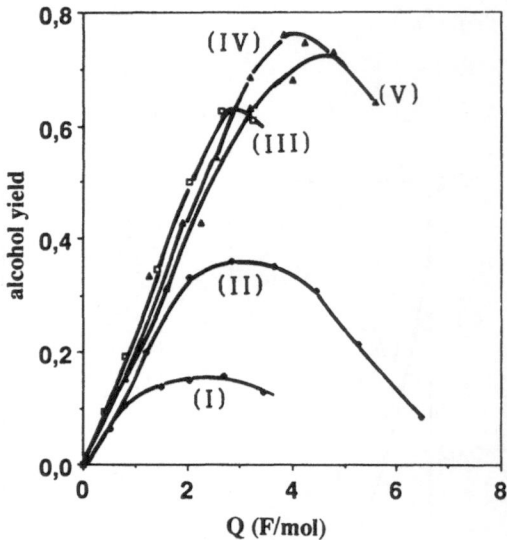

Figure 2. Variation of the yield of alcohol as a function of the charge for different concentrations of H_2SO_4. $C_A° = 0,24$ mol L^{-1}; j = 1071 A m^{-2}; Θ = 50°C; $C_{H_2SO_4}$ in mol L^{-1}: (I) 1; (II) 3; (III) 5; (IV) 7; (V) 9.

Figure 2 shows the variation of the alcohol yield as a function of the electrical charge Q. The variation curves exhibit a maximum which increases by 0.16 to 0.73 whereas the sulfuric acid concentration passes from 1 mol L^{-1} to 7 mol L^{-1}. In parallel, the aldehyde yield decreases and the appearance of picoline is retarded [5]. These two observations can be explained by the existence of two forms in equilibrium for (B): reducible free aldehyde, and the non-reducible hydrated form. Furthermore it is known that dehydration of the aldehyde hydrate is catalyzed by the protons [11] and that the amount of hydrate is greater when the acidity is high [5]. This explains the decrease in aldehyde yield coupled with the increase in alcohol yield when the sulfuric acid concentration increases.

Examining values of both the mass balance and the faradic balance also shows that too high an acidity (9 mol L^{-1}) induces electricity losses lower than 20% for a charge lower than 3 F mol^{-1}. In this case, the charge Q corresponding to the maximum alcohol yield is higher than 4 F mol^{-1}, whereas the picoline has already appeared, and the maximum alcohol yield no longer increases, or even decreases, compared to that obtained in more dilute sulfuric acid media (7 mol L^{-1}, for instance). The losses are probably due to the reduction of protons which results in gaseous hydrogen and principally in the reduction of the aromatic nucleus by the atomic hydrogen adsorbed at the cathode.

Effect of the current density and of the initial ester concentration

These two parameters are shown to play an important role: electrolysis with a high current density makes it possible to reduce the working surface of the electrodes and a high initial substance concentration results in a reduction in the number of batches for a discontinuous process. A current density of 1000 A m^{-2} and an initial ester concentration of 1 mol L^{-1} can be considered to be acceptable values in practice, and they were considered during the experimental study [5].

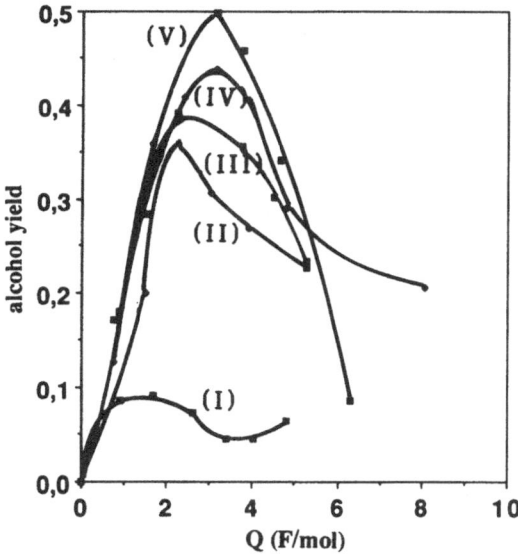

Figure 3. Variation of the yield of alcohol as a function of the charge for different temperatures. $C_A° = 0,24$ mol L^{-1}; $C_{H_2SO_4} = 3$ mol L^{-1}; $j = 1071$ A m^{-2}; Θ (°C): (I) 20; (II) 35; (III) 50; (IV) 65; (V) 80.

The effects of these two parameters are closely linked. The material balance and faradic yield show that a high initial ester concentration (1 mol L^{-1}) favours concurrent polymerization reactions after electrohydrogenation if the medium is only slightly acid (< 3 mol L^{-1}). Nevertheless, in this case a decrease in the current density from 2714 to 1071 A m^{-2} makes it possible to increase the maximum alcohol yield from 23% to 64%. Conversely, for a low value of the initial ester concentration, e. g. 0.24 mol L^{-1}), a moderate current density (571 A m^{-2}) can result in good alcohol yields (77% in a 7 mol L^{-1} sulfuric acid solution), whereas higher current density (1071 A m^{-2}) does not allow a good alcohol yield, probably due to electrohydrogenation induced at the electrode.

Decreasing the current density delays the appearance of picoline for highly acidic media (> 5 mol L^{-1}); the concurrent reductions (reduction of (D) and reduction of the pyridinic nucleus) are of less significance. When the current density increases, the picoline tends to appear before the maximum amount of alcohol is obtained.

Effect of temperature

Figure 3 shows that an increase in temperature from 20°C to 80°C increases the maximum alcohol yield from 10% to 50% for an initial ester concentration of 0.24 mol L^{-1} and a current density of 1071 Am^{-2}. Conversely, the maximum aldehyde yield decreases from 48% to 13% if the temperature rises within the same limits. This can be interpreted, assuming that aldehyde is more easily reduced into alcohol at high temperature, as being the result of a faster dehydration of the hydrate into free aldehyde. Furthermore, since an increase in temperature favours ester hydrolysis, the concentration of less reducible picolinic acid during electrolysis can reach approximately 20% of the initial amount of ester at 80°C and 5% to 10% at 50°C (figure 1). Moreover the appearance of the picoline is observed more rapidly at low temperature; probably a decrease in temperature slows down the reduction of the aldehyde into alcohol to a larger extent than it does that of alcohol into picoline.

THEORETICAL STUDY

A calculation tool was developed for predicting the influence of the operating parameters on the results of electrolysis and to simulate the reactor run for other operation conditions. The goal of the work was to create a numerical model making it possible to predict the variation of the quantities of (A), (B), (C) and (D) as a function of time, on the basis of the set of reactions (1), (2) and (3). However this scheme is highly simplified since it does not take into account either the pyridinic nucleus reduction reactions, nor the reduction of the proton, nor the existence of the aldehyde hydrate or of the picolinic acid.

Hence, the model of reactor operation coupled with a large volume storage tank takes into account the successive reduction stages:

$$A + 2e \rightarrow B \quad (k_{f1}, j_1)$$
$$B + 2e \rightarrow C \quad (k_{f2}, j_2)$$
$$C + 2e \rightarrow D \quad (k_{f3}, j_3)$$

k_{f1}, k_{f2}, k_{f3} represent the charge transfer rate constants and j_1, j_2, j_3 the current densities expressed in A m^{-2} and counted positively. The current density imposed is denoted j and equals $j_1 + j_2 + j_3$.

Since the effects of the variation in the electrode potential on the parameters of the Tafel equation could not be taken into account directly, the charge transfer constants were calculated as functions of time; the resistance to the transport of mass was also introduced. By analogy with flow in electrochemical reactors of similar design it was assumed that the mass transfer correlation used for the model reactor was valid for any reactor scale.

The electronic transfer equations relative to the three reactions envisaged are written:

$$j_1 / 2F = k_{f1} \ C_{AS} \tag{4}$$

$$j_2 / 2F = k_{f2} \ C_{BS} \tag{5}$$

$$j_3 / 2F = k_{f3} \ C_{CS} \tag{6}$$

where C_{AS}, C_{BS}, C_{CS}, C_{DS} are the concentrations of (A), (B), (C) and (D) respectively at the electrode. The mass transfer equations are written:

$$j_1 / 2F = k_{dA} \ (C_A - C_{AS}) \tag{7}$$

$$(j_2 / 2F) - (j_1 / 2F) = k_{dB} \ (C_B - C_{BS}) \tag{8}$$

$$(j_3 / 2F) - (j_2 / 2F) = k_{dC} \ (C_C - C_{CS}) \tag{9}$$

$$j_3 / 2F = k_{dD} \ (C_{DS} - C_D) \tag{10}$$

where C_A, C_B, C_C, C_D are the concentrations of (A), (B), (C) and (D) respectively in the solution and k_{dA}, k_{dB}, k_{dC}, k_{dD} are the mass transfer coefficients of (A), (B), (C) and (D). They are used to express the concentrations at the cathode as a function of the current densities. By carrying over the results to equations (4), (5) and (6), we obtain the relationships between the current densities and the concentrations in the reactor:

$$j_1 / 2F = Z_1 \ C_A \tag{11}$$

$$j_2 / 2F = (Z_2 \ C_A) + (Z_3 \ C_B) \tag{12}$$

$$j_3 / 2F = (Z_4 \ C_A) + (Z_5 \ C_B) + (Z_6 \ C_C) \tag{13}$$

with: $\quad Z_1 = (k_{f1} \ k_{dA}) / (k_{f1} + k_{dA})$

$\qquad Z_2 = (k_{f1} \ k_{f2} \ k_{dA}) / [\ (k_{f1} + k_{dA}) \ (k_{f2} + k_{dB}) \ ]$

$\qquad Z_3 = (k_{f2} \ k_{dB}) / (k_{f2} + k_{dB})$

$\qquad Z_4 = (k_{f1} \ k_{f2} \ k_{f3} \ k_{dA}) / [\ (k_{f1} + k_{dA}) \ (k_{f2} + k_{dB}) \ (k_{f3} + k_{dC}) \ ]$

$\qquad Z_5 = (k_{f2} \ k_{f3} \ k_{dB}) / [\ (k_{f2} + k_{dB}) \ (k_{f3} + k_{dC}) \ ]$

$\qquad Z_6 = (k_{f3} \ k_{dC}) / (k_{f3} + k_{dC})$

Writing the electrochemical reaction rates leads to a link between the current densities and the variations over time in concentrations of (A), (B), (C) and (D) in the reactor:

$$dC_A / dt = - (S / V_S) \ (j_1 / 2F) \tag{14}$$

$$dC_B / dt = - (S / V_S) \ ((j_2 - j_1) / 2F) \tag{15}$$

$$dC_C / dt = - (S / V_S) \ ((j_3 - j_2) / 2F) \tag{16}$$

$$dC_D / dt = (S / V_S) \ (j_3 / 2F) \tag{17}$$

where S represents the surface area of the cathode and V_S the volume of electrolyte in the storage tank. By inserting equations (11), (12), (13) into the above equations, we obtain the following system of differential equations:

$$dC_A / dt = - (S / V_S) \ Z_1 \ C_A \tag{18}$$

$$dC_B / dt = (S / V_S) \ [\ ((Z_1 - Z_2) \ C_A) - (Z_3 \ C_B) \ ] \tag{19}$$

$$dC_C / dt = (S / V_S) \ [\ ((Z_2 - Z_4) \ C_A) + ((Z_3 - Z_5) \ C_B) - (Z_6 \ C_C)] \tag{20}$$

$$dC_D / dt = (S / V_S) \ [\ (Z_4 \ C_A) + (Z_5 \ C_B) + (Z_6 \ C_C) \ ] \tag{21}$$

The calculated mass transfer constants and the time-variable charge transfer constants that occur in the expressions for the coefficients Z_i must be known in order to solve the previous system by the Milne method. The expression for the imposed current density j ($j = j_1 + j_2 + j_3$) leads, after coupling with equations (4), (5) (6), to an equation whose variables are k_{f1}, k_{f2} and k_{f3} for known concentration of products:

$$j / 2F = [(Z_1 + Z_2 + Z_3) \ C_A] + [(Z_3 + Z_5) \ C_B] + [Z_6 \ C_C] \tag{22}$$

The constants are obtained using parameters W_1 and W_2 defined as:

$$k_{f2} = W_1 \ k_{f1} \quad \text{and} \quad k_{f3} = W_2 \ k_{f2}$$

Parameters W_1 and W_2 can be adjusted with a view to obtaining a good correlation between the model and the experiment. Solving equation (22) by the Newton method gives k_{f1} for the concentrations determined by the Milne method in the previous calculation step [12]. We then start solving the equation set (18) - (21) again with the new value of k_{f1} to obtain new concentrations corresponding to the following calculation step, etc. Obviously k_{f1} must be suitably initialized for the first resolution. The initial value is such that the current density j_1 is lower than the limiting diffusion current for the initial concentration of (A).

Estimation of the mass transfer constants requires values for the diffusion coefficients of the various reducible species (A), (B), (C). Neither the case of the aldehyde hydrate, which is not reducible, nor that of the picolinic acid, which is only slightly reducible, are considered. The diffusion coefficients of (A), (B), (C) are calculated using the Wilke and Chang correlation [13]:

$$D_i = [\ 1.173 \times 10^{-13}\ (2.6\ M_i)^{0.5}\ T\ ] / (\ \mu\ V_m^{0.6}\ )$$

taking into account that the solvent is water. M_i designates the molar mass (g mol^{-1}) of species i. The absolute viscosity of the solution μ (mN s m^{-2}) is assumed to be equivalent to that of an aqueous solution of sulfuric acid. Its value is determined at different temperatures on the basis of published data [13,14]. The molar volumes V_m calculated [13] for (A), (B) or (C) are respectively: 0.1743; 0.1153 and 0.1227 m^3 (kmol)$^{-1}$.

Mass transfer coefficients $k_{d,i}$ are calculated by the Carlsson correlation [15], obtained for the type of reactor described above (Electrocell AB) and provided with the same turbulence promoter:

$$k_{d,i} = (D_i / d_h)\ 5.57\ Re^{0.4}\ Sc^{1/3}$$

where d_h designates the hydraulic diameter ($d_h = 5.47 \times 10^{-3}$ m). The cross section of the cathodic compartment is 96 $\times$ 10^{-6} m^2 and the volumic flow rate Q_V is 12.8 $\times$ 10^{-6} m^3 s^{-1}.

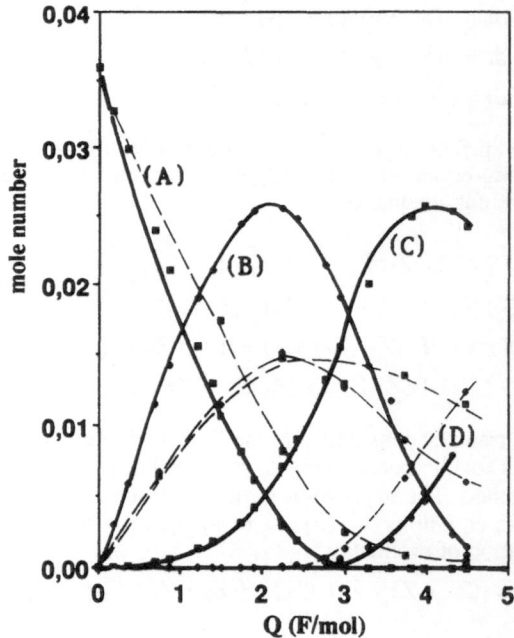

Figure 4. Variation of the mole numbers of reactants (A), (E) and products (B), (C), (D) as functions of the charge. $C_A° = 0,24$ mol L^{-1}; $C_{H_2SO_4} = 3$ mol L^{-1}; j = 1071 A m^{-2}; Θ = 50°C; experimental curves (---); theoretical curves (—).

Table 1. Mass transfer coefficients.

Θ (°C)	$C_{\text{sulfuric acid}}$ (mol L^{-1})	$k_{d,A}$ (ester) (m s^{-1})	$k_{d,B}$ (aldehyde) (m s^{-1})	$k_{d,C}$ (alcohol) (m s^{-1})
50	3.14	2.61 x 10^{-4}	2.74 x 10^{-4}	2.69 x 10^{-4}

Values for the mass transfer coefficients k_d of the products A, B and C are then determined (Table 1).

Parameters W_1 and W_2 are fitted by comparing the time variations of the mole amount of ester, aldehyde, alcohol and picoline with the experimental results; a trial and error procedure was used for this purpose. The compromise chosen ($W_1 = W_2 = 0.1$) for these two parameters leads to figure 4 which shows an example of the variation of the number of moles of the various derivatives as a function of the charge Q.

This model makes it possible to obtain the approximate rate of evolution of the number of moles of derivatives studied as a function of the charge Q (figure 4). Ester has virtually disappeared for charge $Q = 4$ F mol^{-1}. The maximum aldehyde concentration is reached near 4 F mol^{-1}, later than in the experiment. This could be due to the fact that the hydrogenation of the aromatic nucleus was not taken into account. Picoline is formed from 3 F mol^{-1}. However, predicted charges for the maximal productions of A, B and C are slightly shifted from the experimental values. Lastly, the effect of the sulfuric acid concentration has not been introduced into the reaction mechanism considered; taking into account the existence of aldehyde hydrate, whose dehydration is acid-catalyzed, appears to be a promising means of introducing the important role of the medium acidity as shown by experiment. Furthermore, the aromatic nucleus reduction reactions, and the existence of a quantity of less reducible picolinic acid resulting from the hydrolysis of the ester, may affect the validity of the simple model, but to a lower degree.

ACKNOWLEDGEMENTS

Financial support by Isochem Co. (Gennevilliers, France) and Electricité de France (Club Electrochimie Organique) is gratefully acknowledged.

REFERENCES

1. L. Eberson, J.H.P. Utley, Carboxylic acids and derivatives, *in:* "Organic Electrochemistry," M. Baizer ed., M. Dekker Inc., New york (1983).
2. H. Lund, Electroörganic preparations, XI. Reduction of isonicotinic acid in acid solution, *Acta Chem. Scand.* 17:972 (1963).
3. M. Ferles, M. Prystas, Untersuchungen in der pyridinreihe, VI. Betrag zur elektrolytischen reduktion der pyridincarbonsäuren, *Collection Czechoslov. Chem. Commun.* 24:3326 (1959).
4. T. Nonaka, T. Kato, T. Fuchigami, T. Sekine, Electroreduction of substituents on a pyridine ring in aqueous sulfuric acid, *Electrochim. Acta,* 26:887 (1980).
5. A.M. Romulus, Electrosynthèse d'hydroxyméthylpyridines à partir des esters correspondants, Thèse de l'Université Paul Sabatier, Toulouse, 1993.
6. A.M. Romulus, A. Savall, Electrochemical reduction of ethyl-2-picolinate on lead in a sulfuric acid medium, *Electrochim. Acta,* 37:625 (1992).
7. K. Scott, Reactor engineering models of complex electrochemical reaction schemes - I. Potentiostatic operation of parallel and series reactions in ideal reactors, *Electrochim. Acta,* 30:235 (1985).
8. K. Scott, I.F. McConvey, A.N. Haines, Reactor analysis of series and parallel electrochemical reactions during galvanostatic operation, *J. Appl. Electrochem.* 17:925 (1987).

9. A. Savall, J. Quesado, M. Rignon, J. Malafosse, Amino-alcohol electrosynthesis. Modelling of a set-up for producing amino-2-methyl-2-propanediol-1.3, *J. Appl. Electrochem.* 21:805 (1991).
10. K. Scott, Reactor engineering models of complex electrochemical reaction schemes - II. The effect of chemical reaction during batch electrolysis, *Electrochim. Acta,* 30:245 (1985).
11. E. Laviron, Polarographie et études physicochimiques des dérivés carbonylés et halogénés de la pyridine et de bases hétérocycliques, Thèse d'Etat, Dijon (1961).
12. J.P. Nougier, "Méthodes de calcul numérique," 2ème édition, Masson, Paris (1985).
13. J.M. Coulson, J.F. Richardson, "Chemical Engineering," Volume 6, Chapter 8, Pergamon, Oxford (1986).
14. R. C. Weast, "Handbook of Chemistry and Physics," 68th Edition, CRC. Press, Boca Raton (1987).
15. L. Carlsson, B. Sandegren and D. Simonsson, Design and performance of a modular, multi-purpose electrochemical reactor, *J. Electrochem. Soc.* 130:342 (1983).

MODELLING OF A POLYPYRROLE MODIFIED ELECTRODE

Pierre Gros, Alain Bergel, and Maurice Comtat

Laboratoire de Génie Chimique, URA CNRS 192
Laboratoire de Génie Chimique et Electrochimie
Université Paul Sabatier
118 route de Narbonne
31062 TOULOUSE cedex, France

ABSTRACT

Electrochemical polymerisation of organic compounds is a simple and attractive method for immobilizing an enzyme at an electrode surface. An approach that is both experimental and mathematical is presented to explain the behaviour of an electrode modified by entrapment of a glucose oxidase in a polypyrrole film. The model was based on the resolution of the differential mass balance equations involving non-linear homogeneous catalysis. Evaluation of various assumptions such as the variation of the mass transport features inside the film and the occurrence of heterogeneous catalysis, led to the conclusion that the enzyme was non - uniformly distributed throughout the electropolymerised film.

INTRODUCTION

Among the numerous ways of modifying electrode surfaces, electropolymerisation of organic monomers is particularly attractive. The resulting films are generally homogeneous, chemically stable and strongly adherent to the electrode surface. This explains the great number of works devoted to electropolymerisation and the broad field of applications, such as storage batteries, electrochromic displays or protection against corrosion[1-4]. Moreover, a suitable choice of electropolymerisation conditions allows the transport properties of the film to be finely controlled[5-7]. Electropolymerisation has thus been extensively investigated in the area of chemical sensors. The permselective features of the film provides significant improvements in the sensitivity and stability of the response of the sensors. The method has recently been broadened to biosensors[8-10]. The polymer material is used to entrap one or more enzyme(s) in the immediate vicinity of the electrode. The enzyme catalyses a specific reaction involving the substrate to be assayed and consuming or generating an electroactive species. The mass flux variation of this species is most often amperometrically detected, and the current obtained is correlated to the substrate concentration. Many experimental studies have provided efficient biosensors where the enzyme is immobilized in a polymeric matrix, but studies dealing with the modelling of such systems remain very scarce[11-14].

A theoretical approach is therefore necessary with a view to scaling up these devices to a preparative level. Academic examples of combination of an electrosynthesis process with enzyme catalyses are nowadays common, particularly in the area of fine chemical synthesis. Immobilization of the enzymes by electropolymerisation offers promising new opportunities for development in this field. The goal of this work is to improve the understanding of the behaviour of a biochemically modified electrode by both an experimental and a modelling approach.

The experimental support consisted of a glucose oxidase entrapped in a polypyrrole matrix prepared by electropolymerisation on a platinum electrode (Figure 1). The glucose oxidase (GOD) catalysed the oxidation of glucose into gluconic acid:

$$\text{glucose} + O_2 \xrightarrow{\text{GOD}} \text{gluconic acid} + H_2O_2 \qquad (1)$$

The hydrogen peroxide which was generated was oxidized at the electrode surface:

$$H_2O_2 \longrightarrow O_2 + 2\,H^+ + 2\,e \qquad (2)$$

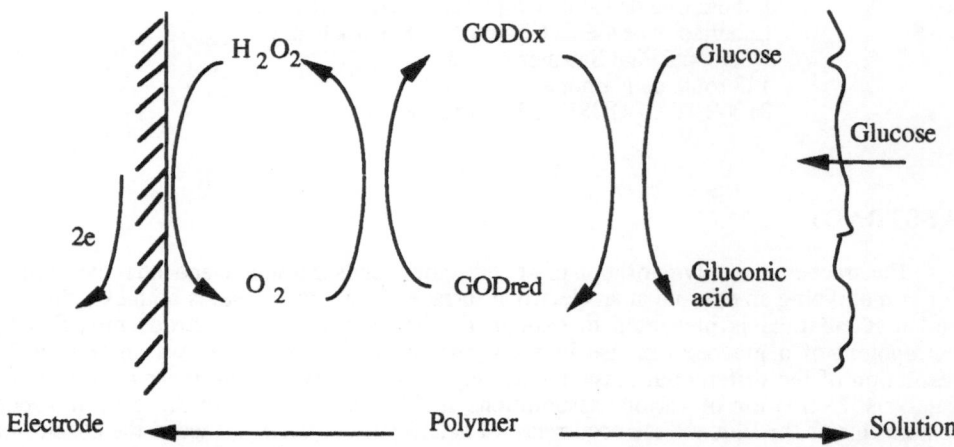

Figure 1. General kinetic scheme for the immobilized enzyme electrode.

It must be mentioned that polypyrrole is in its oxidized and conducting state at the electrode potential where the hydrogen peroxide is oxidized[1] . In this case, the hydrogen peroxide would react at the polymer - bulk interface and the electrons would be conducted through the polymer to the electrode surface. This possibility was not considered because the potential range where the polymer can be used in this way is too narrow, and its stability is too poor, particularly when an oxidant such as hydrogen peroxide is present. In order to obtain an inert, stable and non - conducting material, the polymer was electrochemically overoxidized after its formation at the electrode surface.

EXPERIMENTAL SECTION

General

Glucose oxidase (E.C. 1.1.3.4) type VII from *Aspergillus niger*, β-D-glucose and pyrrole were purchased from Sigma. All the electrochemical experiments were performed with an EGG 362 potentiostat-galvanostat, a satured calomel electrode (SCE) as reference and a large-area platinum-gauze counter electrode. The activity of the glucose oxidase was measured spectrophotometrically according to the protocol from Sigma. The unit is defined as the amount of enzyme necessary to catalyse the oxidation of one micromole of β-D-glucose in one minute at 35°C.

Preparation of the Biochemically Modified Electrode

The pyrrole was galvanostatically polymerized on a 0.03 cm^2 area platinum rotating disk electrode. The solution consisted of a 0.1 M phosphate buffer pH 7.0 containing pyrrole 0.1 M, sodium perchlorate 0.1 M and glucose oxidase 1.5 mg/ml. It was maintained under a nitrogen flux. The polymer thickness was controlled by means of the charge passed according to the correlation[15]:

$$\delta p = \frac{M \; Q}{2 \; F \; \rho \; A} \tag{3}$$

The electrolysis time was 15 s for all the experiments, the charge was controlled by a suitable choice of current. Following its preparation, the film was overoxidized by a constant potential electrolysis carried out at 0.85 V/SCE until the current leveled off at a minimal background value.

Glucose Oxidation Process

The biochemically modified electrode was immersed in a 0.1 M phosphate buffer pH 7.0 containing β-D-glucose 5 mM and dissolved oxygen. The rotation speed was 500 rpm, the potential was held at 0.55 V/SCE to ensure a fast oxidation of the hydrogen peroxide. The current rapidly stabilised to a steady value.

THEORY

Assuming that the kinetics of the enzymatic catalysis is of a "ping-pong" type the steady state mass balance equations in the polymer matrix are:

$$D_g^p \frac{d^2 C_g}{dz^2} - \frac{r_{max}}{1 + \dfrac{K_g}{C_g} + \dfrac{K_o}{C_o}} = 0 \tag{4}$$

$$D_o^p \frac{d^2 C_o}{dz^2} - \frac{r_{max}}{1 + \dfrac{K_g}{C_g} + \dfrac{K_o}{C_o}} = 0 \tag{5}$$

$$D_h^p \frac{d^2 C_h}{dz^2} + \frac{r_{max}}{1 + \dfrac{K_g}{C_g} + \dfrac{K_o}{C_o}} = 0 \tag{6}$$

The boundary conditions at the electrode surface were imposed by the electrochemical conditions:

$$\frac{dC_g}{dz}\Big|_{z=0} = 0 \tag{7}$$

$$D_o^p \frac{dC_o}{dz}\Big|_{z=0} = - D_h^p \frac{dC_h}{dz}\Big|_{z=0} \tag{8}$$

$$C_h|_{z=0} = 0 \tag{9}$$

and the expression for the current was:

$$i = 2 \; F \; A \; D_h^p \frac{dC_h}{dz}\Big|_{z=0} \tag{10}$$

At the polymer - bulk interface, equality of the mass flux densities gives:

$$D_g{}^p \frac{dC_g}{dz} \Big|_{z=\delta p} = \frac{D_g{}^s}{\delta_d} \left(C_g{}^s - \frac{C_g|_{z=\delta p}}{\alpha_g} \right) \tag{11}$$

$$D_o{}^p \frac{dC_o}{dz} \Big|_{z=\delta p} = \frac{D_o{}^s}{\delta_d} \left(C_o{}^s - \frac{C_o|_{z=\delta p}}{\alpha_o} \right) \tag{12}$$

$$D_h{}^p \frac{dC_h}{dz} \Big|_{z=\delta p} = \frac{D_h{}^s}{\delta_d} \left(C_h{}^s - \frac{C_h|_{z=\delta p}}{\alpha_h} \right) \tag{13}$$

After translation into dimensionless form, the numerical process was carried out by a classic finite difference method. An iterative process was necessary because of the non-linearity of the kinetics.

RESULT AND DISCUSSION

Figure 2-a depicts the variation of the current obtained experimentally (discrete points) with a 5 mM glucose solution as a function of the polymer thickness. The current increased rapidly in the first part of the curve and then decreased when the thickness of the film went beyond an optimal value of approximately 90 nm. The continuous line represents the best theoretical curve obtained by adjusting the diffusivities of the three species in the polymer. The values of the diffusivities in the solution can be found in the bibliography[12,16]. All the partition coefficients were estimated at 0.8[12]. The kinetic parameters of the enzymatic catalysis were assumed to have the same value as for the free enzyme in solution. The activity of the glucose oxidase was measured spectrophotometrically in the solution used to prepare the film, and the Michaelis constants were determined from the bibliography[17]. The thickness of the diffusion layer was calculated as a function of the rotation speed according to Levich's equation[18]. The values used for the simulation are given in Table 1. It was not possible to fit the experimental curve satisfactorily, particularly the shape of the peak around 90 nm. A similar discrepancy was noted in work dealing with modelling of a

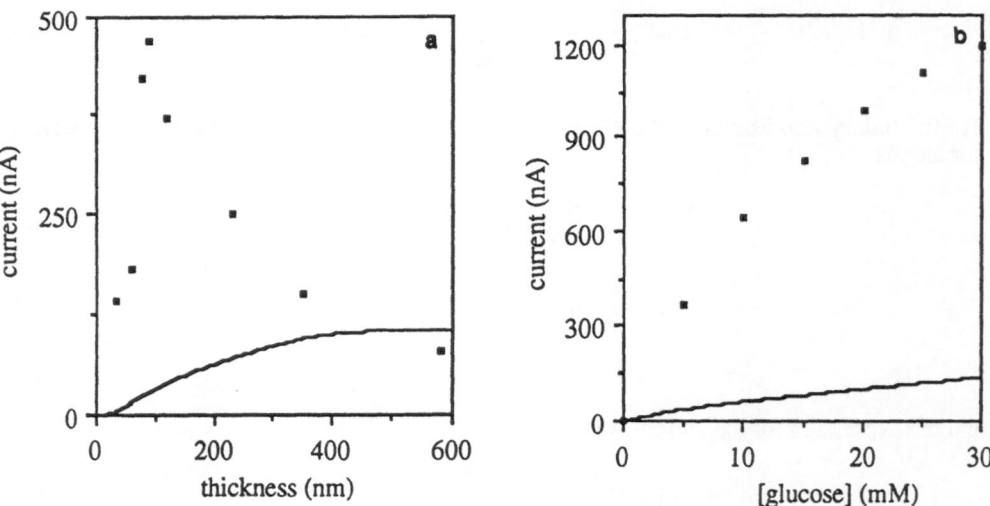

Figure 2. Variation of the modified electrode response as a function of film thickness (2-a) and calibration curve (2-b); theoretical curve (___) and experimental results (•)

glucose oxidase modified polypyrrole electrode [13,14]. The experimental calibration curve, representing the variation of the intensity for a 90 nm thick film as a function of the glucose concentration, was also poorly fitted by the theoretical curve obtained with the same values of the adjusted parameters (Figure 2-b). In both cases the theoretical current obtained was lower than the experimental one.

Table 1. Value of the parameters used in the simulation. One asterisk indicates the value was measured experimentally, two asterisks indicate the value was adjusted.

Variable	Value	Reference
D_g^s	$6.7.10^{-10}$ m^2s^{-1}	16
D_o^s	2.10^{-9} m^2s^{-1}	12
D_h^s	2.10^{-9} m^2s^{-1}	12
r_{max}	13.5 $mol.m^{-3}s^{-1}$	*
K_g	33 $mol.m^{-3}$	17
K_o	0.2 $mol.m^{-3}$	17
δ_d	2.10^{-5} m	*
D_g^p	10^{-14} m^2s^{-1}	**
D_o^p	10^{-12} m^2s^{-1}	**
D_h^p	10^{-12} m^2s^{-1}	**

In order to refine the description of the transport phenomena in the polymer film, the diffusivity of oxygen was measured experimentally for various film thicknesses according to the protocol described by Gough et al [19]. The raw diffusivity values reported in Table 2 were in fact a global estimation of the parameter $\alpha_o D_o^p$. The diffusivity increased as the film thickness increased. During electropolymerisation of the films, the thickness was controlled by the choice of the current of electrolysis whose duration was always the same. The lesser thicknesses were therefore obtained with lower currents The lower values of oxygen diffusivity were consistent with the fact that a slower film growth rate ensures better packing and results in narrower pores[6]. Nevertheless, the introduction of this type of diffusivity variation into the model did not improve the theoretical results.

Table 2. Variation of oxygen diffusivity with film thickness. Electrolysis in 0.1 M phosphate buffer at -0.35 V/SCE

Film thickness (nm)	17	90	230	580
$\alpha_o D_o^p$ (m^2/s)	$4.3\ 10^{-13}$	$1.8\ 10^{-12}$	$3.5\ 10^{-12}$	6.10^{-12}

A previous attempt at modelling attributed the deviation of the theoretical curve from the experimental one to possible adsorption of the glucose oxidase at the electrode surface[14]. The high concentration of the enzyme at this interface could induce an unexpectedly high response for glucose oxidation. The experiments confirmed the occurrence of enzyme adsorption. When electropolymerisation started immediately after the electrode was immersed in the pyrrole solution, the film obtained gave a smaller response to the glucose than when electropolymerisation was started after the electrode was kept 30 s or more at open circuit in the solution. Figure 3 represents the theoretical curve obtained with boundary conditions taking into account the heterogeneous catalysis by the adsorbed glucose oxidase. The boundary condition for the glucose at the electrode surface and the expression of current became:

$$D_g p \frac{dC_g}{dz}\Big|_{z=0} = \frac{r_{max}{}^{ad}}{1 + \frac{K_g}{C_g|_{z=0}} + \frac{K_o}{C_o|_{z=0}}} \qquad (14)$$

$$i = 2FA \left(D_h p \frac{dC_h}{dz}\Big|_{z=0} + \frac{r_{max}{}^{ad}}{1 + \frac{K_g}{C_g|_{z=0}} + \frac{K_o}{C_o|_{z=0}}} \right) \qquad (15)$$

where $r_{max}{}^{ad}$ represents the activity of the enzyme adsorbed per unit of electrode surface. The value of this parameter was adjusted in order to improve the fitting of the greater part of the curve, but it ensured too fast a rate of glucose oxidation for the film of small thickness. This hypothesis improved the shape of the curve for the high thickness values and the reliability of the calibration curve, but the model failed to completely explain the behaviour of the electrode.

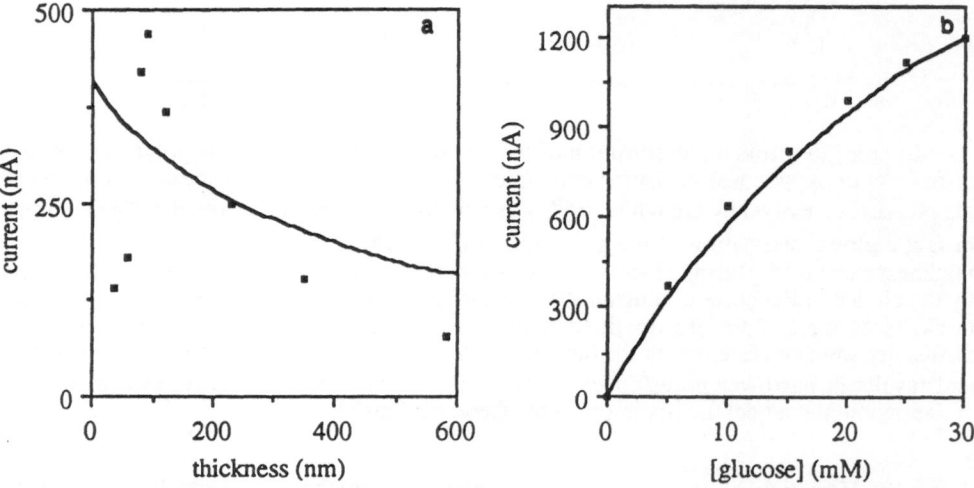

Figure 3. variation of the modified electrode response as a function of film thickness (3-a) and calibration curve (3-b); theoretical curve (___) and experimental results (•). $r_{max}{}^{ad} = 7.10^{-6}$ mol.m^{-2}s^{-1}.

Among the various hypotheses assessed, only one allowed the theoretical curve to approach the experimental results. It was necessary to use a non-uniform distribution of the enzyme. It was assumed that the glucose oxidase was preferentially entrapped inside a small layer in the vicinity of the electrode surface. The polymer grown far from the electrode was not be capable of entrapping a large amount of enzyme. This might be attributed to the occurrence of various successive steps during film formation: an initial nucleation and nucleus growth, allowed a large amount of enzyme to be inserted in the polymeric structure; further polymer growth caused the crosslinkage of the polymer chains, resulting in a more compact form. The theoretical curves obtained with this hypothesis, assuming a 90 nm thick layer with high enzyme load $r_{max} = 140$ mol.m^{-3}s^{-1} and a low enzyme load $r_{max} = 1$mol m^{-3}s^{-1} in the other part of the film, are reported in Figure 4. They clearly exhibit a peak shape as in the experimental evolution of the current with respect to the film thickness. The calibration curve was also satisfactorily fitted. The experimental point represented by an asterisk was the average current obtained with a 90 nm thick film when no enzyme was adsorbed on the electrode surface. Curve 2, plotted with the value $r_{max}^{ad} = 0$, shows that the influence of the enzyme adsorption was correctly taken into account by the model.

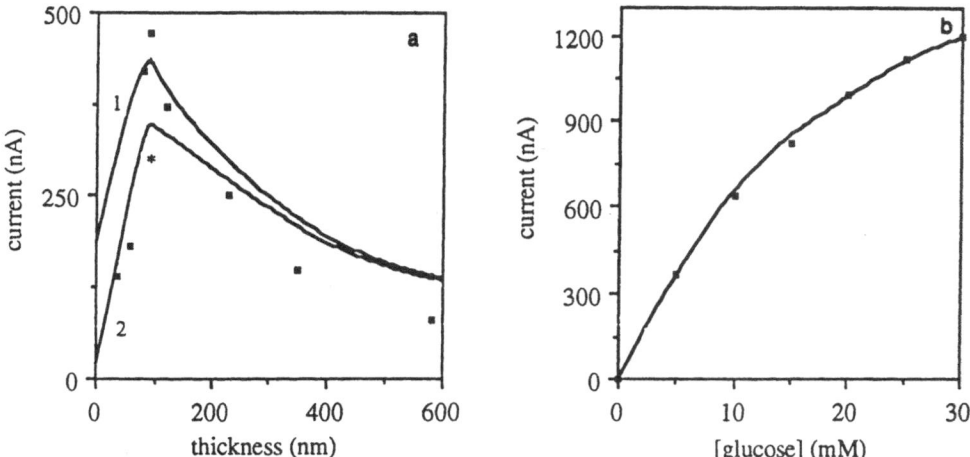

Figure 4. variation of the modified electrode response as a function of film thickness (4-a) and calibration curve (4-b); theorical curve (___) and experimental results (•) and (*).
curve 1: $r_{max}^{ad} = 7.10^{-6}$ mol.m^{-2}s^{-1}; curve 2: without adsorption.

CONCLUSION

The model quantitatively investigated the hypotheses which were generally evoked to explain qualitatively the difference between the experimental results and the small number of theoretical attempts already made. Neither the variation of the diffusivities with respect to the film thickness, nor the occurrence of enzyme adsorption at the electrode surface were sufficient to account for this discrepancy. The present model gives a new insight to the understanding of biochemically modified electrodes by supporting the assumption that the enzyme is non-uniformly distributed throughout the electropolymerised film.

NOTATION

A   electrode area
C_j concentration of the species j
D_j diffusivity of the species j
K_j Michaelis constant relative to the species j
F   Faraday's constant
i   current intensity
M   molar mass of the pyrrole (67.1 g/mole)
Q   electrical charge used in the electropolymerization step
r_{max} activity of the glucose oxidase
r_{max}^{ad} activity of the glucose oxidase adsorbed on the electrode surface
z   distance from the electrode

α_j partition coefficient of the species j

δ_d thickness of the diffusion layer

δ_p thickness of the polymer film

ρ volumic mass of the pyrrole (1500 kg/m^3)[1]

Subscripts

g, o, h   index for glucose, oxygen and hydrogen peroxide species
s   in the solution
p   in the polymeric film

REFERENCES

1. A.F. Diaz, J.I. Castillo, J.A. Logan, and W.Y. Lee, Electrochemistry of conducting polypyrrole films, *J.Electroanal. Chem.*, 129:115 (1981)
2. S. Asavapiriyanont, G.K. Chandler, G.A. Gunawardena, and D. Pletcher, The electrodeposition of polypyrrole films from aqueous solutions, *J. Electroanal. Chem*, 177:229 (1984)
3. E.M. Genies, G. Bidan, and A.F. Diaz, Spectroelectrochemical study of polypyrrole films, *J. Electroanal. Chem.*, 149:101 (1983)
4. G. Tourillon, and F. Garnier, Electrochemical doping of polythiophene in aqueous medium: electrical properties and stability, *J. Electroanal. Chem.*, 161:407 (1984)
5. J. Wang, S.P. Chen, and M.S. Lin, Use of different electropolymerisation conditions for controlling the size-exclusion selectivity of polyaniline, polypyrrole and polyphenol films, *J. Electroanal. Chem.*, 273:231 (1989)
6. A. Witkowski, M.S. Freund, and A. Brajter-Toth, Effect of electrode substrate on the morphology and selectivity of overoxidized polypyrrole films, *Anal. Chem.*, 63:622 (1991)
7. A. Witkowski , and A. Brajter-Toth, Overoxidized polypyrrole films: a model for the design of permselective electrodes, *Anal. Chem.*, 64:635 (1992)
8. N.C. Foulds, and C.R. Lowe, Enzyme entrapment in electrically conducting polymers: Immobilisation of glucose oxidase in polypyrrole and its application in amperometric glucose sensors , *J. Chem. Soc., Faraday Trans.1*, 82:1259 (1986)
9. T. Tatsuma, M. Gondaira, and T. Watanabe, Peroxidase-incorporated polypyrrole membrane electrodes, *Anal. Chem.*, 64:1183 (1992)
10. S. Cosnier,and C. Innocent, A novel biosensor elaboration by electropolymerisation of an adsorbed amphiphilic pyrrole-tyrosinase enzyme layer, *J. Electroanal. Chem.*, 328:361 (1992)
11. J.K. Leypoldt, and D.A. Gough, Model of a two-substrate enzyme electrode for glucose, *Anal. Chem.*, 56:2896 (1984)
12. J.Y. Lucisano, and D.A. Gough, Transient response of the two-dimensional glucose sensor, *Anal. Chem.*, 60:1272 (1988)
13. P.N. Bartlett, and R.G. Whitaker, Electrochemical immobilisation of enzymes; part I. theory, *J. Electroanal. Chem.*, 224:27 (1987)
14. P.N. Bartlett, and R.G. Whitaker, Electrochemical immobilisation of enzymes; part II. glucose oxidase immobilised in poly-N-methylpyrrole, *J. Electroanal. Chem.*, 224:37 (1987)
15. S. Holdcroft, and B.L. Funt, Preparation and electrocatalytic properties of conducting films of polypyrrole containing platinum microparticulates, *J. Electroanal. Chem.*, 240:89 (1988)
16. R.C. Weast, Handbook of Chemistry and Physics, 68th ed, CRC Press Inc., Boca Raton (1987)
17. T.E. Barman, Enzyme Handbook, Springer-Verlag, New York (1969)
18. A.J. Bard, and L.R. Faulkner, Electrochimie, Principes, Méthodes et Applications, Masson, Paris (1983)
19. D.A. Gough, and J.K. Leypoldt, Membrane-covered, rotated disc electrode, *Anal. Chem.*, 51:439 (1979)

ESTIMATION OF CURRENT BYPASS IN A BIPOLAR ELECTRODE STACK

Ch. Comninellis

Institute of Chemical Engineering
Swiss Federal Institute of Technology
CH-1015 Lausanne, Switzerland

ABSTRACT

A simple method is proposed for the estimation of the current bypass from experimental current potential (i-U) curves measured for a "bipolar reactor" and with a one element cell of similar geometry. The model is valid only in the region where a linear i-U relation is obtained.

For the scale-up, a relation between the current bypass (ψ) and two dimensionless numbers (G_b and B_n) has been derived.

$$\psi = G_b (B_n + 1)$$

where G_b (Geometry number) depends on the geometry and B_n (Bipolar number) on the electrochemical parameters of the system.

INTRODUCTION

Direct and indirect electroorganic synthesis have been widely studied on the laboratory scale but only few processes have been applied on an industrial scale [1]. The main reason is the complex nature of the electrochemical reactor compared with the chemical reactor, which is usually a simple stirred-tank.

A new industrial undivided "bipolar electrochemical reactor" of channel type with flat parallel electrodes has been developed in our laboratory and is now commercialized. Its design is very simple, it consists of a bipolar stack of 20-30 metallic sheets (electrodes) separated from each other by insulating plastic spacers at a fixed distance (2-8 mm). The electrode stack is situated in the reactor body which is made of polymer coated steel. (Figure 1).

Electrochemical Engineering and Energy, Edited by
F. Lapicque *et al.*, Plenum Press, New York, 1995

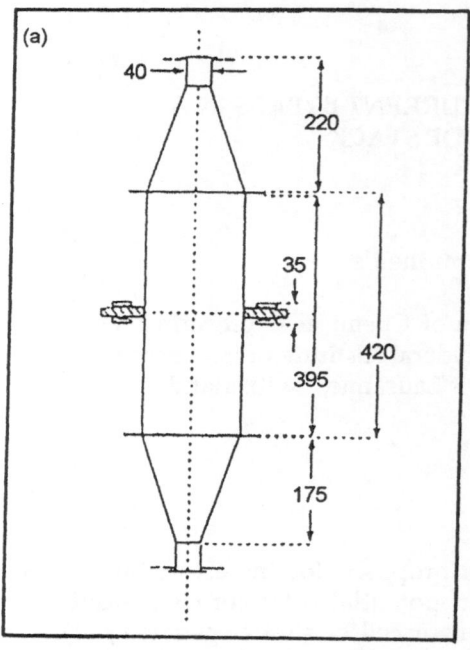

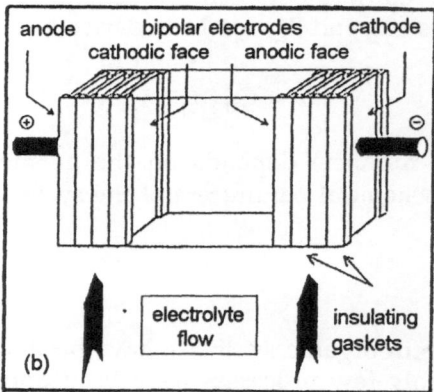

Figure 1. Schematic representation of (a) the electrochemical reactor and (b) the bipolar electrode stack. Dimensions are given in mm.

The main advantages of this electrochemical reactor are: Simple and cheap construction, compact design with high specific electrode area and possibility of using multiphase electrolytes (solid-liquid).

The main disadvantage of the reactor is the presence of parasitic electrical currents, or current bypass, in the lower and upper parts of the electrode stack. This results in an electrical efficiency loss, a non-uniform potential and current distribution, and consequently, a poor selectivity, in addition to an increase in electrode corrosion.

Many investigators have treated the subject of "bipolar electrochemical reactors" using an electrical analogy [2-5], or by direct solution of the Laplace equation with the finite element method or the finite difference method [6,7]. Yet, in spite of the large number of papers in this field, the experimental verification of the proposed models is relatively limited and the studied cases contain relatively few experimental results.

In the first part of this paper, a simple method is proposed for the estimation of the current bypass from experimental current-potential (i-U) curves measured for a bipolar reactor and with a one element cell of similar geometry [8]. In the second part of the paper, a simple relation is derived for scale-up of a bipolar electrochemical reactor. In this relation, the current bypass is related to two dimensionless numbers: B_n (Bipolar number) and G_b (Geometry number) [9].

THEORETICAL TREATMENT

Estimation of the bypass from experimental I-V curves

For a bipolar electrode stack with N cells the applied potential difference between the two electrode feeders (U_N) is equal to the algebraic sum of the individual potential differences (U_j) for each cell in the stack.

$$U_N = \sum_{j=1}^{j=N} U_j \tag{1}$$

Supposing a linear current-potential relation for each cell in the stack we obtain

$$U_N = \sum_{j=1}^{j=N} (U_0 + R_e i_j) \tag{2}$$

As U_0 and R_e are constant for a given electrode reaction and electrolyte conductivity

$$U_N = NU_0 + R_e \sum_{j=1}^{j=N} i_j \tag{3}$$

This relation is compared with that theoretically obtained when the current bypass is almost zero ($\psi = 0$)

$$NU_1 = NU_0 + N R_e i_0 \tag{4}$$

From equation (3), (4) and the definition of the current bypass (5).

$$\psi = \frac{N i_0 - \sum\limits_{j=1}^{j=N} i_j}{N i_0} \tag{5}$$

We thus obtain the relation (6) which allows the estimation of current bypass from the current-potential curve.

$$\psi = \frac{NU_1 - U_N}{NU_1 - NU_0} \tag{6}$$

The graphical determination of current bypass is illustrated in Figure 2.

Derivation of the scale-up relation

For the derivation of the scale-up relation, the model proposed by Burnett and Danly [10] for the filter press electrochemical reactor has been adapted for the configuration used to give the relation (7).

$$\psi = \frac{R_c}{6\,R}\,(N^2 - 1) + \frac{U_0}{6\,R\,A\,i_0}\,(N^2 - 1) \tag{7}$$

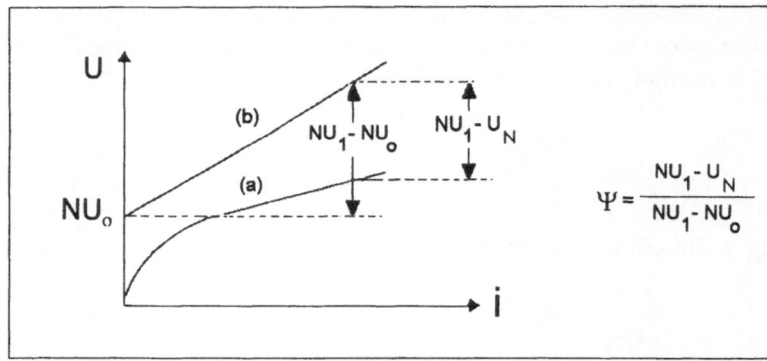

Figure 2. Graphical determination of current bypass from current-potential curves: (a) for the bipolar stack with N cells, and (b) [(i-U) for one element cell] x N.

where R_c is the ohmic resistance in the electrolyte between the electrodes and R is the ohmic resistance in the electrolyte feeder (or collector).

The definition of two dimensionless numbers

$$G_b = \frac{R_c}{6\,R}\,(N^2 - 1) \tag{8}$$

$$B_n = \frac{U_o}{A\,R_c\,i_o} \qquad (9)$$

gives the scale-up relation for the bipolar reactor

$$\boxed{\psi = G_b\,(B_n + 1)} \qquad (10)$$

The dimensionless number G_b (Geometry number) depends only on the geometry of the bipolar reactor, in contrast to the dimensionless number B_n (Bipolar number) which depends on the electrochemical system (U_o), on the electrolyte resistance between electrodes (R_c) and on the operating current density. (i_o).

Relation 10 can be considered in two limiting cases

i) $B_n \leq 1$ or $U_o \leq A\,R_c\,i_o$

In this limiting case, B_n can be neglected relative to unity in equation 10 and the current bypass is given by

$$\psi = G_b \qquad (11)$$

This relation shows that the current by-bass depends only on the geometry of the electrochemical reactor.

ii) $B_n \geq 1$ or $U_o \geq A\,R_c\,i_o$

In this case, unity can be neglected relative to B_n in equation 10 and the current bypass is given by

$$\psi = G_b\,B_n \qquad (12)$$

EXPERIMENTAL

The electrode stack (Fig. 1) contains 4 to 25 bipolar nickel electrodes (396 x 17 x 2 mm) and 2 nickel current feeders; the electrodes are separated from each other by insulating spacers at a fixed distance (3,3 mm) giving a total anode (cathode) area between 0.034 and 0.175 m^2. Figure 3 is a schematic representation of the equipment used.

The electrochemical reaction used for the determination of current bypass of the "bipolar reactor" is water electrolysis in alkaline medium (0,325, 0,50 and 1,0 M KOH). More detail of the experimental determination of current bypass is given elsewhere [8]. The i-U curves were obtained by sweeping the potential (1mV s^{-1}) and recording the resulting current. An electrical rectifier (Electronic Measurements Inc., Type SCR 100 A/100 V) commanded by a computer was used for the scanning of potential and X-Y recorder (Gould 6000) for recording the i-U curves.

RESULTS AND DISCUSSION

Figure 4 presents the effect of KOH concentration on the current-potential curves of a one element cell. For the three KOH concentrations studied the current density can be related to the potential by a linear relation of the type

$$U_1 = U_0 + R_e \, i_0 \tag{13}$$

where U_0 is a constant independent of KOH concentration ($U_0 = 1{,}7$ V) and R_e decreases with increasing KOH concentration.

The current-potential curves obtained for a "bipolar reactor" with 15 cells are given in figure 5 for three KOH concentrations. Transition from a non-linear to a linear i-U relation is observed; the critical current density at which transition occurs depends on the conductivity of the electrolyte (KOH concentration). The non linear i-U behaviour observed at low current densities and high KOH concentration is due to the fact that, under these conditions the electric field inside the reactor is not sufficiently high to polarize all the electrodes and, consequently, some of them remain inactive.

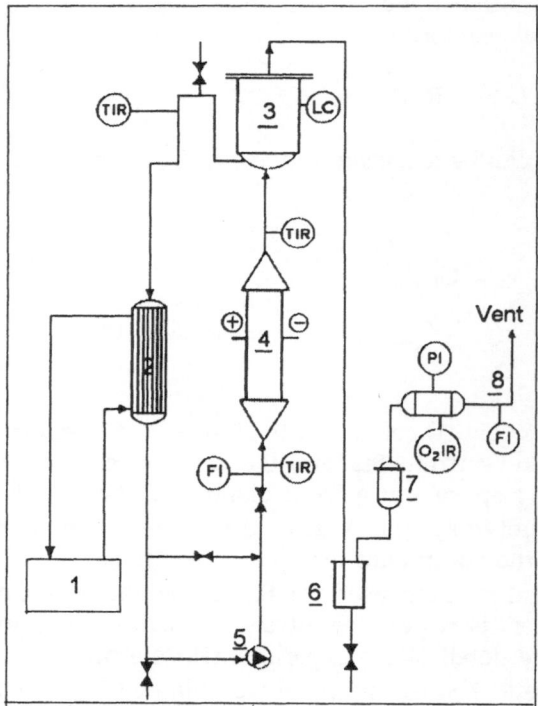

Figure 3. Schematic presentation of the equipment used. Key: (1) thermostat (2) heat exchanger, (3) gas-liquid separator (4) electrochemical reactor, (5) circulation pump, (6) liquid separator (7) alumina bed and (8) gas analysis.

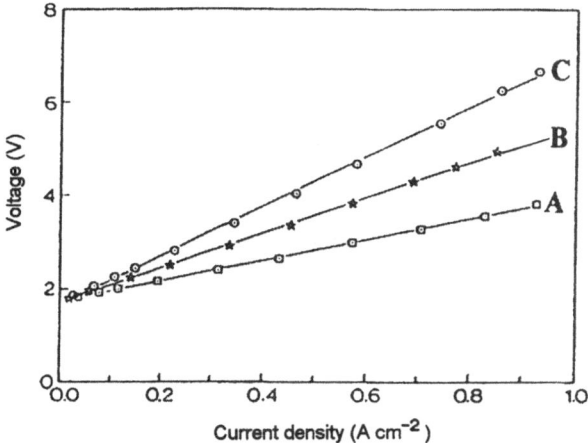

Figure 4. Current potential curves for one element cell at 50° C obtained at different KOH concentrations: (A) 1.0, (B) 0.5 and (C) 0.325 M.

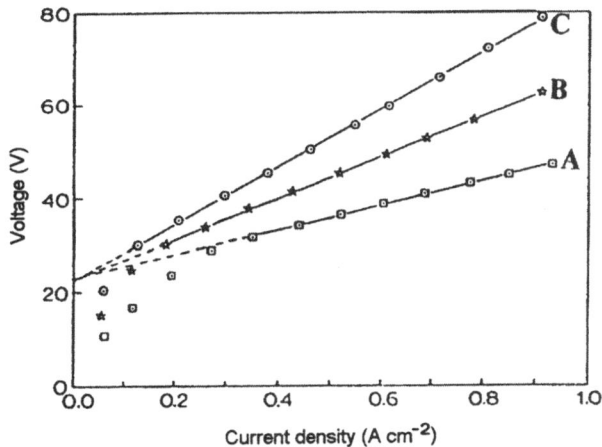

Figure 5. Current-potential curves for a bipolar stack with 15 cells obtained at different KOH concentrations: (A) 1.000   (B) 0.5000 and (C) 0.325 M.

As the proposed model supposes that all cells of the bipolar reactor are active, the linear i-U region only is considered.

The theoretical current bypass calculated from Figs. 4 and 5 using relation (6) and those determined experimentally are given in Figs. 6 and 7 for two different KOH concentrations. A good agreement between experimental and estimated bypass current is observed, especially at high current density.

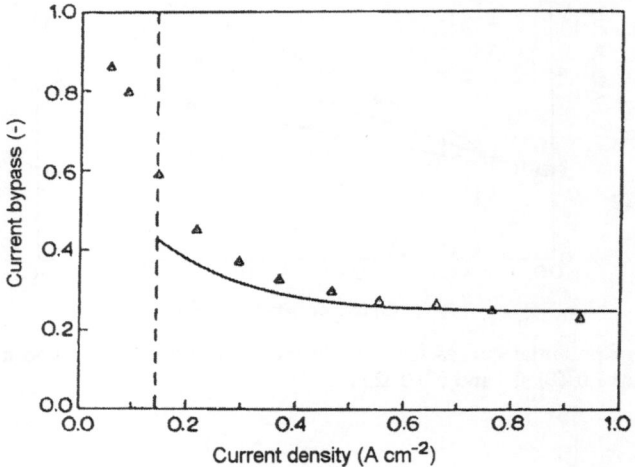

Figure 6. Current bypass as a function of current density with electrolyte: 0.325 M KOH at 50° C (Δ) experimental, (———) estimated from i-U curves (for the linear i-U region only).

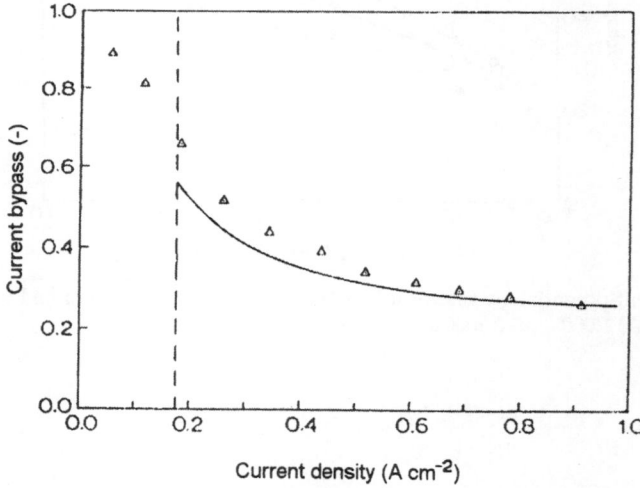

Figure 7. As for Fig. 6 with electrolyte: 0.50 M KOH at 50° C.

The validity of the scale-up relation (equation 10) has been verified (Figure 8) over a wide range of B_n (equation 9) and G_b (equation 8) values by changing the process parameters, (electrolyte conductivity σ and the feeding current density i_o), the number of elements N and by modifying the electrolyte manifolds. Much more detail concerning the experimental verification of the scale-up relation (equation 10) are given elsewhere [9].

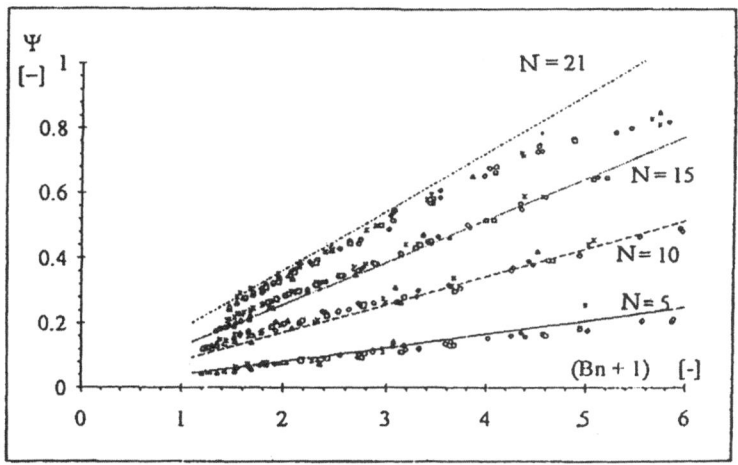

Figure 8. Experimental verification of equation 10, $\psi = G_b (B_n + 1)$, Electrolyte conductivity: 3,1 - 29,2 Ω^{-1} m^{-1}, Current density: 0,6 - 10 kA m^{-2}, Number of elements: 5 - 21

REFERENCES

1 D. E. Danly, Industrial Electroorganic Chemistry, in "Organic Electrochemistry", M. Baizer ed, Marcel Dekker inc., p. 907-943, (1973)

2 A.T. Kuhn and J.S. Booth, Electrical leakage currents in bipolar cell stacks, *J. Appl. Electrochem.*, 10, 233 (1980)

3 J.C. Burnett and D.E. Danly, Current bypass in electrochemical cell assemblies, *AICHE Symposium series*, 185, 75, 8-13 (1979)

4 M. Zahn, P.G. Grimes and R.J. Bellows, *US Patent* No. 4,197, 169 (April 8,1980)

5 E. Kaminski and R. Savinell, A technique for calculating shunt leakage and cell currents in bipolar stacks having divided or undivided cells, *J. Electrochem.Soc.*, 130, 1103 (1983)

6 J.A. Holmes and R.E. White, A finite element model of bipolar plate cells, in "Electrochemical Cell Design", R.E. White ed, Plenum Publishing Co., New York, p. 311 (1984)

7 J.B. Riggs, Evolution of the analog circuit model for the prediction of leakage currents in bipolar electrochemical systems, The Electrochemical Society Extended Abstracts, No. 123, vol. 80-82, 333 (1980)

8 Ch. Comninellis, E. Plattner and P. Bolomey, Estimation of current bypass in a bipolar electrode stack from current - potential curves, *J.Appl. Electrochem.* , 21, 415-418 (1991)

9 G. Bonvin, Ch. Comninellis, E. Plattner, Scale - up of bipolar electrode stack, dimensionless numbers for current bypass estimation, *J. Appl. Electrochem.* (submitted)

Notation

A   Electrode area (m^2)

i_o electrical feed current density ($A\ m^{-2}$)

i_j current density in cell j ($A\ m^{-2}$)

I_o current (A)

N   number of cells

R   Electrolyte resistance between electrodes (Ω)

R_e slope of the linear part of the i-U relation for one element cell ($\Omega\ m^2$)

R_c Resistance in the electrolyte feeder (Ω)

U_o intercept of the linear i-U relation with U axis for one element cell (V)

U_1 potential difference for one element cell (V)

U_N potential difference for a bipolar electrode stack with N cells (V)

U_j potential difference for cell j in the stack (V)

$\dot{V}$ experimental gas flow rate ($m^3\ s^{-1}$)

ψ current bypass

G_b dimensionless number (eq. 8)

B_n dimensionless number (eq. 9)

PROTON EXCHANGE MEMBRANE FUEL CELL MODEL

J. Divisek,[*] J. Mosig,[*] B. Steffen,[†] and U. Stimming[*]

[*]Institute of Energy Process Engineering (IEV)
[†]Institute of Applied Mathematics (ZAM)
Research Center Jülich (KFA)
D-52425 Jülich, Germany

ABSTRACT

A computer model was developed which permits a three-dimensional calculation of mass, charge and heat fluxes in a Proton Exchange Membrane Fuel Cell (PEM). The mathematical basis of the model is the method of finite volume elements. This computation technique also enables balances to be set up for very complex electrochemical processes. In this way, micromodelling of the complete PEM-unit including the membrane/electrodes assembly, the bipolar plates and the current collectors will be possible in principle. The computational results will be described in a second part of the study.

GENERAL DEFINITION OF THE PHYSICAL PROBLEM

The aim of this study is the three-dimensional mathematical modelling of the electronic and ionic charge, mass and heat fluxes in the different regions of the proton exchange membrane fuel cell. With reference to the model concept of Bernardi and Verbrugge[1] we have subdivided the fuel cell into seven regions (Figure 1). For each region we establish the respective charge, mass and heat balances. Figure 1 could be extended with respect to the exact geometry of the current collector system.

The PEM fuel cell is subdivided into the anode, membrane and cathode regions. The electrodes, which are assumed to be porous, are again divided into three regions: the gas compartment, the gas diffuser and the active catalyst layer. In the gas compartment, the wet reaction gases, considered as ideal and well intermixed (e.g. wetted hydrogen on the anode side and wetted air on the cathode side), are fed to the porous gas diffuser.

The porous electrode structure consists of electrically conducting material, carbon and platinum. Some pores have a hydrophobic character. They ensure the transport of the reaction gases into the electrochemically active cell regions, whereas the electrochemically produced water is transported through the hydrophillic pores. The pore water is in thermodynamic equilibrium with the water vapour in the wet reaction gas. The diffusion of the

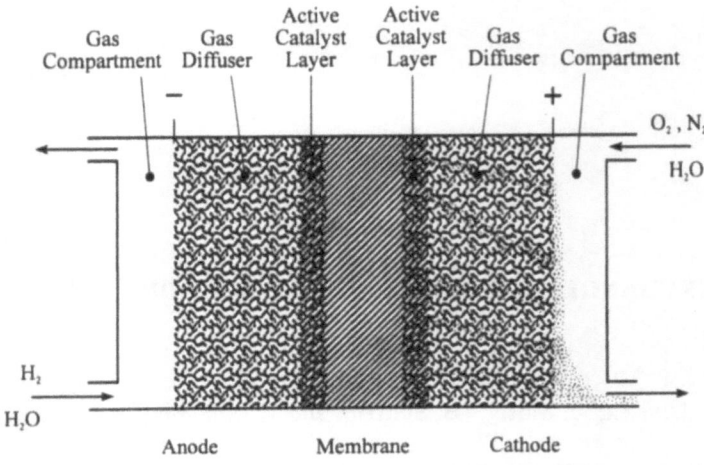

Figure 1. Schematic illustration of a Proton Exchange Membrane Fuel Cell.

multicomponent gas mixtures through the hydrophobic pores of the gas diffusers is described by the Stefan-Maxwell equations for ideal gas mixtures.

The pattern of overall species flow is shown in Figure 2.

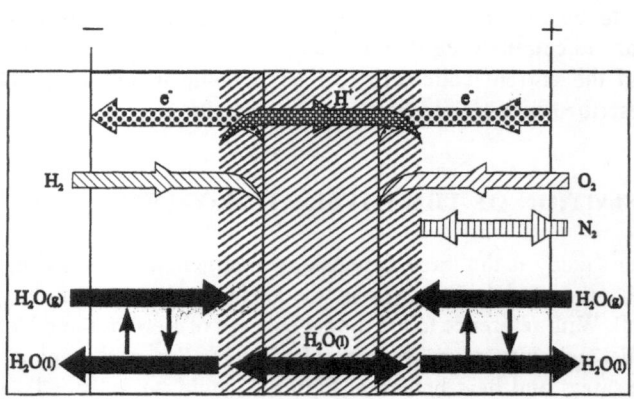

Figure 2. Schematic illustration of the overall species flow in a Proton Exchange Membrane Fuel Cell.

GENERAL MATHEMATICAL FORMULATION OF THE PROBLEM

We base our equations on the conservation quantities of the model: electric charge, thermal energy and mass. For each of these three quantities the flux F_n out of a given control volume V equals the integral over the sources Q contained in that volume (the Gaussian law):

$$\iint_{\partial V} F_n \, dA = \iiint_V Q \, dV \tag{1}$$

The actual physics of the problem is described by the equations which link the respective flow to the associated scalar quantity. This is the voltage Φ for the current flux $\mathbf{i}$, the temperature $\mathbf{T}$ for the heat flux $\dot{\mathbf{q}}''$ and the concentration $\mathbf{c}$ for the species flux $\mathbf{J}$. For the current flux outside surfaces on which electrochemical reactions take place we have the relation

$$\int_{x_1}^{x_2} (\rho \, i, \, ds) = \phi'(x_2) - \phi'(x_1) \tag{2}$$

where ρ is the local resistivity. An analogous relation is valid for the diffusive fraction of the heat flux $\dot{\mathbf{q}}''_D$ (or species flux)

$$\int_{x_1}^{x_2} \left(\frac{1}{\lambda} \, \dot{q}''_D, \, ds \right) = T'(x_2) - T'(x_1) \tag{3}$$

where λ is the heat conductivity.

The flow of electric current across electrochemically active surfaces is described in the simplest case by the symmetrical Butler-Volmer equation:

$$i_n = i_0 \, \sinh\left[\left(\phi'_s - \phi' - \Delta\phi'_0 \right) \frac{\alpha \, F}{R \, T} \right] \tag{4}$$

The convective fraction of the heat flux $\dot{\mathbf{q}}''_C$ (or species flux) is expressed by the relation

$$\dot{q}''_C = T \, c_p \, \vec{v} \tag{5}$$

where $\vec{v}$ is the mass flow vector and c_p the specific heat. The heat sources result from the ohmic losses of electrical conduction and from the chemical reactions between the individual chemical components of the system.

The finite integration technique (FIT)[2] is used for the discretization of the equations. In order to obtain discrete equations, we represent the region by a rectangular grid (Figure 3) of the form

$$x_1 < \ldots < x_{i+1} \quad , \quad y_1 < \ldots < y_{j+1} \quad , \quad z_1 < \ldots < z_{k+1} \tag{6}$$

A grid cell is then defined by

$$GC_{i,j,k} = \left[x_i, x_{i+1} \right] \times \left[y_j, y_{j+1} \right] \times \left[z_k, z_{k+1} \right] \tag{7}$$

Such grid cells are regarded as admissible volumes for the Gaussian equation (1). All conservation quantities are defined in the grid cells. Moreover, we define a dual point grid

$$x_i' = \frac{x_i + x_{i+1}}{2} \quad , \quad y_j' = \frac{y_j + y_{j+1}}{2} \quad , \quad z_k' = \frac{z_k + z_{k+1}}{2} \tag{8}$$

on which the scalar quantities Φ, T and c are defined. The discrete values of the flux quantities are defined on the surfaces of the grid cells, i.e. $(F_x)_{i,j,k}$ is defined at point (x_i, y_j', z_k'). At these points we also define the scalar quantities of the conductivity, concentrations and densities averaged over the respective integration area. If we now replace the integrals in equation (1) to (3) by their discrete approximations with single-point integration schemes, we obtain the discrete equations which are for the grid cell $GC_{i,j,k}$ transformed into:

$$\begin{aligned}
\left[(F_x)_{i+1,j,k} - (F_x)_{i,j,k} \right] (y_{j+1} - y_j) (z_{k+1} - z_k) + \\
\left[(F_y)_{i,j+1,k} - (F_y)_{i,j,k} \right] (x_{i+1} - x_i) (z_{k+1} - z_k) + \\
\left[(F_z)_{i,j,k+1} - (F_z)_{i,j,k} \right] (x_{i+1} - x_i) (y_{j+1} - y_j) = Q\,V
\end{aligned} \tag{9}$$

The corresponding expressions for the source densities are given in Table 1 and those for the fluxes in Table 2 at the end of the paper.

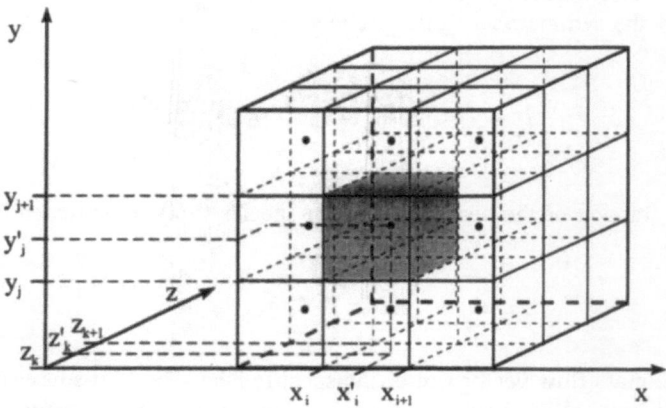

Figure 3. Schematic illustration of the rectangular grid used for the discretization of the conservation equations.

EXPLANATION OF THE MODEL

The coarse subdivision of the model distinguishes between the two electrodes and the proton exchange membrane.

Proton Exchange Membrane

The following assumptions were made for the proton exchange membrane:

190

- the transport of the positively charged water in the membrane pores is effected convectively due to the potential and pressure gradient according to Schlögl[3] and without sources;
- non-oxidized hydrogen and non-reduced oxygen can be transported both diffusively and convectively through the membrane pores;
- the hydrogen ions are transported source-free by migration and convection.

The principle of mass conservation according to equation (1) prevails in the membrane. Discretization is according to the general equation (9). The water velocity in the membrane pores can be determined by an equation of motion according to Schlögl, where the material quantities of density and viscosity are averaged at the surfaces of the grid cells.

The oxygen mole flux is composed of a diffusive and a convective portion and is defined as mole flow per surface area. The diffusion coefficient of oxygen at the surfaces of the grid cells is also averaged. The discretization of the convective fraction of the oxygen flow requires a modification to ensure the resolvability of the resultant equation matrix with efficient iteration methods. The oxygen concentration on the grid surface is determined by a suitable broken linear function which linearly interpolates in the case of vanishing convection, but eliminates coupling of the concentration $c'_{O2,i,j,k}$ to $c'_{O2,i+1,j,k}$ in the extreme case of infinitely strong outflow. The oxygen concentrations at the grid surfaces are determined by equations of the type:

$$\overline{c_{O2,i+1}} = \mu_{i+1}\, c'_{O2,i+1,j,k} + \left(1 - \mu_{i+1}\right) c'_{O2,i,j,k} \tag{10}$$

The function rule for the factor μ reads:

$$\mu_{i+1} = \frac{1}{2}\left(1 - \frac{const.\ v_{i+1,j,k}}{1 + const.\ |v_{i+1,j,k}|}\right) \tag{11}$$

The mass balance for hydrogen in the given control volume V can be calculated by analogy to that of oxygen.

The current flux in the membrane is considered to be source-free. Thus, the same balance equation (1) as for mass transport is applicable. Charge transport is effected by migration and convection, assuming constant hydrogen ion concentration in the membrane.

The general equation (1) also applies to the heat balance. The heat flux is composed of a diffusive and a convective portion and is defined as an energy flow per surface area. The discretization of the equation of conservation can be represented with the ohmic losses in the electrolyte as the source term. The computational treatment of the heat balance is formally analogous to that of the mass balance, i.e. the heat conductivity of the membrane electrolyte is formed as a (harmonic) average, whereas the cell temperature on the grid surface is determined by a broken linear function similar to the method already described for the concentration.

Electrodes

The electrochemical reactions are defined as follows:

cathodic: $O_2 + 4\ H^+ + 4\ e^- \rightleftharpoons 2\ H_2O$

anodic: $2\ H_2 \rightleftharpoons 4\ H^+ + 4\ e^-$

The electrodes are subdivided into the gas diffusion layer and the active catalyst layer. The following assumptions apply to the cathode gas diffuser:

- the electric charge transport in the carbon matrix is effected without sources and sinks due to the potential gradient;
- the flow of oxygen and nitrogen in the ternary gas mixture of oxygen/nitrogen/water vapour is purely diffusive (according to the Stefan-Maxwell equation) without sources and sinks;
- water vapour is in equilibrium with liquid water in the hydrophillic pores according to Raoult's law, so that its flux contains source terms;
- the convective mole flux of the liquid pore water is solely determined by the hydraulic pressure gradient in the hydrophillic pores.

The balance equation of the current flux in the solid matrix of the diffuser is given in the form of equation (1). The current flux is determined by the flow of electrons per cross-sectional area. It is given by Ohm's law and the specific conductivities at the surfaces of the grid cells are obtained by harmonic averaging.

The mass balances for the gases are obtained according to the general equation (1). The respective species flux in the ideal ternary gas mixture takes place diffusively and can be represented for all gases by analogous relations. The relation between the effective diffusion coefficients in a ternary mixture and the binary diffusion coefficients was defined according to Curtiss, Hirschfelder and Bird[4]. The change in the liquid water mole flux due to condensation constitutes its source or sink. The vapour-liquid equilibrium for water is expressed using Raoult's law, the temperature dependence of the saturation pressure of water being described by the vapour pressure equation of Antoine[5]. The anode gas diffuser contains the binary hydrogen/water vapour gas mixture which is regarded as ideal. The basic assumptions for this gas mixture were made by analogy to the case of the cathode gas diffuser. The mole flux of the liquid pore water is solely determined by the hydraulic pressure gradient in the hydrophillic diffuser pores.

The electrode furthermore consists of the active catalyst layer. The following assumptions are made for the cathode side:

- the electric charge transport in the carbon matrix is described by Ohm's law in which the electrochemical reaction represents an electron sink;
- the proton transport is effected by migration and convection assuming a constant ion concentration in the membrane phase (electroneutrality) with the electrode reaction being an ion sink;
- the oxygen mole flux is composed of a convective and a diffusive portion, the cathode reaction being an oxygen sink;
- the nitrogen is not dissolved in the membrane phase (concentration equalling zero);
- the hydrogen concentration equals zero;
- the convective transport of the positively charged water in the membrane phase of the catalyst layer is affected due to the potential and pressure gradient (Schlögl) with the electrochemical cathode reaction as the water source.

The balance equation of charge conservation in the solid matrix of the catalyst layer is expressed according to the general equation (1). A source for the electron flux in the grid cell volume is the electrochemical reaction described in the simplest case by the Butler-Volmer equation (4). This equation is used in the model in the form

$$\phi_s' - \phi' - \Delta\phi_0' = \frac{R\,T}{\alpha\,F} \ln\left[\left(\frac{i_n}{2\,i_0}\right) + \sqrt{\left(\frac{i_n}{2\,i_0}\right)^2 + 1}\right] \qquad (12)$$

in order to determine the approximate working point on the Butler-Volmer curve in an

Table 1. Source densities in the different regions of the proton exchange membrane fuel cell.

Source Density	Anode Gas Diffuser	Anode Catalyst Layer	Membrane	Cathode Catalyst Layer	Cathode Gas Diffuser
$Q_{ts} =$	–	$a^{C^-} i_n^{C^-}$	–	$a^{C^+} i_n^{C^+}$	–
$Q_i =$	–	$a^{C^-} i_n^{C^-}$	–	$a^{C^+} i_n^{C^+}$	–
$Q_{H2} =$	–	$\dfrac{s_{H2}}{n^- F} a^{C^-} i_n^{C^-}$	–	–	–
$Q_{O2} =$	–	–	–	$\left(-\dfrac{s_{O2}}{n^+ F} + \dfrac{s_{H2O}}{n^+ F}\dfrac{c'_{O2}}{c'_{H2O}} \right) a^{C^+} i_n^{C^+}$	–
$Q_{N2} =$	–	–	–	–	–
$Q_{H2O}^{gas} =$	$-\dfrac{1}{V^{D^-}} \displaystyle\iint_{\partial V} J_{H2O}^{liq.}\, dA$	–	–	–	$-\dfrac{1}{V^{D^+}} \displaystyle\iint_{\partial V} J_{H2O}^{liq.}\, dA$
$Q_{H2O}^{liq.} =$	$-\dfrac{1}{V^{D^-}} \displaystyle\iint_{\partial V} J_{H2O}^{gas}\, dA$	–	–	$-\dfrac{s_{H2O}}{n^+ F} a^{C^+} i_n^{C^+}$	$-\dfrac{1}{V^{D^+}} \displaystyle\iint_{\partial V} J_{H2O}^{gas}\, dA$
$Q_{q''} =$	$\dfrac{\Delta H_V^-}{V^{D^-}} \displaystyle\iint_{\partial V} J_{H2O}^{liq.}\, dA - \dfrac{i_s^2}{\sigma_{eff}^{D^-}}$	$\dfrac{i^2}{\kappa_{eff}^{C^-}} + \dfrac{i_s^2}{\sigma_{eff}^{C^-}} - \dfrac{a^{C^-} i_n^{C^-}}{V^{C^-}}\left[\dfrac{T' \Delta S^-}{n^- F}\right] + \dfrac{a^{C^-} i_n^{C^-}}{V^{C^-}}(\phi'_s - \phi')$	$\dfrac{i^2}{\kappa_{eff}^M}$	$\dfrac{i^2}{\kappa_{eff}^{C^+}} + \dfrac{i_s^2}{\sigma_{eff}^{C^+}} + \dfrac{a^{C^+} i_n^{C^+}}{V^{C^+}}\left[\dfrac{T' \Delta S^+}{n^+ F}\right] + \dfrac{a^{C^+} i_n^{C^+}}{V^{C^+}}(\phi'_s - \phi' - \Delta\Phi_0)$	$\dfrac{\Delta H_V^+}{V^{D^+}} \displaystyle\iint_{\partial V} J_{H2O}^{liq.}\, dA - \dfrac{i_s^2}{\sigma_{eff}^{D^+}}$

Table 2. Fluxes in the different regions of the proton exchange membrane fuel cell.

Flux	Anode Gas Diffuser	Anode Catalyst Layer	Membrane	Cathode Catalyst Layer	Cathode Gas Diffuser
$i_s =$	$-\overline{\sigma}_{eff}^{D^-} \nabla\phi_s'$	$-\overline{\sigma}_{eff}^{C^-} \nabla\phi_s'$	–	$-\overline{\sigma}_{eff}^{C^+} \nabla\phi_s'$	$-\overline{\sigma}_{eff}^{D^+} \nabla\phi_s'$
$i =$	–		$-\overline{\kappa}_{eff}^{R} \nabla\phi' + F c_f\, \nu^R$		–
$J_{H2} =$	$-\dfrac{\left(c_{tot}^{D^-}\right)^2}{\overline{\rho}_{tot}^{D^-}}\, M_{H2O}\, \overline{D}_{eff}^{H2-H2O}\, \nabla x_{H2}'$		$-\overline{D}_{eff}^{H2,R} \nabla c_{H2}' + c_{H2}'\, \nu^R$	–	–
$J_{O2} =$	–	–	$-\overline{D}_{eff}^{O2,R} \nabla c_{O2}' + c_{O2}'\, \nu^R$	–	$-\dfrac{\left(c_{tot}^{D^+}\right)^2}{\overline{\rho}_{tot}^{D^+}}\left[M_{N2}\, \overline{D}_{eff}^{O2-N2}\left(\nabla x_{O2}'+\nabla x_{H2O}'\right) + M_{H2O}\, \overline{D}_{eff}^{O2-H2O}\left(\nabla x_{O2}'+\nabla x_{N2}'\right)\right]$
$J_{N2} =$	–	–	–	–	$-\dfrac{\left(c_{tot}^{D^+}\right)^2}{\overline{\rho}_{tot}^{D^+}}\left[M_{O2}\, \overline{D}_{eff}^{N2-O2}\left(\nabla x_{N2}'+\nabla x_{H2O}'\right) + M_{H2O}\, \overline{D}_{eff}^{N2-H2O}\left(\nabla x_{N2}'+\nabla x_{O2}'\right)\right]$
$J_{H2O}^{gas} =$	$-\dfrac{\left(c_{tot}^{D^-}\right)^2}{\overline{\rho}_{tot}^{D^-}}\, M_{H2}\, \overline{D}_{eff}^{H2O-H2}\, \nabla x_{H2O}'$	–	–	–	$-\dfrac{\left(c_{tot}^{D^+}\right)^2}{\overline{\rho}_{tot}^{D^+}}\left[M_{O2}\, \overline{D}_{eff}^{H2O-O2}\left(\nabla x_{N2}'+\nabla x_{H2O}'\right) + M_{N2}\, \overline{D}_{eff}^{H2O-N2}\left(\nabla x_{O2}'+\nabla x_{H2O}'\right)\right]$
$J_{H2O}^{liq.} =$	$c_{H2O}'\, \nu_s^{D^-}$		$c_{H2O}'\, \nu^R$		$c_{H2O}'\, \nu_s^{D^+}$
$\dot{q}'' =$	$-\overline{\lambda}_{eff}^{D^-} \nabla T' + \overline{\rho}_{H2O}^{D^-}\, c_p^{H2O,D^-}\, \nu_s^{D^-}\, T'$		$-\overline{\lambda}_{eff}^{R} \nabla T' + \overline{\rho}_{tot}^{R}\, c_p^{R}\, \nu^R\, T'$		$-\overline{\lambda}_{eff}^{D^+} \nabla T' + \overline{\rho}_{H2O}^{D^+}\, c_p^{H2O,D^+}\, \nu_s^{D^+}\, T'$

external iteration with the approximate values for the electrode and ionic potentials. The electrochemical reaction represents a sink for the hydrogen ion flux. The mass balance in the active catalyst layer is formed for water and for oxygen.

The assumptions for the anode side of the active catalyst layer are analogous to those for the cathode side except that the balance equation for water does not contain any source terms and analogous relations for hydrogen are valid instead of the balance equations for oxygen.

The heat balances in the two electrodes, anode and cathode, are set up separately for all relevant regions, i.e. for the active catalyst layer and the gas diffuser. As in the case of the membrane region, it is assumed that the heat flux is composed of a diffusive and a convective portion, the heat radiation being negligible due to the relatively low working temperature of the fuel cell. The discretization of the general equation (1) is carried out according to the same pattern as already shown in treating the electrolyte. The entropy production of the electrode reactions represents the source terms for the heat flux in the individual electrodes. These source terms are calculated using the corresponding thermodynamic data for hydrogen, oxygen and water as well as those for protons and electrons as functions of cell temperature and pressure.

NOMENCLATURE

a	contact area catalyst/membrane per unit volume, cm^3/cm^2
A	grid cell area, cm^2
c_{f-}	concentration of the membrane fixed-charge-site species, mol/cm^3
c'_i	concentration of species i in the membrane phase, mol/cm^3
$\bar{c}_p$	average heat capacity at constant pressure, J/mol K
$\bar{c}_{tot}$	average concentration of a gas mixture, mol/cm^3
$\bar{D}^{i,j}$	average diffusivity of the gas pair i,j in a mixture, cm^2/s
$\bar{D}^i$	average diffusion coefficient of species i dissolved in the membrane, cm^2/s
F	Faraday's constant, 96487 C/mol
F_n	flux out of a grid cell volume
ΔH_V	enthalpy of vaporization, J/mol
i	current density in the membrane phase, A/cm^2
i_0	exchange current density, A/cm^2
i_n	current density of charge transfer, A/cm^2
i_s	current density in the electronically conductive material, A/cm^2
M_i	molecular weight of species i, g/mol
n	number of electrons participating in a reaction
Q	source density
R	gas constant, 8.3143 J/mol K
s_i	stoichiometric coefficient of species i
T	absolute temperature, K
v^i	water velocity in the membrane pores of cell region i, cm/s
V	grid cell volume, cm^3
x'_B	gas-phase mole fraction of species B
α	charge transfer coefficient
$\bar{\kappa}_{eff}$	average effective membrane conductivity, mho/cm
$\bar{\lambda}_{eff}$	average effective heat conductivity, W/cm K
μ	coupling factor for convective flux
$\bar{\rho}_{H2O}$	average mass density of liquid water in the pores of the gas diffuser, g/cm^3
$\bar{\rho}_{tot}$	average mass density of a gas mixture, g/cm^3
$\bar{\sigma}_{eff}$	average effective conductivity of the electronically conductive phase, mho/cm

ϕ' local potential value in the ionic phase, V
ϕ'_s local potential value in the electronically conductive phase, V
$\Delta\phi'_0$ reversible potential difference, V

REFERENCES

1. Bernardi D.M., Verbrugge M.W., A mathematical model of the solid-polymer-electrolyte fuel cell, *J. Electrochem. Soc.*, 139, 9, 2477-2491 (1992)
2. Weiland T., On the numerical solution of Maxwell's equations and applications in the field of accelerator physics, *Particle Accelerators*, 15, 245-292 (1984)
3. Schlögl R., Zur Theorie der Anomalen Osmose, *Z. Physik. Chem.*, 3, 73, Frankfurt (1955)
   Schlögl R., Membrane permeation in systems far from equilibrium, *Ber. Bunsenges. Physik. Chem.*, 70, 400 (1966)
4. Curtiss C.F., Hirschfelder J.O., Bird R.B., "Molecular Theory of Gases and Liquids", John Wiley & Sons Inc., New York (1960)
5. Gmehling J., Kolbe B., "Thermodynamik", Lehrbuchreihe Chemieingenieurwesen/Verfahrenstechnik, Georg Thieme Verlag, Stuttgart (1988)

A MODEL OF A PHASE TRANSFER CATALYST LIQUID/LIQUID ELECTRO-CHEMICAL MEMBRANE REACTOR

K. Scott

Department of Chemical and Process Engineering
University of Newcastle
Newcastle upon Tyne, NE1 7RU, England

ABSTRACT

Problems of phase contacting with reaction between aqueous based reagents and immiscible organics using mixer settlers or staged contactors can be overcome by membrane reactors. The membrane with high specific area, provides a stable interface for reagent transfer and reaction. After reaction between reagent (typically a redox species) and organic, regeneration of the reagent is achieved electrochemically. A general mathematical model is developed for a membrane based reaction between an organic species, of low solubility in the aqueous phase with a redox agent. The reaction in the organic phase employs a phase transfer catalyst to facilitate transfer of the aqueous based reactant into the organic phase. The important parameters in the model(s) are identified and their influence described through simulations of the model. The effect of using either a hydrophobic or hydrophilic membrane is discussed.

INTRODUCTION

The promotion of reactions between aqueous and organic reacting phases which are mutually insoluble can be achieved by using phase transfer catalysis. Typical catalysts, as employed in industrial applications, are positively charged agents which facilitate the transfer of the aqueous phase reactants (typically present as an anion) into the organic phase. The usual mode of carrying out these reactions is by dispersing the two phases in a well stirred reactor to facilitate phase transfer and reaction and then separation of the mixture into the two phases for product (and catalyst) recovery. The method is used for the preparation of several, pharmaceuticals, agricultural chemicals and speciality chemicals[1].

There are several disadvantages in the use of mixer-settler based reactions, which include emulsification, limitations in, and poor definition of, interfacial area for phase transfer. The use of membrane based reactors has been proposed in several applications[2] and recently

Electrochemical Engineering and Energy, Edited by
F. Lapicque *et al.*, Plenum Press, New York, 1995

in phase transfer catalysis[3]. The use of a porous membrane enables a well defined interface to be formed between the two phases for transfer of the catalyst/reactant species. These two phases flow on opposite sides of the membrane polymer which enables a stable interface to be formed for reaction over a wide range of operating conditions. The membrane can also give control over the overall rate of the reaction process. This paper deals with the application of membrane reactors in phase transfer catalysis where the active aqueous based reactant is generated by electrochemical reaction. The electrochemical production of mediators for the oxidation of organic compounds is of considerable interest and is rapidly developing in an industrial context. Oxidants such as cerium ions, hyprobromite, hypochlorite, manganese ions and dichromate are useful species[4]. The use of phase transfer catalysts with such redox mediators has been applied in several syntheses[5] to try to overcome the low solubility of these inorganic reactants in the organic phase.

A model of an electrochemical phase transfer membrane reactor is described which considers the electrochemical generation of the mediator, the formation of the phase transfer ion pair and its transport across the membrane based liquid/liquid interface. The influence of the type of membrane, whether hydrophobic or hydrophilic, may be an important factor. A hydrophilic membrane will have its pores filled with the aqueous electrolyte phase, while a hydrophobic membrane will be filled by the organic phase. Thus a hydrophobic membrane will present a greater resistance to diffusional mass transfer in the organic phase. For operation as a continuous reactor it provides a greater active volume for reaction, but for batch recycle operation this is not a significant factor. A hydrophilic membrane offers a purely diffusional resistance to mass transfer in the aqueous phase. If the chemical reaction is relatively slow the latter system will tend to be preferable. However regardless of the type of membrane the overall model of the system will have a common format of hydrodynamic mass transport resistance, membrane diffusion and chemical and equilibrium reaction.

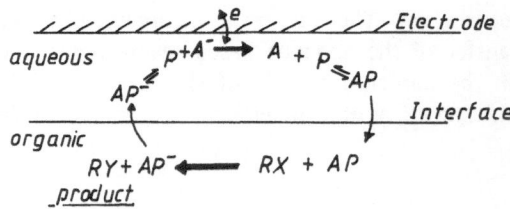

Figure 1   Mechanism of Phase Transfer Catalysis

MATHEMATICAL MODEL ANALYSIS

The theoretical model of the use of phase transfer catalysts in two phase electrochemical reactions considers that chemical reaction occurs solely in the organic phase and that there is a negligible quantity of organic phase in the aqueous electrolyte. The reaction mechanism is depicted in Figure 1.

The phase transfer catalyst, P, effectively shuttles between the organic and aqueous phase, carrying with it the active redox species, A. The ion pair PA undergoes reaction with the organic phase which liberates the "spent" catalyst, PA⁻, which is readily rejected back into the aqueous phase. The transport characteristics of the phase transfer ion pairs in relation to the rate of reaction in the organic phase and the rate of the electrode process will have a significant bearing on the overall system performance. It is possible to consider three cases;

1.   When chemical reaction is very fast and is located at or near the organic side of the aqueous, organic interface. The transport of species in the aqueous phase will then be a significant factor.

2.   When chemical reaction is relatively slow and occurs mainly in the bulk organic phase.

3.   When chemical reaction occurs immediately the phase transfer ion pair enters the organic phase and proceeds simultaneously during diffusional transport of the former. This will be more significant with a hydrophobic membrane with the organic phase filling the pores, due to a reduction in the effective diffusivity.

This work will focus on case 2 where the rate of chemical reaction is a major factor.

Slow Chemical Reaction

The general mechanism adopted is one of ten discrete reaction or rate processes. These are, with reference to Figure 1, as follows, starting with the electrode process:-

electrode reaction,	$A^- \rightarrow A + e$	(1)
phase transfer ion pair formation,	$A + P \Leftrightarrow PA$	(2)
mass transfer to the liquid/liquid interface,	$PA \rightarrow PA_{iaqu}$	(3)
interfacial equilibrium,	$PA_{iaqu} \Leftrightarrow PA_{io}$	(4)
mass transfer in the organic phase,	$PA_{io} \rightarrow PA_o$	(5)
chemical reaction,	$PA_o + RX \Leftrightarrow PA^-_o + RY$	(6)
mass transport of spent agent,	$PA_o^- \rightarrow PA^-_{io}$	(7)
interfacial equilibrium,	$PA^-_{io} \Leftrightarrow PA^-_{iaqu}$	(8)
dissociation of spent ion pair,	$PA_{iaqu} \Leftrightarrow A_i^- + P_i$	(9)
mass transfer in the aqueous phase,	$A^-_i \rightarrow A^-$	(10)
	$P_i \rightarrow P$	(11)
	$PA_i^- \rightarrow PA^-$	(12)

In the above the species are identified as, $A^- \equiv$ redox ion, $A \equiv$ redox mediator, $P \equiv$ phase transfer catalyst, $PA \equiv$ phase transfer ion pair, $RX \equiv$ organic reagent, $RY \equiv$ organic product, $PA^- \equiv$ spent PTC ion pair. The suffices, i, o and aqu indicate the interface, organic phase and aqueous phase respectively. It is assumed that the dissociation of the spent PTC ion pair is negligible in the organic phase and that ion pair association/dissociation reactions are fast equilibrium processes. Thus dissociation of the species PA^- at the aqueous interface occurs before transport of the dissociated species. The relevant model equations for this system are given in Table 1 where r represents the reaction rate in mol $dm^{-3}s^{-1}$, a_e and a_o are the specific surface area of the electrode and the liquid/liquid interface respectively and k_L represent mass transfer coefficients in the membrane model. j is the current density for the generation of mediator. k_L^* is the effective mass transfer coefficient for the boundary layer and the membrane.

Table 1 **Rate and Equilibrium Equations for the Phase Transfer Catalyst Membrane Reactor**

Reaction Number	Equation
1	$r_1 = {a_e j}/{nF}$
2	$[PA] = K_2 [A] [P]$
3	$r_3 = a\, k_L^* ([PA] - [PA]_{i,aq})$
4	$[PA]_{io} = K_4 [PA]_{i,aq}$
5	$r_5 = k_{Lo}\, a_o ([PA]_{io} - [PA]_o)$
6	$r_6 = k_c [PA]_o[RX] - k_{-c}[PA^-{}_o][RY]$
7	$r_7 = k_{Lo}\, a_o ([PA^-]_o - [PA^-]_{oi})$
8	$[PA^-]_{oi} = K_8 [PA^-]_{iaqu}$
9	$[PA^-]_{iaqu} = K_9 [A^-]_i[P]_i$
10	$r_{10} = k_{LA}\, a ([A^-]_i - [A^-])$
11	$r_{11} = k_{LP}\, a ([P]_i - [P])$
12	$r_{12} = k_{LDA}^-\, a ([PA^-]_i - [PA^-])$

Reactor Model

The reactor system is depicted schematically in Figure 2. It consists of an electrochemical cell supplying mediator in a flowing aqueous stream to a membrane module which provides a relatively high surface area for contact between the aqueous phase and the organic reactant phase. The unit is in practice a batch recycle reactor but for convenience is modelled here as a simple batch unit, with appropriate fluid flows. The material balances for appropriate species in the batch reactor are:

In the organic phase, for the reaction of organic reagent (and PTC Species)

$$\frac{dN_{RX}}{dt} = -V_o r_6 \tag{13}$$

$$\frac{dN_{PA^-}}{dt} = V_o (r_6 - r_7) \tag{14}$$

$$\frac{dN_{PA_o}}{dt} = V_o (r_3 - r_6) \tag{15}$$

In the aqueous phase, the generation and consumption of mediator A,

$$\frac{dN_A}{dt} = V_{aqu}(r_1 - Rr_3) \qquad (16)$$

In the aqueous phase, the consumption and regeneration of redox agent A^-,

$$\frac{dN_{A^-}}{dt} = V_{aqu}(Rr_7 - r_1) \qquad (17)$$

V_O and V_{aqu} are the volumes of the aqueous electrolyte and organic phase respectively. R is the ratio V_O/V_{aqu} and N_j the number of mols of species j.

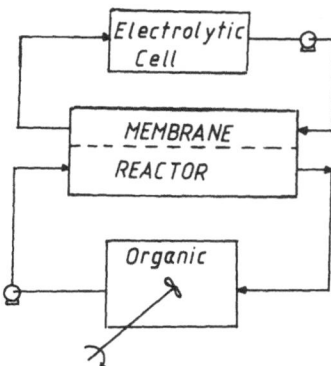

Figure 2   Electrochemical membrane reactor

Equations (1) to (11) of Table 1, which couple the above material balance equations, can be simplified by applying the steady state approximation at the membrane interface for species PA i.e. $r_3 = r_5$. This gives

$$r_3 = k_{Lo}a_o\left(K_4[PA] - [PA]_o\right) / \left[1 + \frac{K_4 k_{Lo}a_o}{ak_L^*}\right] \qquad (18)$$

Equations (7), (8) and (9) combine to give

$$r_7 = k_{Lo}a_o\left([PA^-]_o - K_8 K_9 [A^-]_i [P]_i\right) \qquad (19)$$

Material balances at the interface for A^-, P and PA^- are

$$\frac{d[A^-]_i}{dt} = Rr_7 - r_{10} \qquad (20)$$

$$\frac{d[P]_i}{dt} = Rr_7 - r_{11} \qquad (21)$$

$$\frac{d[PA^-]}{dt}i = Rr_7 - r_{12} \tag{22}$$

The set of differential equations describes the variation in concentration of species RX, A, A^-, PA^-_o, A_i, and P_i and are solved (numerically) in conjunction with overall material balances which define the concentrations of P and PA. These are

$$N_p = V_{aqu} ([P] + [PA^-] + [PA]) + V_o ([PA]_o + [PA^-]_o) \tag{23}$$

$$N_A^o - N^o_p = V_{aqu} ([A] - [P] + [A^-]) \tag{24}$$

where N^o_p and N^o_A represent the initial mols of P and A in the reactor.

Membrane Diffusion with Reaction

The mathematical model discussed does not consider chemical reaction in the organic phase while the P.T.C. diffuses from the liquid/liquid interface. This diffusion with reaction will occur in the membrane if it is hydrohobic and also possibly in the organic phase hydrodynamic mass transfer layer. The mathematical equations describing this process are a set of 4 second order O.D.E.'s for the species PA_o, PA_o^-, RX and RY, with appropriate boundary conditions for the interfacial and bulk organic phase concentrations.

$$\frac{d^2C^*}{dz} = \overline{D}D_a r_6^* \tag{25}$$

where z = x/d, with d, the membrane thickness, $\overline{D}$ is the ratio of diffusion coefficients based on the concentration RX_o initially present. D_a is a modified Dämköhler number

$$D_a = \frac{k_6 RX^o d^2}{D_{RX}}$$

This dimensionless group is a governing factor in determining performance. Low values, indicate only a relatively small amount of reaction in the membrane, because either the kinetic rate is low or the effective membrane diffusion mass transfer rate is small e.g. a thin membrane. High values would indicate a large amount of reaction in the membrane phase. In the limit when the reaction between the phase transfer ion pair reagent and organic is fast, then mass transport in the organic phase will not be important. Reaction will occur at the liquid/liquid interface where the reagent concentration is the interfacial concentration in the organic phase ($[PA]_{io}$). Thus the mass transport rate r_3 is given by

$$r_3 = ak_L^* \left([PA]^- - \frac{[PA]o}{K_4} \right) \tag{26}$$

Treatment of this model is beyond the scope of this work and is the subject of a later publication.

Results

The following results consider a model in which diffusional mass transfer through the membrane (assumed thin) is not rate limiting. Figure 3 shows the influence on the electrochemical reaction rate on the components of this reaction system. The effective electrochemical rate constant is given generally by

$$\frac{j}{nF} = \frac{k_{f1}C_{A^-}}{1 + \frac{k_{f1}}{k_L}} = k_1 C_{A^-} \tag{27}$$

where k_{f1} is an experimental function of potential and k_L is the mass transfer coefficient at the electrode. The volumes of both liquid phases and initial concentrations of reactants are fixed and chemical reaction rate constants are given in Table 2 and are fixed in all simulations.

Table 2   Values of Parameters for Simulations

k_c	$= 10^{-3}$ dm^{3}kmol^{-1}s^{-1}, $k_{-c} = 10^{-4}$ dm^{3}kmol^{-1}s^{-1}
K_4	$= 0.1$, $K_2 = 0.01$ (kmol m^{-3})
$k_L{}^*a$	$= k_{Lo} a_o = 10^{-2}$ s^{-1}, $K_9 = 4.10^{-3}$ dm^{3}kmol^{-1}
K_8	$= 8$
V_o	$= V_{aqu}$, Pure organic reagent
C_{A^-} initially	$= 1.0$ kmol/m^3 of total solution
C_p initially	$= 1.0$ kmol/m^3 total solution

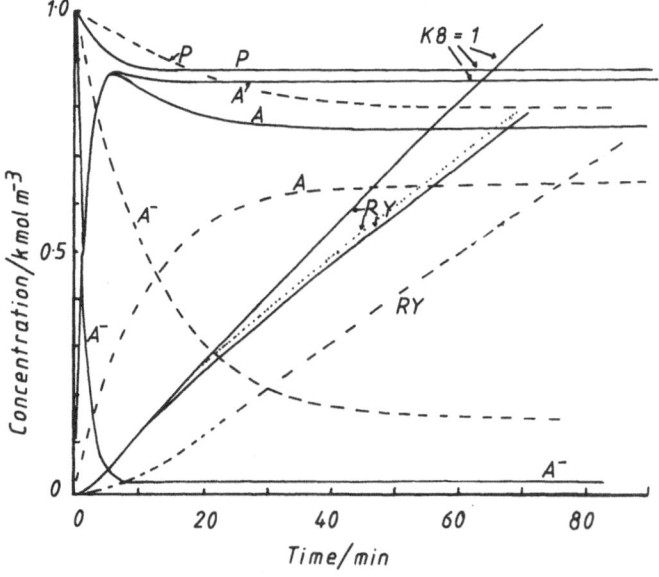

Figure 3   Effect of electrochemical rate constant on product distributions

(————) : $k_1 = 0.01$, (-----) : $k_1 = 0.001$, (·····) : $k_1 = 0.025$ (s^{-1})

Generally the reactant concentrations of species originating in the aqueous phase approach steady values at short reaction times of less than one hour, whilst the organic product RY increases steadily. A higher electrochemical rate constant increases the rate of RY formation, which is reflected in a higher concentration of A. A decrease in the value of interfacial equilibrium constant for species PA$^-$, makes for a greater concentration of PTC and reagent A$^-$ in the aqueous phase. This is also shown in Figure 3, (with $K_8 = 1.0$) to increase the production of RY and A$^-$.

Figure 4 shows the influence of a variation in interfacial specific mass transport rate on the species distribution. As expected lower values of $k_L a$ decrease the rate of organic product generation. In practice variations in the value of $k_L a$ will arise from changes in hydrodynamic conditions and/or changes in the membrane (interfacial area). Figure 5 shows the effect of a significant mass transport resistance in the membrane, which as expected decreases the rate of product generation. This again is seen in lower concentrations of reagent A$^-$ in the aqueous phase.

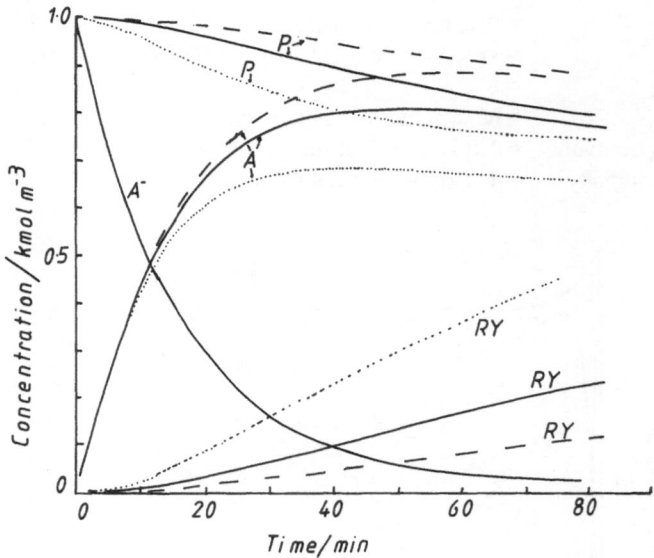

Figure 4   Effect of mass transport on the production of organic product . $k_1 = 0.001$ s^{-1},
(———) : $k_{Lo}a_o = 0.001$ s^{-1}, (-----) : $k_{Lo}a_o = 0.004$ s^{-1} (·····) : $k_{Lo}a_o = 0.004$ s^{-1}

Figure 6 shows the effect of imposing a constant current in the rate of generation of species A, ie, the reaction is independent of the concentration of ionic species A$^-$. As expected the rate of production of product RY increases with increasing current density. Also the concentration of species A$^-$ falls rapidly and in practice will cause a reduction in the current efficiency for generation of the reagent A.

204

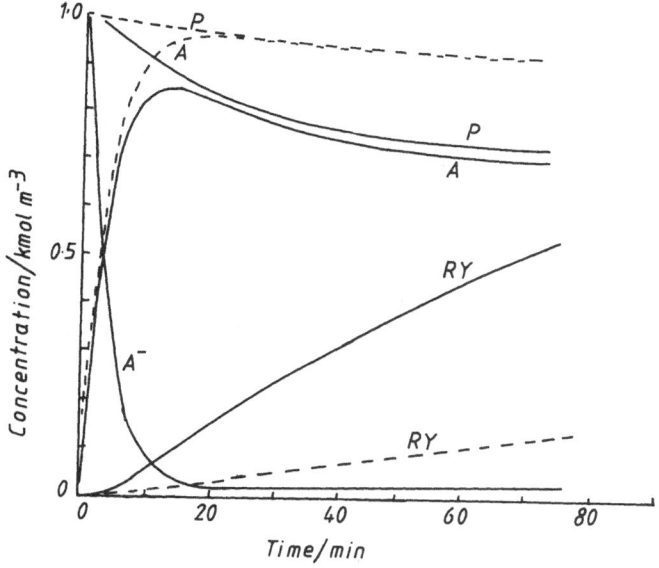

Figure 5   The effect of membrane diffusion on the production of organic product $k_1 = 0.004$ s^{-1}, (———) : $k_L^*a = 0.004$ s^{-1}, (-----) : $k_L^* = 0.0004$ s^{-1}

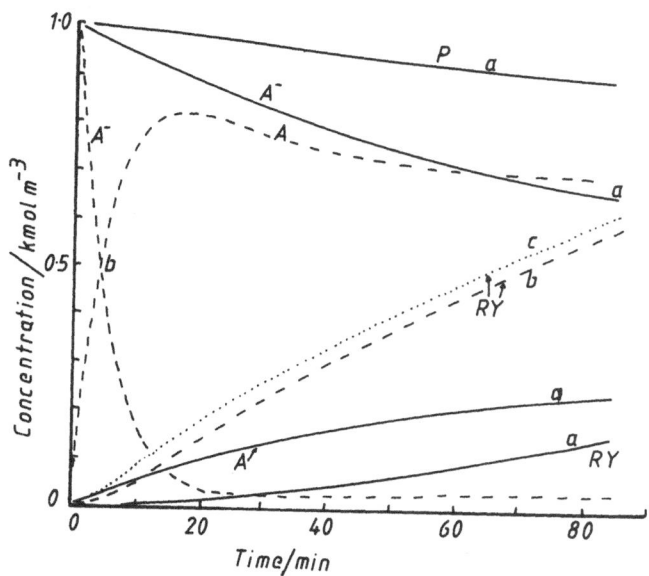

Figure 6   Performance at a constant rate of generation (current density) of redox reagent. $k_{Lo}a_o = 0.004$ s^{-1}. Values of j/nF are (a) : 0.0001, (b) : 0.0003, (c) : 0.001.

CONCLUSIONS

The importance of determining the influence of interfacial mass transport on the operation of membrane reactors has been demonstrated. Both hydrodynamics, membrane properties and interfacial equilibria play a crucial role. The models presented are sufficiently general to enable comparisons of performance to be made with a l/l interface formed by mixing and dispersion. The influence of more complex reaction chemistry and electrochemistry will be addressed in subsequent work as will the case of the electrolysis of organic liquid films.

REFERENCES

1. C.M. Starks and C. Liotta. "Phase Transfer Catalysis, Principles and Techniques," Academic Press, New York (1978).
2. S.L. Matson and J.A. Quinn. "Indirect Electrolysis Involving Phase Transfer Catalysis," Ann. N.Y. Acad. Sci., 469, 152, (1986).
3. T.J. Stanley and J.A. Quinn. "Phase-Transfer catalysis in a Membrane Reactor", Chem. Eng. Sci. 42 (10), 2313, (1987).
4. K. Scott. "Electrochemistry, Clean Technology for Synthesis and the Environment", Royal Society of Chemistry. To be published (1994).
5. D. Pletcher."Indirect electrolysis involving phase transfer", in "Electro-organic Synthesis", Ed. R.D. Little and N.L. Weinberg, Marcel Dekker Inc., N.Y. (1992).

ADVANCED DYNAMIC OPTIMIZATION OF BATCH ELECTROCHEMICAL REACTORS

M.A. Latifi

Laboratoire des Sciences du Génie Chimique
CNRS-ENSIC, BP 451, 1-rue Grandville
54001, Nancy Cedex, France

ABSTRACT

The maximum principle was used to find the control optimal policy in a batch electrochemical reactor where a simple cathodic reaction A<—>B or a series of cathodic reactions A<—>B<—>C takes place. In the first case the optimal current required to reduce the concentration of A from C_0 to C_F in a given time t_F, while minimizing the overall consumption of electrical energy and while imposing a 100% electrical yield at all times is determined for different values of electrical resistance of the solution. In the last case, the optimal electrode potential, which maximizes the final concentration of the intermediate product B in a given time t_F, is determined for different values of liquid/solid mass transfer coefficient of the desired product B.

INTRODUCTION

The optimization of chemical and electrochemical processes is usually carried out by studying the influence of different operating parameters such as current density, cell voltage, flowrates, temperature, concentrations, etc. Naturally, this kind of optimization provides uniform profiles of these control variables. This is what is called static optimization and it is not always optimal.

In many electrochemical operations the control variables may also be a function of some independant variables such as time and/or space. When only one independent variable is involved, i.e., time or space, the problem is termed a *lumped parameter problem*. Problems where more than one independent variable is involved, for example, both space and time, are referred to as *distributed parameter problems* [1].

In the following, only lumped parameter problems will be investigated, where the time is the only independent variable, since only batch electrochemical reactors will be considered.

Though this type of optimization is widely encountered in chemical systems[2], in electrochemical systems, few works have been devoted to this form of advanced optimization[3,4,5,6]

In this communication, we consider a batch electrochemical reactor where a simple cathodic reaction A<—>B or a series of cathodic reactions A<—>B<—>C takes place. The objective is :

(i) the determination, for the simple cathodic reaction, of the optimal profiles of operating current required to reduce the concentration of A from C_0 to C_F in a given time t_F, while minimizing the overall consumption of electrical energy and imposing a 100% electrical yield at all times.

(ii) the determination, for the series of cathodic reactions, of the optimal electrode potential policy, which maximizes the final concentration of the intermediate product B, in a given time t_F under given constraints.

OPTIMIZATION METHOD

Maximum principle [1,7,8]

The general class of problems we wish to optimize can be represented by the modeling equation :

$$\frac{dx(t)}{dt} = f(x(t), u(t)) \quad ; \quad 0 \le t \le t_F \tag{1}$$

where $x(t)$ is an n dimensional vector of the state variables and $u(t)$ is an m dimensional vector of control variables which we wish to choose optimally. The boundary conditions will depend on the physical nature of the problem.
If only the initial state $x(0)=x_0$ is specified, then the result will be a straightforward initial value problem. If the final state $x(t_F)=x_F$ is also fixed, then we have a two-point boundary value problem.

The problem is to find an optimal control $u(t) \in [u_-, u^+]$ which transfers x_0 to x_F.

In order to specify the meaning of optimal, we define an objective functional :

$$J[u(t)] = \int_0^{t_F} f^0(x(t), u(t))dt \tag{2}$$

which we wish to maximize or minimize.

Optimal controls can be determined using a variety of techniques. In the following, only the Pontryaguin maximum principle will be used.
The application of this principle consists in defining of the Hamiltonian of the system as :

$$H[\varphi(t), x(t), u(t)] = \varphi_0(t)f^0(x(t), u(t)) + \sum_{i=1}^{n} \varphi_i(t)f^i(x(t), u(t)) \tag{3}$$

where $\left(\varphi_0(t), \varphi^T(t)\right)$ is an n+1 dimensional vector of adjoint variables.

The components of this vector are given by :

$$\frac{d\varphi_0(t)}{dt} = 0 \tag{4}$$

and :

$$\frac{d\varphi_i(t)}{dt} = -\frac{\partial H}{\partial x_i} = -\sum_{i=1}^{n} \varphi_i(t)\frac{\partial f^i}{\partial x_i}(x(t), u(t)) \quad i = 1,...,n \tag{5}$$

According to Eq.4, φ_0 is a constant. As the vector $\varphi(t)$ is non-trivial, φ_0 is non-zero and its value is normalized to -1.

The boundary values of the functions $\varphi_i(t)$ are given by the transversality conditions :
if only some of the components of $x(t_F)$ are unspecified then $\varphi_i(t_F) = 0$ holds for those components. Similarly, if some components of $x(0)$ were not to be specified, then $\varphi_i(0) = 0$ would hold for those components.

The necessary condition for optimal control is that :

$$\frac{\partial H}{\partial u} = 0 \tag{6}$$

on the unconstrained portion of the path and H is maximum on the constrained portion of the path.

Computational technique

If one can solve, analytically or numerically, Eqs.(6) and then substitute the solution into Eqs.(1) and (5) the result is a set of 2n equations with split boundary conditions. $\varphi(0)$ and $\varphi(t_F)$ assume that all $x(0)$ are specified and all $x(t_F)$ are not specified. In some practical systems, we may end up with two sets of boundary conditions for Eqs.(1) and none for Eqs.(5), as will be illustrated by the case of a simple cathodic reaction. This two-point boundary value problem has a solution which provides the optimal values of u(t), x(t) and $\varphi(t)$. Many computational approaches can be used, e.g., control vector iteration, boundary condition iteration, invariant imbedding, quasi-linearization, control vector parameterisation, etc. In this paper only the boundary condition iteration will be used. This approach attempts to find the missing boundary conditions $x(t_F)$ or $\varphi(0)$ by minimizing the error in the boundary conditions by a direct search so that Eqs.(1) and (5) can be integrated together in the same time direction.

The optimization algorithm is then :
(i) Guess $\varphi(0)$,
(ii) u(t) are deduced from Eqs.(6), $u(t) \in [u_-, u^+]$)
(iii) Integrate Eqs.(1) and (5) forward in time to provide x(t) and $\varphi(t)$. Repeat steps (ii) and (iii) from t = 0 until t = t_F,
(iv) Repeat steps (i) to (iii) until the precision on the error in the boundary conditions is reached.

REACTOR MODEL and OPTIMIZATION

One considers a batch electrochemical reactor in which a simple cathodic reaction or a series of cathodic reactions takes place. The following simplifying assumptions are made:

(i) the reactor is perfectly stirred,

(ii) the mass transfer resistance occurs by material diffusion through a Nernst diffusion layer and is characterized by a mass transfer coefficient,

(iii) no other reactions in addition to those considered occur,

(iv) the capacitance of the double layer is negligible,

(v) the current distribution is uniform.

Case of a simple cathodic reaction A<—>B [3]

The objective is to determine the time-varying operating current which minimizes the overall consumption of electrical energy under given constraints.

Optimization criterion. The overall consumption of electrical energy, J, in the reactor during the time t_F is :

$$J[I(t)] = \int_0^{t_F} U_c I dt \tag{7}$$

where I is the operating current. The overall voltage between anode and cathode, U_c, is given by the following expression :

$$U_c = U_0 + \eta_a + |\eta_c| + R_e I \tag{8}$$

where U_0 is the reversible voltage, η_a the overall anodic overpotential, η_c the overall cathodic overpotential and R_e the electrical resistance of the solution.

Constraints. In this problem, we have an equality constraint and an inequality constraint : The first one is that the mass balance equation must be satisfied and, subject to the listed assumptions, may be written as :

$$V\frac{dC_{As}}{dt} = -\frac{I}{v_e F} \tag{9}$$

with : $C_{As} = C_{A0}$ at $t = 0$ and $C_{As} = C_{AF}$ at $t = t_F$ (10)

The second constraint requires that in order to maintain an electrical yield equal to 100% at all times, the instantaneous operating current must be less than the limiting current I_L :

$$0 < I(t) < I_L(t) = v_e F k_d A_e C_{As}(t) \tag{11}$$

where $v_e F$ is the Faraday number, k_d the liquid/solid mass transfer coefficient and A_e is the electrode area.

Minimal operating time. The minimal operating time required to reduce the concentration of A from C_{A0} to C_{AF} is obtained by operating at all times at the limiting current. Thus, I is replaced by I_L in Eq.(9) and after integration, we obtain :

$$t_L = -t_C \ln\left(\frac{C_{AF}}{C_{A0}}\right) \quad \text{with} \quad t_C = \frac{V}{k_d A_e} \tag{12}$$

Since the constraint (11) must be satisfied, the operating time t_F must be greater than the minimum time t_L.

The cathodic reaction studied here is the electrodeposition of copper. Its kinetics are given by the following equation [9]:

$$I = A_e i_0^0 \left(C - \frac{I}{v_e F k_d A_e} \right)^\beta \left(\exp\left\{ \alpha_1 \frac{v_e F}{RT} \left(\eta_c + \ln\left(1 - \frac{I}{I_L} \right) \right) \right\} - \exp\left\{ \alpha_2 \frac{v_e F}{RT} \left(\eta_c + \ln\left(1 - \frac{I}{I_L} \right) \right) \right\} \right) \qquad (13)$$

The anodic reaction is assumed to be described by a Tafel law [10] :

$$\eta_a = \frac{RT}{\alpha v_e F} \ln\left(\frac{I}{I_0} \right) \qquad (14)$$

Optimization. Here $n = m = 1$ so that $x = C_{As}$ and $u = I$. The Hamiltonian of this system is :

$$H = U_c I + \varphi \frac{I}{v_e F V} \qquad (15)$$

The time-varying adjoint variable is given by :

$$\frac{d\varphi}{dt} = - \frac{\partial H}{\partial C_{As}} \qquad (16)$$

and the necessary condition for optimality is :

$$\frac{\partial H}{\partial I} = 0 \qquad (17)$$

The error in the boundary conditions is :

$$\left(C_{As}(t_F) - C_{AF} \right)^2 + \left(Q_{real} - Q_{cal} \right)^2 \qquad (18)$$

$$\text{where :} \quad Q_{real} = v_e F V (C_{A0} - C_{AF}) \quad \text{and} \quad Q_{cal} = \int_0^{t_F} I dt \qquad (19)$$

Case of a series of cathodic reactions A<—>B<—>C

Here the objective is to determine the electrode potential-time policy which maximizes the concentration of the desired product B at the end of a specified batch period t_F under given constraints.

Optimization criterion. When only reactant A is present at $t=0$, the concentration of the desired product B at the end of a specified batch period t_F is :

$$J[E(t)] = \int_0^{t_F} \frac{dC_{Bs}}{dt} dt \qquad (20)$$

where E is the electrode potential.

Constraints. Only equality constraints are involved in this problem. This means that the mass balance equations must be satisfied and, subject to the listed assumptions, may be written as :

$$V\frac{dC_{As}}{dt} = -\frac{i_1 A_e}{v_1 F} \tag{21}$$

$$V\frac{dC_{Bs}}{dt} = \frac{i_1 A_e}{v_1 F} - \frac{i_2 A_e}{v_2 F} \tag{22}$$

with : $C_{As}(0) = C_{A0}$; $C_{Bs}(0) = 0$ and $C_{Cs}(0) = 0$ $\tag{23}$

$$C_{Cs} = C_{A0} - C_{As} - C_{Bs} \tag{24}$$

Kinetic equations. Let us consider the following reaction scheme[11] :

$$A + v_1.e \overset{k_{f1}}{\underset{k_{b1}}{\longleftrightarrow}} B + v_2.e \overset{k_{f2}}{\underset{k_{b2}}{\longleftrightarrow}} C$$

The reactions are assumed to be first order with respect to the reactants. Thus, the reaction rates for both steps are :

$$\frac{i_1}{v_1 F} = k_{f1} C_{Ae} - k_{b1} C_{Be} \tag{25}$$

$$\frac{i_2}{v_2 F} = k_{f2} C_{Be} - k_{b2} C_{Ce} \tag{26}$$

which can be written in terms of mass transfer coefficients as :

$$\frac{i_1}{v_1 F} = k_{dA}(C_{As} - C_{Ae}) \tag{27}$$

$$\frac{i_2}{v_2 F} = k_{dC}(C_{Ce} - C_{Cs}) \tag{28}$$

$$\frac{i_1}{v_1 F} - \frac{i_2}{v_2 F} = k_{dB}(C_{Be} - C_{Bs}) \tag{29}$$

By combining these equations to eliminate the surface concentrations, the reaction rates for both steps become :

$$\frac{i_1}{v_1 F} = a_1 C_{As} + a_2 C_{Bs} + a_3 C_{Cs} \tag{30}$$

$$\frac{i_2}{v_2 F} = b_1 C_{As} + b_2 C_{Bs} + b_3 C_{Cs} \tag{31}$$

where :

$$a_1 = \frac{k_{f1}Y}{\Delta} \; ; \qquad a_2 = \frac{k_{b1}(k_{f2} - Yk_{dB})}{k_{dB}\Delta} \; ; \qquad a_3 = -\frac{k_{b1}k_{b2}}{k_{dB}\Delta} \; ;$$

$$b_1 = \frac{k_{f1}k_{f2}}{k_{dB}\Delta} \; ; \qquad b_2 = \frac{k_{f2}(Xk_{dB} - k_{b1})}{k_{dB}\Delta} \; ; \qquad b_3 = -\frac{X.k_{b2}}{\Delta} \; ;$$

$$X = 1 + \frac{k_{f1}}{k_{dA}} + \frac{k_{b1}}{k_{dB}} \; ; \qquad Y = 1 + \frac{k_{f2}}{k_{dB}} + \frac{k_{b2}}{k_{dC}} \; ; \qquad \Delta = XY - \frac{k_{b1}k_{f2}}{k_{dB}^2}$$

The rate constants are defined as :

$$k_{fi} = k_{fi0}\exp(\alpha_i E) \quad i = 1, 2 \tag{32}$$

$$k_{bi} = k_{bi0}\exp(-\beta_i E) \quad i = 1, 2 \tag{33}$$

The reaction considered here is the electroreduction of oxalic acid to glyoxilic acid followed by the reduction of glyoxilic acid to glycollic acid[11,12].

Optimization. Here $n = 2$ and $m = 1$ so that $x(1) = C_{As}$, $x(2) = C_{Bs}$ and $u = E$. The Hamiltonian of this system is :

$$H = -(1 + \phi_1 - \phi_2)\frac{A_e i_1}{v_1 FV} - (\phi_2 - 1)\frac{A_e i_2}{v_2 FV} \tag{34}$$

The time-varying adjoint variables are given by :

$$\frac{d\phi_1}{dt} = (1 + \phi_1 - \phi_2)\frac{A_e}{v_1 FV}\frac{\partial i_1}{\partial C_{As}} + (\phi_2 - 1)\frac{A_e}{v_2 FV}\frac{\partial i_2}{\partial C_{As}} \tag{35}$$

$$\frac{d\phi_2}{dt} = (1 + \phi_1 - \phi_2)\frac{A_e}{v_1 FV}\frac{\partial i_1}{\partial C_{Bs}} + (\phi_2 - 1)\frac{A_e}{v_2 FV}\frac{\partial i_2}{\partial C_{Bs}} \tag{36}$$

and the necessary condition for optimality is :

$$\frac{\partial H}{\partial E} = 0 \tag{37}$$

The error in the boundary conditions is :

$$\left(Q_{1,real} - Q_{1,cal}\right)^2 + \left(Q_{2,real} - Q_{2,cal}\right)^2 + \left(\phi_1(t_F)\right)^2 + \left(\phi_2(t_F)\right)^2 \tag{38}$$

where :

$$Q_{1,real} = v_1 FV(C_{A0} - C_{AF}) \text{ and } Q_{1,cal} = \int_0^{t_F} I_1 dt \tag{39}$$

$$Q_{2,real} = v_2 FV C_{CF} \qquad \text{and } Q_{2,cal} = \int_0^{t_F} I_2 dt \tag{40}$$

RESULTS and DISCUSSIONS

Case of the reaction A<—>B

The influence of several parameters on the optimal control I(t) can be analysed. In this work, our interest is focused on the study of the influence of the electrical resistance Re. Data used in this case are the following :

$V = 10^{-3}$ m^3; $A_e = 8 \times 10^{-4}$ m^2; $k_d = 5 \times 10^{-5}$ m/s; $T = 293.15$ K; $\nu_e = 2$; $F = 96500$ C/equi; $R = 8.31$ J/mol.K; $U_0 = 0.9$ V; $\alpha = 0.5$; $I_0 = 8 \times 10^{-7}$ A; $\alpha_1 = 0.84$; $\alpha_2 = 1.16$; $\beta = 0.42$; $i_0^0 = 1.445$ SI; $t_F/t_L = 21.5$; $C_{A0} = 100$ mol/m^3; $C_{AF} = 1$ mol/m^3.

Case of large R_e. In this case, the charge transfer law is linear :

$$I \qquad = \qquad \frac{1}{R_e} \qquad x \qquad U_c \qquad\qquad (41)$$

flux(C/s)   conductance(Ω^{-1})   driving force (V)

Then, the optimal profile of operating current is uniform and given by the following expression :

$$\overline{I} = \frac{\nu_e F V (C_{A0} - C_{AF})}{t_F} \qquad\qquad (42)$$

Tondeur [13,14] has demonstrated that in separation processes, when the heat transfer law is linear, the opimal profile which minimizes the overall consumption of thermal energy is also uniform.

Case of small R_e. Figure 1 presents the optimal operating current profiles for different values of R_e.

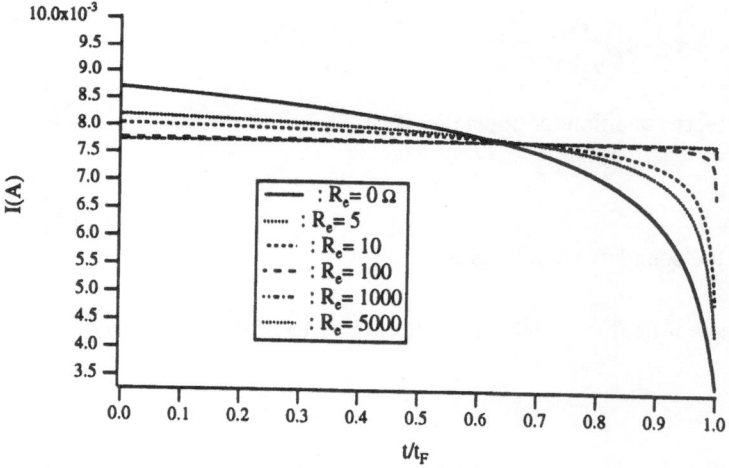

Figure 1. Optimal profiles of the current for different values of R_e

The uniform profile is no longer optimal because of the non-linearity of the charge transfer law. However, for resistances larger than 1kΩ, the transfer law becomes linear and the optimal profile is uniform. The non-uniform profiles obtained for resistances smaller than 1kΩ are optimal provided that the power U_cI is a convex function of I. This is true in the working domain, i.e., $I < I_L$.

Case of the series of reactions A<—>B<—>C

Only the influence of the liquid/solid mass transfer coefficient of the desired product B is studied. In this case the physical data used are :

$A_e = 22$ cm^2, $V = 157$ cm^3, $C_{A0} = 680$ mol/m^3, $C_{B0} = 0$, $C_{C0} = 0$, $k_{f10} = 10^{-13}$ m/s, $k_{f20} = 3.3 \times 10^{-14}$ m/s, $k_{b10} = k_{b20} = 0$, $k_{dA} = k_{dB} = k_{dC} = k_d$, $\alpha_1 = \alpha_2 = 12.45$ V^{-1}, $t_F = 10000$ s.

In figure 2, the optimal-time profiles of the electrode potential, for different values of k_d, are presented.

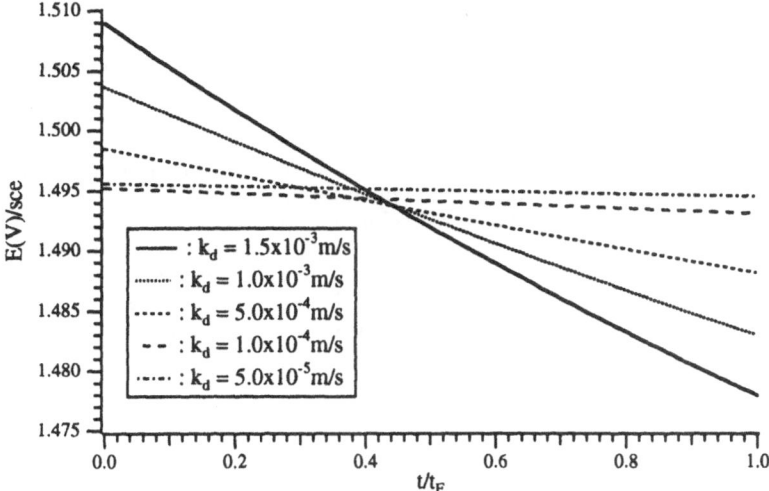

Figure 2. Optimal profiles of the electrode potential

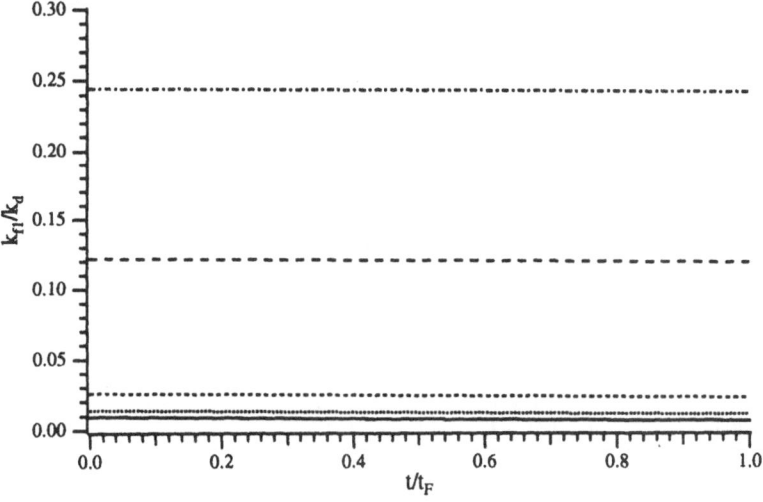

Figure 3. Ratio of the rate constant and the mass transfer coefficient

In this reaction scheme, the optimal procedure required to maximize $C_{Bs}(t_F)$ is to produce the maximum of B by the first reaction (A—>B) at the beginning of the operation and to transform the minimum of B by the second reaction (B—>C) at the end of the operation. This optimal procedure explains the decrease of the electrode potential with time.

The competition between diffusional and electrochemical regimes of the first reaction is the origin of the significant decrease of the electrode potential with time when the mass transfer coefficient k_d is significant (Figure 3). For small values of k_d ($<1.0 \times 10^{-4}$ m/s), the electrochemical reaction is limited by diffusion and plays an insignificant role in the optimization : the optimal profile of the potential is practically constant. In contrast, for high values of the mass transfer coefficient ($k_d > 1.0 \times 10^{-4}$ m/s), the reaction regime is purely electrochemical and the reaction plays the main role in the optimization which becomes more interesting than before.

CONCLUSION

The maximum principle was used to determine the optimal policy of control of batch reactors for two electrochemical examples. In the first example, the optimal profiles of current which minimize the overall consumption of electrical energy have been determined in a batch reactor where a simple reaction A<—>B takes place. The optimal profiles of electrode potential which also maximize the final concentration of the intermediate product in a batch reactor involving a series of reactions A—>B—>C are determined in the last example. All these optimal profiles are determined in the open loop, i.e., possible perturbations could not be considered. Further studies will deal with closed loops and experiments.

Beyond the minimization of the electrical energy or the maximization of the final concentration, the most important is the introduction of this advanced optimization into electrochemical engineering. To our knowledge, this kind of optimization is not available in the literature for electrochemical processes and we think that it is interesting to do further work in this direction.

REFERENCES

1. W.H.Ray and J.Szekely, "Process Optimization", John Wiley and Inc., NY(1973).
2. J.De Morant, "Contrôle en temps minimal des réacteurs chimiques discontinus", Thèse de l'université de Rouen, France, (1992).
3. M.A.Latifi, S.Risson and A.Storck, "Optimisation énergétique d'un réacteur électrochimique discontinu : Application à l'électrodéposition du cuivre", *Entropie* n° 163 : 37(1991).
4. M.A.Latifi, "Dynamic optimization of a batch electrochemical reactor with respect to the overall consumption of electrical energy", *Hung.J.Ind.Chem.* 20 : 213(1992).
5. M.A.Latifi and A.Storck, "Optimisation dynamique de la sélectivité dans un réacteur électrochimique discontinu", Tec. & Doc. Lavoisier (Paris), Récents Progrès en Génie des Procédés, Vol.7, n°29 : 95(1993).
6. R.Bakshi and P.S.Fedkiw, "Optimal time-varying potential control", *J.Appl.Electrochem.* 23 : 715(1993).
7. E.B Lee and L.Markus, "Foundations of optimal control theory", John Wiley and Inc., NY(1967).
8. L.Pontriaguine, V.Boltianski, R.Gamkrélidzé and E.Michtchenko, "Théorie mathématique des processus optimaux", French translation, Edition Mir., Moscou(1974).
9. J.Newman, "Electrochemical systems", Prentice Hall Inc., Engelwood Cliffs, NJ(1973).
10. A.Storck and F. Coeuret, "Eléments de Génie Electrochimique", Tec.&Doc., Lavoisier, Paris(1984).
11. K.Scott, "Reactor engineering models for complex electrochemical reaction schemes- I.Potentiostatic operation of parallel and series reactions in ideal reactors", *Electrochimica Acta* 30 : 235(1985).
12. D.J.Pickett and K.S.Yap, "A study of the production of glyoxilic acid by the electrochemical reduction of oxalic acid solutions", *J. Appl. Electrochem.* 4 : 17(1974).
13. D.Tondeur and E.Kvaalen, "Equipartition of entropy production. An optimality criterion for transfer and separation processes", *Ind. Eng. Chem. Res.* 26 : 50(1987).
14. D.Tondeur, "Un principe variationnel pour la conception des échangeurs de chaleur et de matière : l'équipartition des sources d'entropie", Proceedings de Rencontre S.F.T.89 "Thermique et Génie des Procédés", Nancy 23-25 mai, 285 (1989).

MODELLING AND DESIGN OF INDUSTRIAL FLUORINE ELECTROLYZERS

Jean-Pierre Caire,[1] Laroussi Ben Abdallah,[1] Patrick Ozil[1], and François Nicolas[2]

[1]Centre de Recherche en Electrochimie Minérale et en Génie des Procédés,
URA C.N.R.S. 1212, Institut National Polytechnique de Grenoble,
Ecole Nationale Supérieure d'Electrochimie et d'Electrométallurgie de
Grenoble,Domaine Universitaire, BP.75, 38402 Saint Martin d'Hères (France)
[2]COMURHEX, Usine de Pierrelatte, BP 29, 26701 Pierrelatte (France)

ABSTRACT

A general purpose software package is presented for predicting the primary and secondary current distributions in an electrolyser viewed as an equivalent network of discrete resistors. This easy-to-use package was made for designing industrial fluorine cells characterised by a complex geometry. Numerical simulation shows that the primary distribution is irrelevant to the understanding of the local corrosion problems observed in such cells. On the other hand, the computed secondary distribution is able to explain the unexpected high local current densities in some regions of the cell where corrosion predominately acts. Such software is an efficient tool in the initial design of an electrochemical cell devoted to a given industrial working target.

INTRODUCTION

Considerable production of uranium hexafluoride has been required since the 1950's for military needs. This has been increased since the 1960's for nuclear power plants but production has been conducted by only a few companies in the world. All fluorine cells have been designed, like the Union Carbide cell, from empirical criteria and technological restrictions.

The geometry of an industrial fluorine cell is complex due to mechanical, thermal, electrochemical and electrical constraints. In this kind of molten salt device, large overvoltages and high current density distribution induce strong heat transfer problems. Thus, poor thermal control of the working cell may lead to partial solidification of the molten salt electrolyte, while local hot spots may occur in another part of the cell. Such events result in disastrous consequences in terms of mechanical stress and cell lifetime.

Designing such electrolysers is a cumbersome, time consuming and expensive task. The cost of a fully instrumented pilot plant may be prohibitive, particularly due to the highly corrosive molten salt to be used. Up to the present the common industrial practice for design has been a trial and error procedure based on empirical rules and the experience gained from

Electrochemical Engineering and Energy, Edited by
F. Lapicque *et al.*, Plenum Press, New York, 1995

the very few feasible laboratory tests and measurements. Now numerical modelling promises to be a fast and money saving approach to comprehensive design.

SPECIFIC PROBLEMS RELATED TO FLUORINE CELLS

The aim of this paper is not to discuss the technology for manufacturing fluorine and its specific problems which have been reported elsewhere.[1,2] Here we are interested in the Allied Chemical Co. cell,[3] which will be used as a cell pattern for testing our software. This cell works at a medium temperature, around 90 °C, using an electrolyte based on a HF-2KF mixture. As shown on figure 1, a fluorine cell has a wide anode-cathode gap, large enough

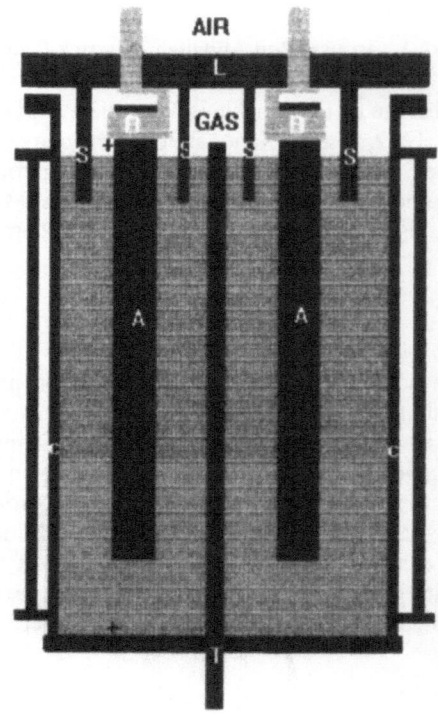

Figure 1. Schematic diagram of the Allied Chemical Co Cell

Legend: A: Anode B: Bus Bar C: Cathode L: Lid S: skirts T: Tank

to prevent the recombination of small hydrogen bubbles evolving at the cathode and large fluorine bubbles crawling up the anode. The final separation of the two gases is achieved by skirts immersed in the electrolyte.

The operation is characterized by a relatively poor conducting electrolyte and high overvoltages, especially at the anode. All these distinctive features lead to large thermal effects.

Furthermore, the materials usually used for the cell components (carbon for anodes, stainless steel for tank and cathodes and copper for the bus bar) have different thermal and electrical conductivities. As an example, we may compare the respective electrical conductivity of electrolyte, carbon and copper (0.18, 0.25×10^5 and 0.57×10^8 ohm^{-1}.m^{-1}). This large range of values, spanning over eight decades, generates thermal problems leading to mechanical stresses in the electrolyzer and also induces convergence troubles for numerical computation when simulating the cell.

MODELLING OF CURRENT DISTRIBUTION IN ELECTROCHEMICAL CELLS

The prediction of current and potential distributions in an electrochemical cell is an important step in the design of industrial devices. Solving such a problem may be performed through experiments, but this is both expensive and time consuming and modelling generally appears to provide a better approach. Some analytical evaluations of the current distribution are available for simple cell geometries. However, numerical computation is obviously required for determining potential and current density distributions in electrolyzers characterised by both complex geometry and non-linear boundary conditions at the electrolyte-electrode interfaces.

THE MAIN NUMERICAL METHODS

As pointed out by Prentice and Tobias[4], in a general survey of the numerical methods devoted to current distribution problems in electrolyzers, most methods have been known since the 1960's : Green's functions, perturbation, straight lines, coordinate inversion, variational methods, orthogonal collocation, finite elements and finite difference methods. However, only the two last methods are able to treat realistic current distribution problems in electrolyzers having a complex geometry like most industrial cells.[2] These methods are used by the few commercial codes including CAD tools which were developed these last few years for treating such modelling. However, such sophisticated software is not only expensive but, moreover, requires both a powerful computer and specific training for the user. In contrast, the purpose of the present work was to propose simple software designed to be used on a standard personal computer by a designer without any special preparation.[2]

THE METHOD OF RESISTANCES

The numerical method chosen here for computing the current and potential distributions is the method of resistances which was previously employed by several authors for different purposes.[5-8] This method is derived from the finite difference method and uses a two-dimensional resistive network.

The main idea consists in dividing the cell into a mesh of parallelepipedic dX.dY elements centred on a node related by four crosswise resistors to its four neighbouring nodes, as shown on figure 2.

The definition of each resistive element implies the specification of both horizontal resistors R_h and the vertical ones R_v as:

$$R_h[i,j] = \frac{\rho_h[k].dX}{e.dY} \qquad R_v[i,j] = \frac{\rho_v[k].dY}{e.dX} \tag{1}$$

from the horizontal and vertical resistivities $\rho_h[k]$, $\rho_v[k]$ of a given medium k and the width e of the differential element dX.dY. It is easy to demonstrate that the choice of the width has no influence on the calculations of current densities and potentials.

For determining the primary current and potential distribution in an electrochemical cell, such a description enables us to take into account the non-isotropic resistive properties of materials in the cell. In particular it permits consideration of transitions between different media: for example, electrolyte-electrode interfaces or electrolyte-cell wall transitions.

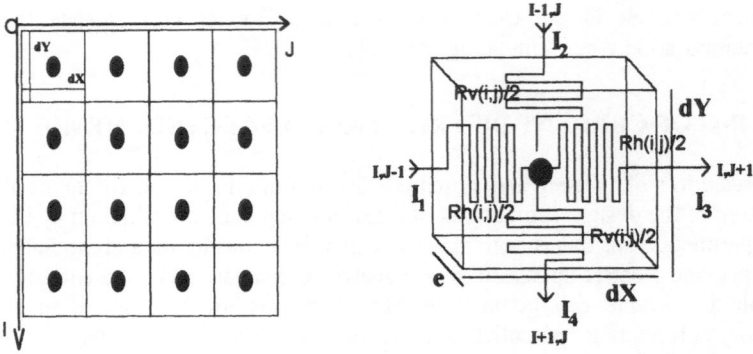

Figure 2. The resistive network and the detail of the mesh (I,J)

Hence the whole electrolyzer can be replaced by a single network of resistances only differing in their values. There are no boundary conditions except the Dirichlet conditions which are simply given as the experimental potential imposed at the edge of cathodes and anodes along linear segments indicated with the mouse. These segments are used at the end of the computation to add up the currents flowing through each individual edge electrode resistor to compute the global current entering and leaving the cell. It must be noted that the insulators are simulated by a mesh of infinite resistors where the current density is zero, so the Neumann conditions are automatically verified along the electrolyte-gas interface.

After defining the boundary conditions of Dirichlet, the calculations leading to the local primary current and potential values in a cell are reduced to the determination of electrical characteristics in an electrical network for which the classical Ohm's law can be applied for any node (I,J):

$$I_1 = \frac{V_{i,j-1} - V_{i,j}}{\left[R_h(i,j) + R_h(i,j-1) \right]/2} = \frac{V_{i,j-1} - V_{i,j}}{R_1} \tag{2}$$

$$I_2 = \frac{V_{i-1,j} - V_{i,j}}{\left[R_v(i,j) + R_v(i-1,j) \right]/2} = \frac{V_{i-1,j} - V_{i,j}}{R_2} \tag{3}$$

$$I_3 = \frac{V_{i,j} - V_{i,j+1}}{\left[R_h(i,j) + R_h(i,j+1) \right]/2} = \frac{V_{i,j} - V_{i,j+1}}{R_3} \tag{4}$$

$$I_4 = \frac{V_{i,j} - V_{i+1,j}}{\left[R_v(i,j) + R_v(i+1,j) \right]/2} = \frac{V_{i,j} - V_{i+1,j}}{R_4} \tag{5}$$

as well as the Kirchoff law:

$$I_1 + I_2 = I_3 + I_4 \tag{6}$$

Practical calculations are performed according the following iterative steps:
- Step 1: Initialisation of current and potential for all the nodes of the network,
- Step 2: Calculation of currents for all the nodes from Ohm's law [eqns (2) to (5)],
- Step 3: Determination of the deviation from the Kirchhoff law at each node from the residue [RES = (I1+I2) - (I3+I4)],
- Step 4: Correction of potentials by a relaxation method at each node except the points of the electrode edges where the conditions of Dirichlet apply, from:

$$V[i,j] = V[i,j] + \mathrm{Re}\,s/[R_1{}^{-1} + R_2{}^{-1} + R_3{}^{-1} + R_4{}^{-1}] \tag{7}$$

- Step 5: Iteration from step 2 until

$$\left|Max\left[\mathrm{Re}\,s/[R_1{}^{-1} + R_2{}^{-1} + R_3{}^{-1} + R_4{}^{-1}]\right]\right| \le \varepsilon \quad \forall i,j \tag{8}$$

ε being the chosen calculation precision for voltages.

For determining the secondary current distribution, the resistance method is able to solve this problem by defining, between the electrolyte and an electrode, an interfacial resistance depending on local current density:

$$R_{equivalent} = \eta[i]/i \tag{9}$$

For this example, the activation overvoltages $\eta[i]$ have been taken into account by using an approximated linear law:

$$\eta_{anode}[i] = 3.77 + 2.12\,10^{-4} * i \quad \text{and} \quad \eta_{cathode}[i] = 0.25 + 1.88\,10^{-4} * i \tag{10}$$

corresponding to the resistive scheme shown in figure 3.

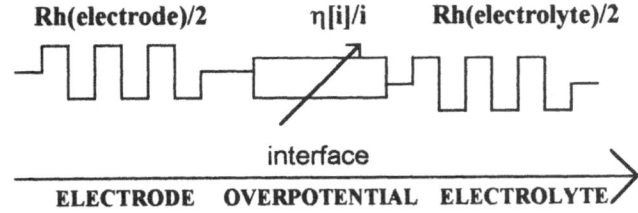

Figure 3. Equivalent circuit for electrode overvoltage

The calculations imply an extra iterative loop in the previous computational scheme because of the unknown values of the interfacial resistance, except for the special case where $\eta(i)/i$ is constant.

THE SOFTWARE

The software presented here extensively uses mouse capabilities and runs on simple PC micro computers only equipped with 2 Mo RAM and VGA graphical output.

The acquisition of the cell geometry is realised from a simple two-dimensional plan by using an A4 IBM 3119 scanner. The different parts of the cell (electrolyte, anodes, cathodes, cell, insulating gas, bus bar etc.) appearing on the screen are identified by clicking the mouse and painting them with specific colours. Then the resistivity to be attributed to each colour is extracted from an internal data file which may be extended if necessary. Lastly the user has to define, with the mouse, the lines where the potential is imposed (the so-called boundary conditions of Dirichlet).

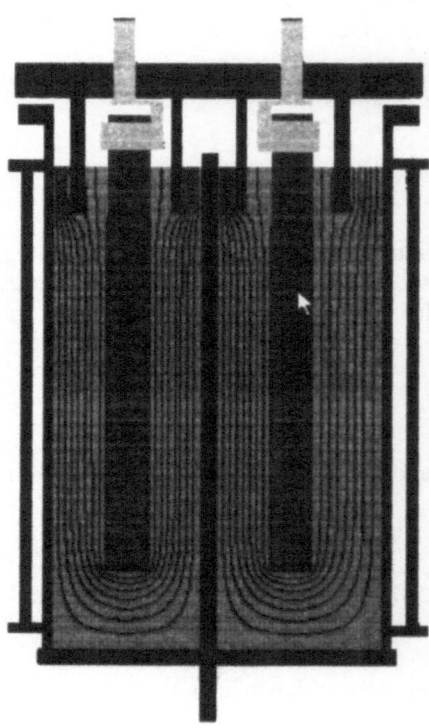

Figure 4. Primary distribution of potential in the Allied Chemical Co. cell

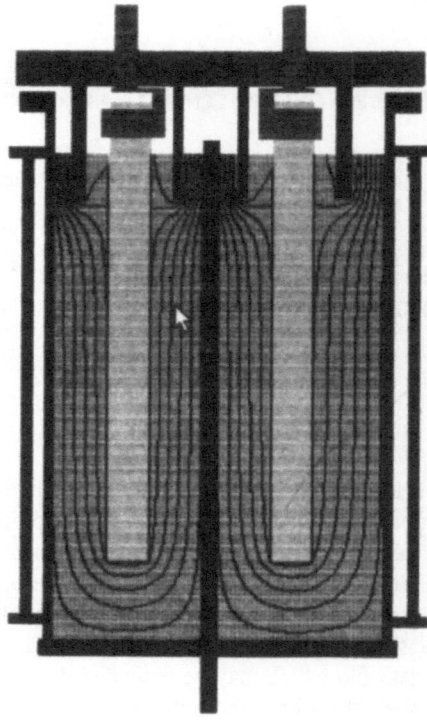

Figure 5. Secondary distribution of potential in the Allied Chemical Co. cell

In this example, potentials are imposed at the top of the electrodes, with an operating voltage of 9 Volts. From this step, all operations are performed in an automatic mode. Calculations concerning either the primary or the secondary current distribution in an industrial cell are short enough to allow rapid checking of numerous geometrical configurations and electrical working conditions.

Typically - when using an IBM PS/2 equipped with 80386 SX and 80387 processors - computation times, for a cell such as the one presented here and described by a mesh of 100x50 nodes, are about 3 minutes for a final 0.1 % relative precision on the potential. This becomes 47 minutes for a 0.001 % numerical precision.

At the end of these computations, the corresponding equipotential curves and current lines can be drawn on the screen and printed or plotted on external outputs. Moreover, the user is able to explore any particular region of interest in the cell (including bus bar and electrodes) with the mouse and obtain the local potential and current. Moreover, the code gives an estimation of anodic and cathodic cell currents obtained by summing the individual currents on every point where a potential was imposed.

SOME SIGNIFICANT RESULTS FOR THE FLUORINE CELL

Figure 4 and 5 present a comparison of primary and secondary equipotential lines calculated for an Allied Chemical Co. fluorine cell containing an electrolyte having a composition close to that used by Comurhex for fluorine production.

For the same working voltage, the primary and secondary equipotential curves have the same global shape (Figures 6 and 7), but as expected, the secondary equipotential curves in the electrolyte are more sparse than the primary ones. This effect is due to the significant anodic overvoltage as seen in equation (9).

In such a fluorine cell, the conductive tank is electrically connected to the cathodes, so many current lines linking the tank and the anodes can be observed, some of them flowing through the gas separation conducting skirts which are immersed in the electrolyte.

It is also noticeable that there are numerous current lines between the anodes and the bottom of the cell. In fact, the secondary current distribution differs here from the primary one, essentially in the regions between the cathodes and the skirts.

Figure 8 shows the distribution of the secondary current density along the left anode in the cell, from the bottom of the cell (left side of the figure) to its top, as defined by two mouse clicks appearing as the two crosses "+" on Figure 1.

This figure clearly shows that the current density is quite uniform over nearly 70 % of the anode length but presents large variations close to the ends of the anode. The left peak is obviously related to the concentration of leakage current lines flowing from the corner of the anode to the tank, as mentioned above, while the other inverse peak is due to the shadow effect of the skirts. It must be noticed that the skirts are not connected to the electrical source and are, in fact, at a floating potential.

More surprisingly, the simulation predicts a very high and unexpected current density (about ten times the average value) passing from the anode to the cathodic gas-separation skirt.

This current line flows along the gas-electrolyte interface, and may generate a high thermal spot on the anode. This effect is amplified by the proximity of the thermally insulating gas and may explain some corrosion problems appearing in the upper part of such cells.

A similar effect was suspected for the Comurhex cells and led us to undertake a study of the thermal effects generated by currents. An initial thermal modelling has already confirmed the first numerical observations.[9]

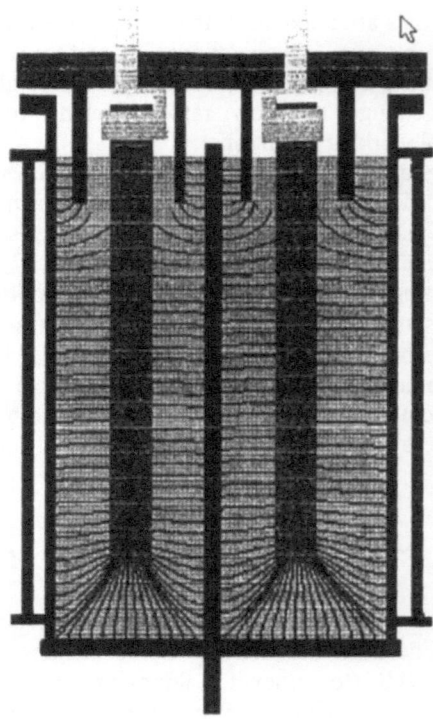

Figure 6. Primary current lines

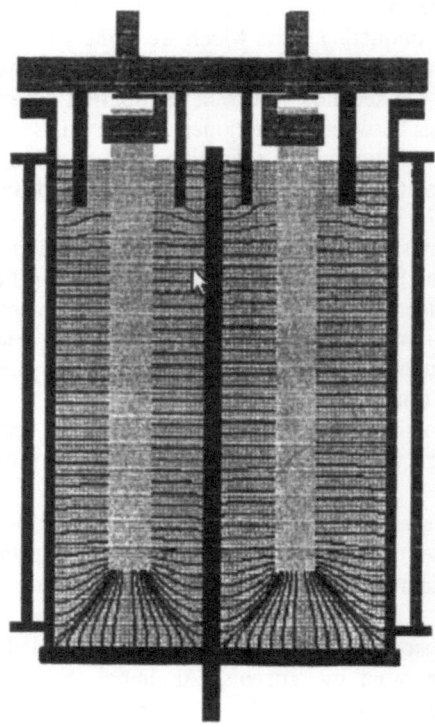

Figure 7. Secondary current lines

A similar study was performed for different types of cell used by Comurhex. The numerical values obtained for the secondary potential distribution fitted perfectly the few available working data, i.e. the skirt-anode voltage, the potential-current curve and the global currents observed.

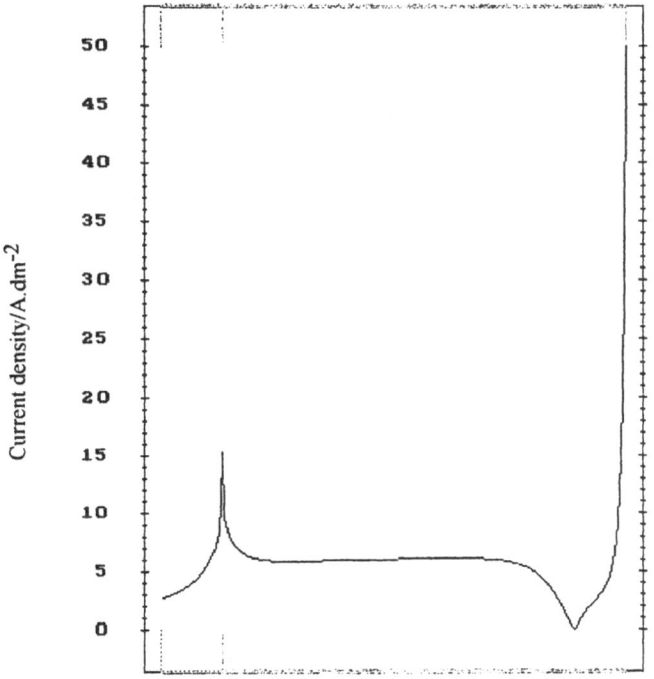

Figure 8. Screen hardcopy of current density along an anode

CONCLUSIONS

The software presented for determining two-dimensional primary and secondary current and potential distributions in electrolysers can easily produce helpful information for designing industrial cells in which electrolyte flow has no significant effect.

Coupling of electrode kinetics and thermal phenomena by exploitation of the direct analogy between thermal and electrical phenomena is now possible by use of this software. However, the study of the influence of hydrodynamics on the cell behaviour requires interfacing of this software to codes for laminar or turbulent flow simulation. Such a feasibility is under study at the present time.

A worthwhile advantage of this software has been shown when studying a real working fluorine cell. Indeed, simulation highlighted regions showing high local current densities for which unexpected corrosion problems were experimentally observed.

Moreover, the modelling capabilities of such software allow easy analysis of the impact of geometric modifications of a cell design in order to decrease the overall cell voltage, to obtain a suitable current distribution and to ensure a proper electrode operation.

REFERENCES

1. M. Jaccaud, F. Nicolas. Procédés Industriels in: "Techniques de l'Ingénieur", Lavoisier ed., Paris (1990).
2. L. Ben Abdallah, Modélisation de la répartition du courant et du potentiel dans une cellule d'électrolyse industrielle. Thesis. Institut National Polytechnique de Grenoble (1993).
3. Kirk - Othmer, Electrochemical Processing-Inorganic, in : "Encyclopedia of Chemical Technology", Wiley ed., N.Y. (1985).
4. G. A. Prentice, C.W. Tobias, A survey of numerical methods and solutions for current distribution problems . *J. Electrochem. Soc.*, 129: 72 (1982).
5. C.W. Walton, R.E. White, Utility of an empirical method of modelling combined zero gap / attached electrode membrane chlor-alkali cells, *J. Electrochem. Soc.* 134 : 565 (1987).
6. J.P. Bessaguet, J.P. Caire, J.C. Delachaume, P. Ozil , Modélisation des lignes de courant dans un composite Electrode-Membrane, in : "Récents Progrès en Génie des Procédés", 2:347, Lavoisier ed., Paris (1987).
7. M.Z. Yang, H. Wu, J.R. Selman , A model for bipolar current leakage in cell stacks with separate electrolyte loops. *J. Appl. Electrochem.*, 19 : 247 (1989).
8. J. Divisek, R. Jung, D. Britz , Potential distribution and electrode stability in a bipolar cell. *J. Appl. Electrochem.*, 20:247 (1990).
9. S. Carpentier, Modélisation de la répartition thermique dans un électrolyseur industriel, DEA Génie des Procédés, Institut National Polytechnique de Grenoble (1993).

POTENTIAL DISTRIBUTION IN A POROUS PERCOLATED REACTOR WITH RADIAL FIELD MODELING AND SIMULATION

Laurent Foucher and Germain Lacoste

Laboratoire de Génie Electrochimique et d'Energétique des Réacteurs
18, chemin de la Loge
31078 Toulouse Cédex, France

ABSTRACT

A theoretical study is described for a fixed bed electrochemical reactor with perpendicular current and electrolyte flows. A two dimensional model with different sets of boundary conditions was developed for limiting current conditions. The model was solved by a finite element method.

INTRODUCTION

Pollution due to heavy metals, arising mostly from industrial activities, constitutes a severe environmental problem. Conventional methods for the removal and recovery of these metals can operate economically only with relatively concentrated solutions and are unable to reduce pollutants to low levels. Electrochemical techniques are an alternative method for the treatment of heavy metal contaminated liquid effluents, and in particular, three dimensional electrode technology[1,2]. Such electrodes are characterized by a very large specific area and high mass transfer coefficient so that they can operate at low current density but with a processing time compatible with industrial constraints[3].

There are two principal kinds of configuration for such three dimensional reactors : radial and axial field electrodes. The latter have been widely studied[4,5] because solution of the model for the potential distribution does not present real difficulty.

The second configuration, the radial field flow-through porous electrode, has a potential distribution more complex than in the case of the axial field and a two-dimensional analysis of the problem must be developed[6]. This kind of reactor has undergone significant technical development in recent years, notably with the use of a sinusoidal type pulsation to the fluid percolating the cell. The geometry of such electrochemical reactors becomes complex with, for example, the implantation of anodes inside the granular matrix[7].

The purpose of this work is to develop a better understanding of the potential distribution inside a porous electrode with radial field operating under limiting conditions, which would be useful for engineering design and scale-up calculations. In the general case, the two-dimensional models describing this type of reactor are based on the mass transfer equation over a differential element of electrode and the conservation of charge in the electrolyte phase[8,9], which is considered as a pseudo-continuous media with an electrical equivalent conductivity σ_s.

LOCAL METAL-SOLUTION POTENTIAL DIFFERENCE

A radial field flow through porous electrode can be designed in two different configurations. The first one is the internal counter electrode cell and corresponds to a working porous electrode located in the annular region between the cylindrical counter electrode and the insulating outer wall. The second one, the external counter electrode cell, is shown in figure 1; here the working electrode is surrounded by the cylindrical counter electrode[9].

Both cell types can be treated theoretically in the same way, and only the boundary conditions are different. In this paper, we will develop an external counter electrode model and indicate only the boundary conditions for an external counter electrode cell.

As can be seen in figure 1, the considered porous electrode is a cylindrical fixed bed of copper particles which is separated by a cylindrical diaphragm from the cylindrical counter electrode. Several assumptions have been made in order to simplify the mathematical treatment :
- electrolysis proceeds under isothermal steady-state conditions, with steady plug-flow of electrolyte in the axial direction,
- a single reversible electrode reaction occurs,
- the porous electrode is highly conductive and has a uniform porosity and a specific surface area which do not change during electrolysis,
- the presence of a supporting electrolyte suppresses the migration of reacting species,
- axial dispersion is neglected.

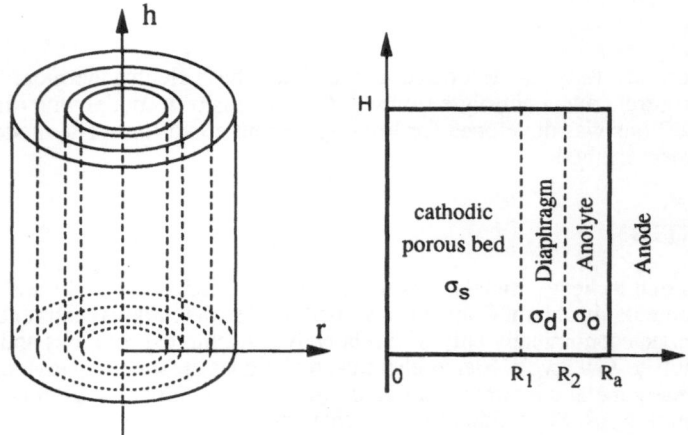

Figure 1. External anode cell configuration.

With these assumptions, a simple mass balance for a system operating at limiting current with an upward flow, gives the variation of the electroactive species concentration as[10]

$$C = C_0 \exp(-\alpha h), \quad \text{with } \alpha = \frac{k \, S_p}{v} \tag{1}$$

The general equation of the local metal-solution potential difference within a porous electrode can be found by a charge-balance. In the case of a highly conductive porous electrode[11] :

$$\Delta E = -\frac{n \, F \, r^*}{\sigma_s} \tag{2}$$

where r^* is the rate of reaction for the limiting reactant. However, in the case of the radial flow configuration, the cell is composed of three different parts.

The cathodic compartment, with a solution conductivity of σ_s, is the only part of the cell containing the limiting reactant. The reaction rate r^* in this part is given by :

$$r^* = - k \, S_p \, C \tag{3}$$

The combination of (1), (2) and (3) gives the potential distribution :

$$\frac{\partial^2 E}{\partial r^2} + \frac{1}{r} \frac{\partial E}{\partial r} + \frac{\partial^2 E}{\partial h^2} = - \frac{n \, F \, k \, Sp \, C_0}{\sigma_s} e^{-\alpha h} = A \, e^{-\alpha h} \tag{4}$$

The separator, which is satured with electrolyte, has a conductivity σ_d. In this part, the solution-potential distribution is given by ($r^* = 0$) :

$$\frac{\partial^2 E}{\partial r^2} + \frac{1}{r} \frac{\partial E}{\partial r} + \frac{\partial^2 E}{\partial h^2} = 0 \tag{5}$$

When the current lines pass through the separator, because of conductivity changes in different parts of the reactor, the solution potential gradient changes and can be mathematically represented by the following expressions written over the left and right hand faces of the separator :

$$\text{at } r = R_1, \qquad \sigma_s \left(\frac{\partial E}{\partial r} \right)_{R_1^-} = \sigma_d \left(\frac{\partial E}{\partial r} \right)_{R_1^+}$$

$$\text{at } r = R_2, \qquad \sigma_d \left(\frac{\partial E}{\partial r} \right)_{R_2^-} = \sigma_0 \left(\frac{\partial E}{\partial r} \right)_{R_2^+} \tag{6}$$

In these expressions, $(\partial E/\partial r)_{R^-}$ and $(\partial E/\partial r)_{R^+}$ represent the radial potential variations at both sides of a surface situated at r.

The anodic compartment contains supporting electrolyte with a conductivity σ_0. In this part also, the reaction rate for the limiting reactant is zero and the solution potential distribution is given by equation (5). The oxidation of water at the anode surface is a very fast reaction and as a result the solution-potential distribution over the anodic surface is uniform.

BOUNDARY CONDITIONS FOR EXTERNAL ANODE CELL

In order to integrate the model equations (4) and (5), four boundary conditions are required. We have chosen the three following boundary conditions, corresponding to experimental results[8,11] :

$$\text{at } h=0, \qquad \frac{\partial E}{\partial h} = 0 \qquad \text{for all } r$$

$$\text{at } r=0, \qquad \frac{\partial E}{\partial r} = 0 \qquad \text{for all } h \tag{7}$$

$$\text{at } h = H, \qquad E = E^* \text{ for } 0 < r < R_1 \text{ and } E = E_a \text{ for } R_1 < r < R_a$$

Acccording to the choice of the fourth boundary condition, two different models have been proposed.

Simplified Model

Under some conditions such as small bed length, high separator conductivity and small radial thickness of the anodic compartment, the separator wall may be assumed equipotential, i.e. :

$$\text{at } r = R_1, \qquad E = E^* \qquad \text{for all } h \tag{8}$$

In this model, only the cathodic compartment is considered.

Complete Model

To generalise the problem, a more realistic boundary condition may be used. Effectively, the solution-potential on the anodic surface is approximately uniform, i.e.

$$\text{at } r = R_a, \qquad E = E_a \qquad \text{for all } h \tag{9}$$

Using this boundary condition, the whole of the reactor is considered and influence of the anodic compartment and separator are taken into account .

MODEL INTEGRATION

To integrate the model, the finite element method was chosen. Its main advantages over other methods are that the element can be easily deformed, small elements can be used in regions of interest and irregular domains are easily handled. The discretized model equation was obtained by the method of weighted residuals using the Galerkin approach[12,13,14] By making use of integration by parts and inserting the boundary conditions, the resulting system of linear algebric equations was solved by a band-matrix method[14].

Finite Element Formulation

First, we used the Galerkin approach in the finite element formulation. Detailed explanation of the use of this method is given in Dhatt & Thouzot[12]. Using a cylindrical system cordinate (r, θ, z), the potential distribution equation has the following volume form :

$$\iiint_v N \left[\frac{\partial^2 E}{\partial r^2} + \frac{1}{r} \frac{\partial E}{\partial r} + \frac{\partial^2 E}{\partial h^2} - g(h) \right] dv = 0 \tag{10}$$

The elemental volume is axisymetric, i.e. $dv = 2 \pi r \, dr \, dh$ and Green's theorem is used to transform the second derivate and incorporate the boundary conditions :

$$I = -2\pi \int_0^H \int_0^R \left\{ \frac{\partial N}{\partial r}\frac{\partial E}{\partial r} + \frac{\partial N}{\partial h}\frac{\partial E}{\partial h} \right\} r\, dr\, dh$$

$$+ 2\pi \int_0^H \left[r\, N \frac{\partial E}{\partial r} \right]_{r=0}^{r=R} dh + 2\pi \int_0^R \left[N \frac{\partial E}{\partial h} \right]_{h=0}^{h=H} r\, dr \qquad (11)$$

$$- 2\pi \int_0^H \int_0^R N\, A\, e^{-\alpha h}\, r\, dr\, dh = 0$$

In order to simplify the calculations, isoparametric elements are used, based on a local coordinate system with independent variables ξ and η which usually vary from -1 to +1. In such elements we can write the following expressions :

$$E = \sum_{j=1}^{nn} N_j\,(\xi,\eta)\, E_j$$

$$r = \sum_{j=1}^{nn} N_j\,(\xi,\eta)\, r_j \qquad (12)$$

$$h = \sum_{j=1}^{nn} N_j\,(\xi,\eta)\, h_j$$

Since we are dealing with Cartesian derivates in the integral equation, we transform the derivates of $N_i\,(\xi,\eta)$ using the chain rule:

$$\frac{\partial N_i}{\partial \xi} = \frac{\partial N_i}{\partial r}\frac{\partial r}{\partial \xi} + \frac{\partial N_i}{\partial h}\frac{\partial h}{\partial \xi}$$

$$\frac{\partial N_i}{\partial \eta} = \frac{\partial N_i}{\partial r}\frac{\partial r}{\partial \eta} + \frac{\partial N_i}{\partial h}\frac{\partial h}{\partial \eta} \qquad (13)$$

These equations can be rewritten using the jacobian matrix expression :

$$\begin{bmatrix} \dfrac{\partial N_i}{\partial \xi} \\[2mm] \dfrac{\partial N_i}{\partial \eta} \end{bmatrix} = \begin{bmatrix} \dfrac{\partial r}{\partial \xi} & \dfrac{\partial h}{\partial \xi} \\[2mm] \dfrac{\partial r}{\partial \eta} & \dfrac{\partial h}{\partial \eta} \end{bmatrix} \begin{bmatrix} \dfrac{\partial N_i}{\partial r} \\[2mm] \dfrac{\partial N_i}{\partial h} \end{bmatrix} = J \begin{bmatrix} \dfrac{\partial N_i}{\partial r} \\[2mm] \dfrac{\partial N_i}{\partial h} \end{bmatrix} \qquad (14)$$

To find the Cartesian derivates of N_i, we simply write :

$$\begin{bmatrix} \dfrac{\partial N_i}{\partial r} \\[2mm] \dfrac{\partial N_i}{\partial h} \end{bmatrix} = J^{-1} \begin{bmatrix} \dfrac{\partial N_i}{\partial \xi} \\[2mm] \dfrac{\partial N_i}{\partial \eta} \end{bmatrix} = \frac{1}{|J|} \begin{bmatrix} \dfrac{\partial h}{\partial \eta} & -\dfrac{\partial h}{\partial \xi} \\[2mm] -\dfrac{\partial r}{\partial \eta} & \dfrac{\partial r}{\partial \xi} \end{bmatrix} \begin{bmatrix} \dfrac{\partial N_i}{\partial \xi} \\[2mm] \dfrac{\partial N_i}{\partial \eta} \end{bmatrix} \qquad (15)$$

where $|J|$ is the determinant of J. Using relations (12), all the terms of the jacobian matrix are evaluated from these expressions :

$$J = \begin{bmatrix} \sum_{i=1}^{nn} \dfrac{\partial N_i}{\partial \xi} r_i & \sum_{i=1}^{nn} \dfrac{\partial N_i}{\partial \xi} h_i \\[2mm] \sum_{i=1}^{nn} \dfrac{\partial N_i}{\partial \eta} r_i & \sum_{i=1}^{nn} \dfrac{\partial N_i}{\partial \eta} h_i \end{bmatrix} \qquad (16)$$

The integral in term of ξ and η is defined as

$$\int_{-a}^{a} \int_{-b}^{b} F(x,y)\, dx\, dy = \int_{-1}^{1} \int_{-1}^{1} f(\xi,\eta)\,|J|\, d\xi\, d\eta \qquad (17)$$

Using eq (7), (12), (15) and (17), the integral form becomes:

$$I = \sum_{j=1}^{nn} \left(\int_{-1}^{1} \int_{-1}^{1} \left\{ \frac{\partial N_i}{\partial r} \frac{\partial N_j}{\partial r} + \frac{\partial N_i}{\partial h} \frac{\partial N_j}{\partial h} \right\} r\,|J|\, d\xi\, d\eta \right) E_j$$
$$- \int_{-1}^{1} \int_{-1}^{1} N_i\, A\, e^{-\alpha h}\, r\,|J|\, d\xi\, d\eta = 0 \qquad i=1,nn \qquad (18)$$

Finite Element Algorithm

The reactor is divided into rectangles as shown in figure 2 .

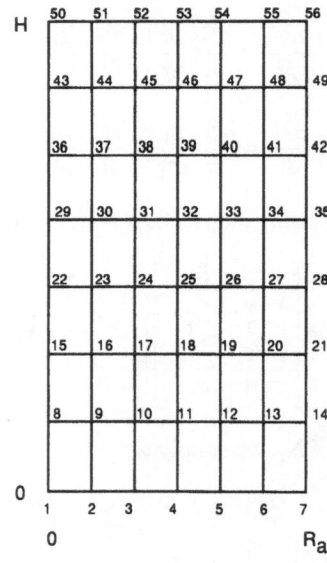

Figure 2. Example of two-dimensional mesh consisting of linear rectangular elements.

Due to the geometry of the electrode, linear rectangular elements have been considered. The interpolation functions for those elements are expressed as:

$$N_1 = \frac{1}{4}\,(1-\xi)\,(1-\eta)$$

$$N_2 = \frac{1}{4}\,(1+\xi)\,(1-\eta)$$

232

$$N_3 = \frac{1}{4} (1+\xi) (1+\eta) \tag{19}$$

$$N_4 = \frac{1}{4} (1-\xi) (1+\eta)$$

The algorithm consists of assembling the linear equation (18) related to each element. In order to obtain the correct assembly of equations, the coupling of the element is carried out by taking into account that the nodal values are identical for a node located on the common boundary of the elements[12,13].

NUMERICAL EXAMPLE

The electrode potential distribution measured by Mowla[11] on a flow-by porous electrode is compared with the simplified model developed previously to validate the method. The working electrode (cathode) was a fixed bed of spherical copper particles and the electrolyte solution was 1N, H_2SO_4, which contained Cu^{++} from 60 to 600 ppm.

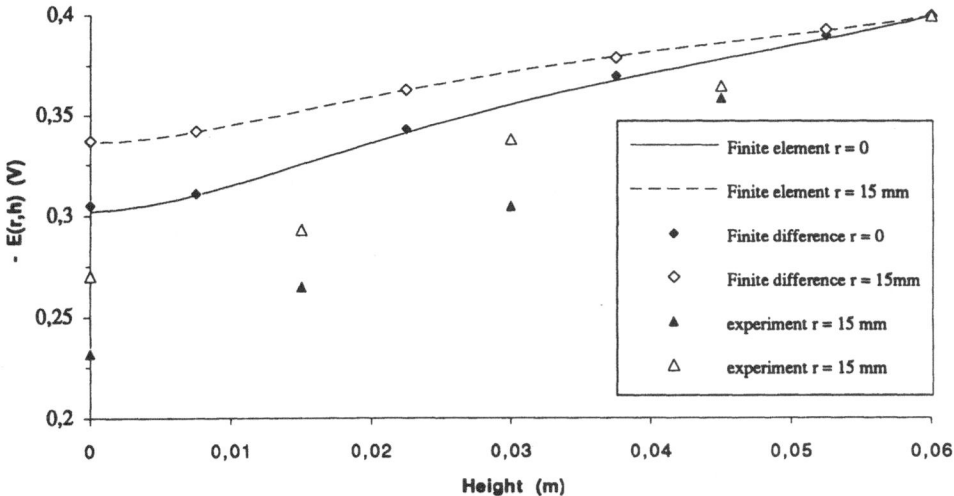

Figure 3. Theoretical and experimental distributions of the local potential difference. Comparison between the finite element and the finite difference methods.

As seen in figure 3, the resolution of the simplified model by the finite element method gives the same results as the finite difference method. Nevertheless, the values obtained by the simplified model are lower than the experimental ones, especially when the bed height is sufficiently high. The simplified model does not take into account the vertical variation of solution potential in the separator and anolyte compartment. Consequently, the separator wall may be considered as an equipotential surface and used as a boundary condition for the charge balance equation. However, the experimental equipotential surfaces in the solution phase show vertical drops occuring in these compartments of the reactor, and demonstrate that only the anodic wall can be assumed equipotential. For these reasons, the complete model is now under development.

NOMENCLATURE

A, α   constant
C   concentration of electroactive species (mol.m^{-3})

C_0   inlet concentration (mol m^{-3})
E   local metal-solution potential difference (V)
E_a   metal-solution potential on anode wall (V)
E*   applied potential at the top of the reactor (V)
E_j   value of local metal-solution potential difference at node j (V)
F   Faraday's constant

k   average liquid-solid mass transfer coefficient (m.s^{-1})
H   height of cathodic porous bed (m)
h   axial distance through porous bed (m)
h_j   axial coordinate of node j (m)
r   radial distance through porous bed (m)
r_j   radial coordinate of node j (m)
n   number of electrons transferred in reaction of limiting reactant
nn   number of nodes per finite element
N_i   linear interpolation function
r*   electrochemical reaction rate (mol.m^{-3}.s^{-1})
Sp   specific area (m^{-1})
v   inlet flow velocity (m.s^{-1})
σ_0 actual conductivity of the solution (Ω^{-1}.m^{-1})
σ_d conductivity of diaphragm when filled with electrolyte (Ω^{-1}.m^{-1})
σ_s effective conductivity of solution within the pore (Ω^{-1}.m^{-1})
ξ, η local coordinate in the finite element

REFERENCES

1. R.E. Sioda, Electrolysis with a flowing solution on graphite packing, *Electrochim. Acta*, 13:1559 (1968)
2. R. Alkire and B. Gracon, Flow-through porous electrode, *J. Electrochem. Soc.*, 122:1594 (1972)
3. J. Van Zee and J. Newman, Electrochemical removal of silver ions from photographic fixing solutions using a porous flow-through electrode, *J. Electrochem. Soc.*, 12:706 (1977)
4. M. Paulin, D. Hutin, and F. Coeuret, Theoretical and experimental study of flow-through porous electrode, *J. Electrochem. Soc.*, 124:180 (1977)
5. J A. Trainham and J. Newman, A flow-through electrode model : application to metal-ion removal from dilute streams, *J. Electrochem. Soc.*, 124:1528 (1977)
6. P.S. Fedkiw, Ohmic potential drop in flow-through and flow-by porous electrode, *J. Electrochem. Soc.*, 128:831 (1981)
7. H. Olive and G. Lacoste, Application of a porous electrode for silver recovery from photographic fixing solution. Part III - Radial field and multi-anode reactor," *J. Photogr. Sci.*, 32:227 (1984)
8. A. Storck, M.A. Enriquez-Granados, and M. Roger, The behaviour of porous electrodes in a flow-by regime. Part I - Theoretical study, *Electrochim. Acta*, 27:293 (1982)
9. G. Kreysa and K. Jüttner, Cylindrical three-dimensional electrodes under limiting current conditions, *J. Appl. Electrochem.*, 23:707 (1993)
10 R.B. Bird, W.E. Steward and E.N. Lightfoot. "Transport Phenomena," John Wiley & Sons, New-York (1960)
11. D. Mowla. "Réacteur à Electrodes Volumique : Flux de Matière et de Charge Croisés," Thesis, I.N.P. Toulouse (1982)
12. G. Dhatt and G. Thouzot. "Une Présentation de la Méthode des Eléments Finis," Presses de l'université Laval, Québec (1981)
13. D.W Pepper and J.C Heindrich. "The Finite Element Method. Basic Concepts and Applications," Hemisphere Publishing Corporation, Washington (1992)
14. O. Axelsson and V.A. Barker. "Finite Element Solution of Boundary Value Problems. Theory and Computation," Academic Press, Inc, Orlando (1984)

CONTINUOUS ELECTRODEPOSITION MODELLING : TERTIARY CURRENT DISTRIBUTION ON A PLANE ELECTRODE - EFFECT OF AXIAL DIFFUSION AT LOW PECLET NUMBER

Jacques Josserand,[1] Patrick Ozil,[2] Antoine Alemany,[1] Serguei A. Martemianov,[3] Bernadette Nguyen,[2] and Pascale Pham[4]

[1] LEGI, Insitut de Mécanique de Grenoble, BP53X, 38041 Grenoble
[2] Centre de Recherche en Electrochimie Minérale et en Génie des Procédés (URA CNRS 1212) ,ENSEEG (INPG), BP75, 38402 Saint Martin d'Hères Cedex
[3] A.N. Frumkin Inst. of Electrochemistry, Académie des Sciences de Russie, Moscou
[4] DT2I, 8 Chemin des Prèles, 38240 Meylan Zirst

INTRODUCTION

The purpose of the work is to develop a general model of the phenomena occurring in continuous electrodepositon processes. This study involves the adaptation of the commercial software FLUX-EXPERT® - based on a finite element method - for determining the current distribution in an electrolyzer, from the primary distribution to the tertiary. This latter case requires a simultaneous solving of electrical and mass balance equations for deriving the electrical potential Φ and the concentration C of the electroactive species, which are strongly coupled at the electrolyte-cathode interface.

Determining the tertiary current distribution in systems with flowing electrolyte is complex. Whereas there are many studies of primary and secondary current distributions in various systems, only a few studies related to tertiary distribution limited to simple configurations have been reported : for example among the basic works are free convection along a vertical plane electrode[1], forced laminar convection for either a rotating disk[2-4] or ring-disk electrode[5], fixed electrodes in rectangular[6-8] and tubular channels[9], a moving sheet electrode[10] and anodic levelling of triangular surface profiles[11].

The present work presents a way of determining the tertiary current distribution by using the capabilities of Galerkine's formulation on which the finite element method is based. This numerical approach is checked for a classical system involving a plane electrode embedded in a rectangular channel in which laminar electrolyte flow occurs, characterised by a constant velocity gradient and for an irreversible electrochemical reduction process partially controlled by diffusion and described by a simplified Butler-Volmer equation.

Electrochemical Engineering and Energy, Edited by
F. Lapicque *et al.*, Plenum Press, New York, 1995

For large Peclet numbers, results are in agreement with those obtained by Parrish and Newman[6,7]. But for more moderate Pe values, axial diffusion appears to be important at the downstream end of the cathode for high limiting diffusion current densities. Then the uneven primary current distribution is not totally reduced by cross diffusion and the Lévêque solution is no longer valid for locally determining the limiting diffusion current.

DEFINITION OF THE PROBLEM

The system under study consists of two short plane parallel electrodes embedded in the opposite walls of an insulating rectangular channel located downstream of a Couette flow (Figure 1). The small size of the cathodic system (electrode length L = 1mm and gap H = 0.5 mm) is a choice which allows the diffusion layer to be described by a fine mesh of finite elements connected by their nodes.

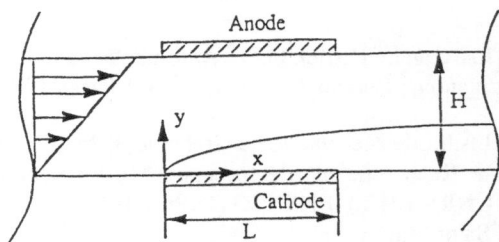

Figure 1. System of short plane electrodes embedded in the wall

The cathode is assumed to be sufficiently short to allow the wall velocity gradient to be considered as a constant, so leading to a linear velocity profile inside the diffusion layer and along the electrode. Further assumptions are : (i) the electrolyte is a dilute solution of an ion in an excess of supporting electrolyte, (ii) the migration effect is neglected and so the transport number of the reacting species is zero, (iii) both the ionic diffusivity D of the electroactive species and the conductivity σ of the electrolyte are constant.

The Lévêque solution specifies a constant wall concentration along the plate (having a zero value for the diffusion limiting case), thus neglecting axial diffusion. This assumption is valid for short distances from the leading edge, particularly for large values of the Peclet number. On the other hand, for small Pe values, the effects of axial diffusion both at the two upstream and downstream discontinuity points between electrode and insulator are no longer negligible. The effects of axial dispersion increase when the current density increases (corresponding to small values of the Wagner number). Indeed, edge effects significantly increase consumption of the electroactive species at both the ends of the electrode. In such a case, axial diffusion can significantly affect the current distribution.

In the present work, only downstream axial diffusion is considered. Thus the concentration at the upstream end of the plate is assumed to be the same as the concentration C_∞ in the bulk. This last condition allows upstream axial diffusion to be neglected and leads to an easier comprehensive comparison between the present results and those of Parrish and Newman who totally neglected axial diffusion.

If considering the system at steady-state without charge accumulation or generation and consumption within the bulk, the local conservation of current density j and mass flux N_i for the electroactive species may be written as:

$$\nabla j = \nabla \left(-\sigma . \nabla \Phi \right) = 0 \tag{1}$$

$$\nabla N_i = \nabla(V.C - D.\nabla C) = 0 \tag{2}$$

where ϕ represents the electrical potential, V the fluid velocity and C the concentration of the electroactive species.

Moreover the reduction rate is assumed to to be sufficiently high to consider the process as irreversible and partially controlled by diffusion. Thus, only the cathodic term appears in the Butler-Volmer equation which is reduced to the following simplified form:

$$j = j_0 \frac{C_0}{C_\infty} e^{\left[-\beta \frac{z\Phi}{RT} \eta \right]} \tag{3}$$

FORMULATION OF THE PROBLEM

First formulation

For implementation of the finite element method, the local equations (1) and (2) are replaced by integral forms for the electrolyte bulk using a projective function α according to Galerkine's method:

$$\iint_{\text{bulk}} \left[\alpha . \nabla(-\sigma \nabla \Phi) \right] . ds = 0 \tag{4}$$

and:

$$\iint_{\text{bulk}} \left[\alpha(V . \nabla C + \nabla(-D.\nabla C)) \right] . ds = 0 \tag{5}$$

This strong integral formulation can be replaced by a weak one by first decomposing the product between the function α and the function divergence:

$$\alpha . \nabla(-\sigma \nabla \Phi) = -(-\sigma \nabla \Phi).\nabla \alpha + \nabla \left[\alpha(-\sigma \nabla \Phi) \right] \tag{6}$$

and, secondly, using the Ostrogradski theorem for rejecting the divergence terms to the boundary where they express the non-homogeneous Neumann boundary conditions for fluxes at the wall:

$$\iint_{\text{bulk}} \nabla \left[\alpha(-\sigma.\nabla \Phi) \right] ds = \int_{\text{cathode}} \alpha(-\sigma.\nabla \Phi).n.dl \tag{7}$$

Then the resulting equation corresponding to (4) is rewritten as:

$$\iint_{\text{bulk}} \left[\sigma.\nabla \alpha.\nabla \Phi \right].ds = \int_{\text{cathode}} \alpha. \left(-\sigma \frac{\partial \Phi}{\partial n} \right) dl \tag{8}$$

and similarly, equation (5) leads to:

$$\iint_{\text{bulk}} \left[\alpha V.\nabla C + D.\nabla \alpha.\nabla C \right].ds = - \int_{\text{cathode}} \alpha. \left(-D \frac{\partial C}{\partial n} \right).dl \tag{9}$$

Hence increasing the derivation order of the projective function α allows a decrease in the derivation order to one for the unknown variables C and Φ.

Boundary conditions

The electrodes are assumed to be made of the same material as the deposit and to be equipotential. The potential Φ is measured versus a reference electrode also made of the same material and so the $\Phi - \Phi_c$ value at equilibrium is zero.

Under this assumption the boundary conditions are as follows:
- for the anode

The overpotential is neglected and the potential of the equipotential surface is imposed (Dirichlet condition) thus allowing control of the current in the cell.

$$\Phi_a = Cte \tag{10}$$

- for the insulating walls and the axis of symmetry

No current or mass flow passes through these elements, and the current density perpendicular to the boundary is zero (homogeneous Neumann condition):

$$\frac{\partial \Phi}{\partial n} = 0 \quad ; \quad \frac{\partial C}{\partial n} = 0 \tag{11}$$

- for the cathode, where electrodeposition occurs

The condition expresses the equality at the wall between the current density of charge transfer j and the ohmic and diffusional current densities (non-homogeneous Neumann condition):

$$j = -\sigma \frac{\partial \Phi}{\partial n} = -zFD \frac{\partial C}{\partial n} \quad \text{(at y=0)} \tag{12}$$

New formulation

Finally, taking into account the boundary condition (12), the two boundary terms may be expressed as functions of electronic current at the wall, i.e. as functions of wall potential change (3), so leading to:

$$\iint_{bulk} [\sigma. \nabla \alpha. \nabla \Phi].ds = - \int_{cathode} \alpha. j_0 \frac{C_0}{C_\infty} e^{-\beta \frac{zF}{RT}(\Phi - \Phi_c)}.dl \tag{13}$$

and:

$$\iint_{bulk} [\alpha V. \nabla C + D. \nabla \alpha. \nabla C].ds = - \int_{cathode} \frac{\alpha}{zF}. j_0 \frac{C_0}{C_\infty} e^{-\beta \frac{zF}{RT}(\Phi - \Phi_c)}.dl \tag{14}$$

Two cases are successively considered here for modelling the current distribution.

In the first step, downstream axial diffusion is neglected, so leading to the following classical simplification to be introduced into the first term of equation (14):

$$D. \nabla \alpha. \nabla C = D \left[\frac{\partial \alpha}{\partial y}. \frac{\partial C}{\partial y} \right] \tag{15}$$

238

then in a second step, this phenomenon is taken into account by using a more complete form:

$$D.\nabla\alpha.\nabla C = D\left[\frac{\partial\alpha}{\partial x}\cdot\frac{\partial C}{\partial x}+\frac{\partial\alpha}{\partial y}\cdot\frac{\partial C}{\partial y}\right] \tag{16}$$

The coupled equations (13) and (14) replacing equations (1)-(3) can be directly implemented into the software FLUX-EXPERT® by using an internal equation generator. Then the classical discretization of the finite element method is performed by approximating the projective function and the unknown variables C and Φ by 1st and 2nd order polynomial functions.

RESULTS AND COMMENTS

For a given geometry (ratio H/L fixed), the value of the average cathodic current density j_{avg} is is a function of the anodic potential ϕ_a which is also the cell potential when defining a zero value for the cathodic potential ϕ_c. It is defined as :

$$j_{avg} = \frac{1}{L}\int_0^L j(x)dx \tag{17}$$

Moreover this density depends on the coefficient β and on five other parameters: σ, j_0, D, S_0 and C_∞ which can be grouped into three current terms:
- the exchange current density j_0,
- the migration current density j_m:

$$j_m = \sigma\frac{RT}{zFL} \tag{18}$$

- the average limiting diffusion current density which has been determined for the Lévêque case as:

$$j_{ld} = \frac{0.807}{1-t_+}zFC_\infty\left[\frac{S_0D^2}{L}\right]^{1/3} \tag{19}$$

From these currents, one can define, following Parrish and Newman[6,7], dimensionless currents relative to the migration current:

$$J = \frac{j_o}{j_m} \qquad N = \frac{j_{ld}}{0.807\,j_m} \qquad \delta = \frac{j_{avg}}{j_m} \tag{20}$$

For all the cases discussed, the following set of common parameters: β= 0.5, z=1, C_∞= 1 mol.m^{-3}, S_0= 10 s^{-1}, D= 10^{-7} m^2.s^{-1} are considered in order to compare the results of the present work with previous ones[6,7].

Results without axial diffusion

The purpose of this preliminary study was to validate the numerical approach from finite elements by comparing the results to those previously obtained by other methods[6,7].

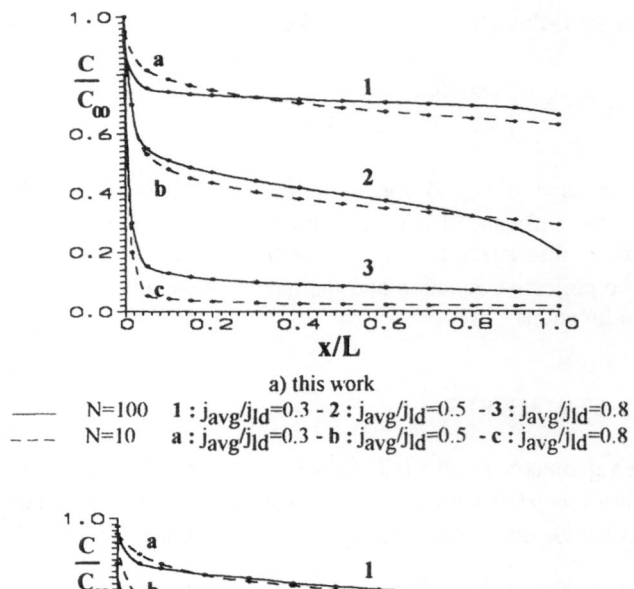

a) this work

——— N=100 **1** : j_{avg}/j_{ld}=0.3 - **2** : j_{avg}/j_{ld}=0.5 - **3** : j_{avg}/j_{ld}=0.8
- - - N=10   **a** : j_{avg}/j_{ld}=0.3 - **b** : j_{avg}/j_{ld}=0.5 - **c** : j_{avg}/j_{ld}=0.8

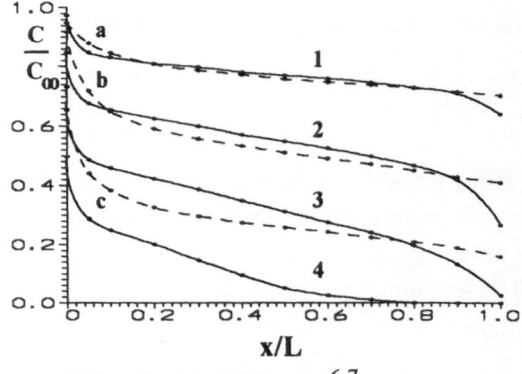

b) from Parrish and Newman[6,7]:

——— N=100 **1** : j_{avg}/j_{ld}=0.25 - **2** : j_{avg}/j_{ld}=0.50 - **3** : j_{avg}/j_{ld}=0.75 - **4** : j_{avg}/j_{ld}=0.95
- - - N=5   **a** : j_{avg}/j_{ld}=0.25 - **b** : j_{avg}/j_{ld}=0.50 - **c** : j_{avg}/j_{ld}=0.75 - **d** : j_{avg}/j_{ld}=0.95

Figure 2. Results without axial diffusion - Wall concentration profiles for high or low N values ($J\cong0$, σ= 0.04 $\Omega^{-1}.m^{-1}$ for N=100, σ= 0.4 $\Omega^{-1}.m^{-1}$ for N=10)

As shown in figures 2 and 3, for J close to zero and N=100 or 10, both the wall concentration and the current density profiles computed by the present method are in good agreement with those presented by Parrish and Newman and allow an edge effect at the upstream end, growing as a function of the average current density, to be identified. Thus, the chosen computing method can be considered appropriate for determining the current distribution.

Results with axial diffusion

Figures 4 and 5 show some computed current density and wall concentration distributions when taking into account downstream axial diffusion.

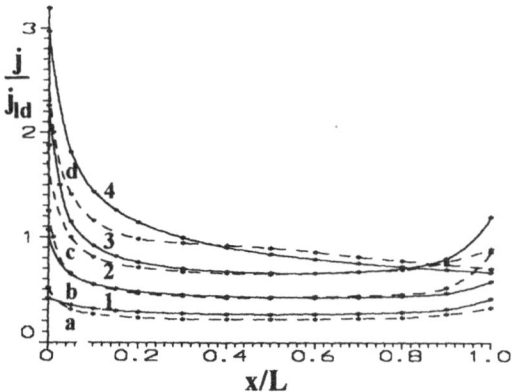

Figure 3. Results without axial diffusion - Current density profiles for high and low N values ($J\cong 0$, $\sigma = 0.04\ \Omega^{-1}.m^{-1}$ for N=100)

—— this work **1**: $j_{avg}/j_{ld}=0.3$ - **2**: $j_{avg}/j_{ld}=0.5$ - **3**: $j_{avg}/j_{ld}=0.8$ - **4**: $j_{avg}/j_{ld}=1$

– – – from Parrish and Newman[6,7] **a**: $j_{avg}/j_{ld}=0.25$ - **b**: $j_{avg}/j_{ld}=0.50$

c: $j_{avg}/j_{ld}=0.75$ - **d**: $j_{avg}/j_{ld}=0.95$

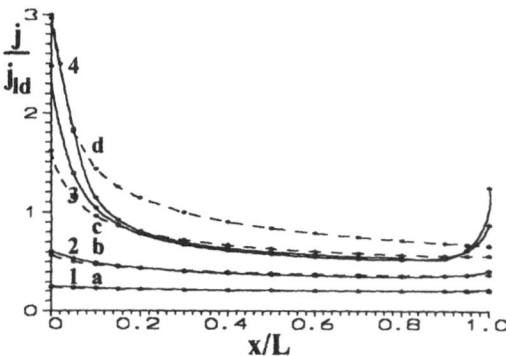

Figure 4. Comparison between current density profiles with and without axial diffusion for low N values ($J=0.01$, N=10, $\sigma = 0.4\ \Omega^{-1}.m^{-1}$)

—— with axial diffusion   **1**: $j_{avg}/j_{ld}=0.2$ - **2**: $j_{avg}/j_{ld}=0.4$

3: $j_{avg}/j_{ld}=0.7$ - **4**: $j_{avg}/j_{ld}=0.85$

– – – without axial diffusion   **a**: $j_{avg}/j_{ld}=0.2$ - **b**: $j_{avg}/j_{ld}=0.4$

c: $j_{avg}/j_{ld}=0.7$ - **d**: $j_{avg}/j_{ld}=1$

241

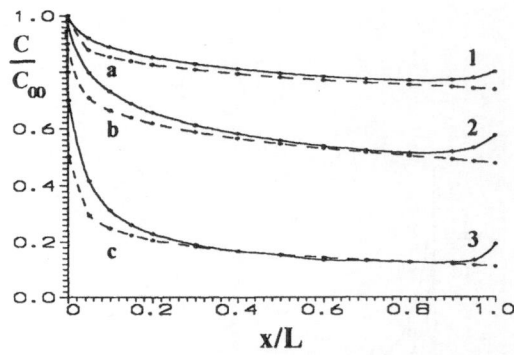

Figure 5. Comparison between concentration profiles with and without axial diffusion for low N values
(J=0.01, N=10, σ= 0.4 Ω⁻¹.m⁻¹)
——— with axial diffusion **1**: j_{avg}/j_{ld}=0.2 - **2** : j_{avg}/j_{ld}=0.4 - **3** : j_{avg}/j_{ld}=0.7 - **4:** j_{avg}/j_{ld}=1
– – – without axial diffusion **a** : j_{avg}/j_{ld}=0.2 - **b** : j_{avg}/j_{ld}=0.4 - **c** : j_{avg}/j_{ld}=0.7

Along the main part of the electrode length , these profiles appear to be in agreeement with those obtained by Parrish and Newman[7] without axial diffusion. However, when axial diffusion acts, a significant edge effect appears close to the downstream electrode end, showing a simultaneous increase of wall concentration and current density which is amplified when the cathodic current increases.

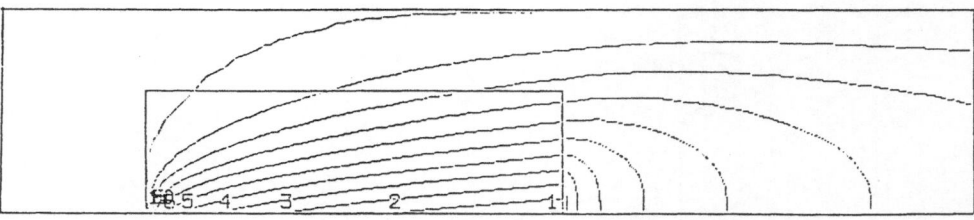

a) without axial diffusion (N=10, J=0.4, C_0/C_∞=0.56)
concentration(kg.m⁻³) 1=0.480 - 2=0.532 - 3=0.584 - 4=0.635 - 5=0.687
- 6=0.739 - 7=0.790 - 8=0.842 - 9=0.894 -10=0.945 - 11=0.997

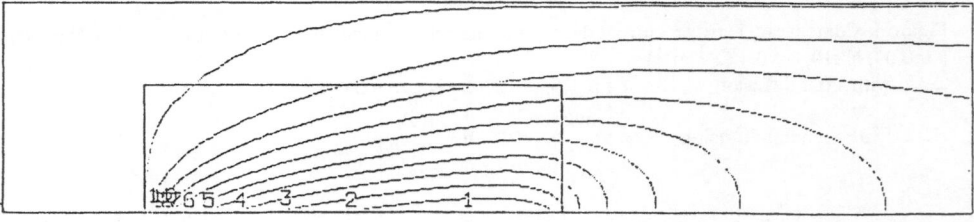

b) with axial diffusion (N=10, J=0.4, C_0/C_∞=0.60)
concentration(kg.m⁻³) 1=0.515 - 2=0.563 - 3=0.611 - 4=0.659 - 5=0.707
- 6=0.755 - 7=0.803 - 8=0.851 - 9=0.899 -10=0.947 - 11=0.995

Figure 6. Concentration distributions in the electrolyte

242

An analogous edge effect was pointed out by Ling[12] for a similar thermal problem concerning the heat transfer from a small isothermal spanwise strip on an insulated boundary. Indeed it was shown that in some conditions the stream-wise conduction term cannot be neglected in accurate predictions of the thermal behaviour at the trailing edge.

Physically the results observed without axial diffusion are due to a diminution of the mass flux because of the gradual thickening of the boundary diffusion layer even downstream of the region of interest.

For the geometry under study, the end of the consumption zone (x > L) causes a break in thickening which induces concentration and flux increase. Figures 6 show the concentration distribution in the electrolyte with and without considerations of downstream axial diffusion. It is it clear that neglecting axial diffusion leads to a doubtful boundary condition at the electrode trailing edge, preventing the prediction of a wall concentration increase which is quite possible in practice.

CONCLUSIONS

This work demonstrates the effect of axial diffusion on the current distribution for systems characterised by low Peclet numbers. In further work, axial diffusion at the upstream end of a plane electrode will be taken into account and larger systems having more complex geometries will be studied. A specific methodology is being developed based on a suitable adjustment of the finite elements mesh inside the boundary layers. Nevertheless complex systems will require a decomposition into sub-systems, equations being treated in some of them in an uncoupled way.

Acknowledgements: Thanks to ANVAR, ANRT, MRT and ADEME for their financial support

NOTATION

C	local concentration of the electroactive species (mole.m^{-3})
C_0	concentration of the electroactive species at the cathode surface (mole.m^{-3})
C_∞	concentration of the electroactive species in the bulk (mole.m^{-3})
D	ionic diffusivity of the electroactive species (m^2.s^{-1})
F	Faraday's constant (96,487 C.mol^{-1})
j	local working current density (A.m^{-2})
j_0	exchange current density (A.m^{-2})
j_{avg}	average working current density (A.m^{-2})
j_m	migration current density (A.m^{-2})
j_{ld}	average limiting current density (A.m^{-2})
J	dimensionless exchange current density (see eqn 16)
L	electrode length (m)
N	dimensionless average limiting current density (see eqn 16)
N_i	mass flux density for the electroactive species i (mol.m^{-2}.s^{-1})
Pe	Peclet number [= $S_0 L^2/D$]
S_0	characteristic velocity gradient (s^{-1})
T	temperature (K)
t_+	transport number of the electroactive species
V	electrolyte velocity (m.s^{-1})
z	number of electrons per reactant ion

Greek symbols

α projective function

β cathodic transfer coefficient

δ dimensionless average current density (see eqn 16)

η overall overpotential (V)

σ electrolyte conductivity ($\Omega^{-1}.m^{-1}$)

Φ local electric potential (V)

Φ_a anodic potential (V)

Φ_c cathodic potential (V)

REFERENCES

1. K. Asada, F. Hine, S. Yoshikawa and S. Okada, Mass transfer and current distribution under free convection conditions, *J. Electrochem. Soc.* 107:242 (1960).
2. J. Newman, Resistance for flow of current to a disk, *J. Electrochem. Soc.* 113:501 (1966).
3. J. Newman, Current distribution on a rotating disk below the limiting current, *J. Electrochem. Soc.* 113:1235 (1966).
4. R.V. Homsy and J. Newman, Current distribution on a plane below a rotating disk, *J. Electrochem. Soc.* 121:1448 (1974).
5. P. Pierini, P. Appel and J. Newman, Current distribution on a disk electrode for redox reactions, *J. Electrochem. Soc.* 123:366 (1976).
6. W.R. Parrish and J. Newman, Current distribution on a plane electrode below the limiting current, *J. Electrochem. Soc.* 116:169 (1969).
7. W.R. Parrish and J. Newman, Current distributions on plane, parallel electrodes in channel flow, *J. Electrochem. Soc.* 117:43 (1969).
8. A. Katagiri, Calculation of steady-state distributions of concentrations and potential controlled by diffusion and migration of ions, *J. Appl. Electrochem.* 21:487 (1991).
9. R. Alkire and A.A. Mirarefi, The current distribution within tubular electrodes under laminar flow, *J. Electrochem. Soc.* 120:1507 (1973).
10. K. Viswanathan and D.T. Chin, Current distribution on a continuous moving sheet electrode, *J. Electrochem. Soc.* 124:709 (1977).
11. R. Sautebin and R. Landolt, Anodic leveling under secondary and tertiary current distribution conditions, *J. Electrochem. Soc.* 129:946 (1982).
12. S.C. Ling, Heat transfer from a small isothermal spanwise strip on an insulated boundary, *J. of Heat Transfer, Trans.ASME.* 230 (1963).

A NEW APPROACH FOR SOLVING MASS AND CHARGE TRANSPORT IN ELECTROCHEMICAL SYSTEMS

Leslie Bortels and Johan Deconinck

Department of Electrical Engineering
Vrije Universiteit Brussel
Pleinlaan 2
1050 Brussel (Belgium)

ABSTRACT

This paper presents a numerical method for the calculation of concentration, potential and current distributions in electrochemical cells controlled by diffusion, convection and migration of ions. The influence of migration effects gives rise to a system of coupled non-linear partial differential equations which describes the transport of mass and charge. A multi-dimensional upwinding method (MDUM) has been adapted in order to solve this non-linear system. The developed model can deal with two dimensional electrochemical cells at steady state involving multiple ions. The electrolyte solutions are considered to be dilute and at a constant temperature. Examples are given for solutions with 3 ions in one and two dimensional geometries and with imposed current density on the electrodes.

INTRODUCTION

Starting from a dilute solution of i species in an unionised solvent at constant pressure and temperature, one can write down six equations describing transport of mass and charge[1,2,3]. The flux of each dissolved species i due to convection, diffusion and migration is given below:

$$\overline{N}_i = -z_i u_i F c_i \overline{\nabla} U - D_i \overline{\nabla} c_i + c_i \overline{v} , \tag{1}$$

with c_i the concentration, $\overline{v}$ the velocity of the solvent, z_i the charge, u_i the mobility, U the potential, D_i the diffusion coefficient and F the Faraday constant. For each species i one can state that at each point in the solution, the change of concentration is equal to the net input plus the local production due to chemical reactions:

$$\frac{\partial c_i}{\partial t} = -\overline{\nabla}.\overline{N}_i + R_i , \tag{2}$$

with:

$\overline{\nabla}.\overline{N}_i$: the divergence of the flux vector,

R_i : the production rate of an ion (positive or negative) due to homogeneous reactions in the bulk of the solution.

In an electrolyte, the conductivity is so large that free charges do not exist. Equation (3) is not valid very close to the electrodes where the double layer exists. The phenomena taking place in the double layer will be encompassed in the non-linear boundary conditions. Hence electroneutrality is expressed as follows:

$$\sum_i z_i c_i = 0. \tag{3}$$

Combination of equations (1) to (3), together with some mathematical manipulations leads to the following well known system of equations describing transport of mass and charge in dilute electrochemical solutions:

$$\frac{\partial c_i}{\partial t} + \overline{v}.\overline{\nabla}c_i = z_i F\overline{\nabla}.(u_i c_i \overline{\nabla}U) + \overline{\nabla}.(D_i \overline{\nabla}c_i) + R_i, \tag{4}$$

$$\sum_i z_i c_i = 0. \tag{5}$$

This set contains $i+1$ equations which, after solution, will give the $i+1$ unknowns, c_i and U. Even in the case of an infinite dilute electrolyte, a complex system of coupled partial differential equations has to be solved.

The velocity field $\overline{v}$ has first to be determined by solving the continuity and the Navier-Stokes equation. This can be done since the fluid dynamic equations are not coupled with the electrochemical ones in the case of an infinitely dilute electrolyte. Therefore the velocity in (4) is considered to be known. Furthermore it is supposed that the reactions are restricted to the electrodes and so R_i in equation (4) equals zero.

THE MULTIDIMENSIONAL UPWINDING METHOD (MDUM)

General

The Multidimensional Upwinding Method (also referred to as the Fluctuation Splitting Method) has been developed in the field of fluid dynamics. It is an alternative approach to both finite element methods and finite volume methods[4,5]. This method has been extended to the equations describing transport of mass and charge in dilute electrochemical systems.

MDUM for the Transport of Mass

In order to simplify the notation the mass transport equation (4) will be written in terms of the unknown concentration c. Introduction of the constant $K = zFu$ yields:

$$\frac{\partial c}{\partial t} = \overline{\nabla}.(D\overline{\nabla}c) - \overline{v}.\overline{\nabla}c + K\overline{\nabla}.(c\overline{\nabla}U) . \tag{6}$$

In order to reduce the number of unknowns, the two-dimensional domain Ω on which this equation holds is subdivided into triangles. On each triangle T, a linear approximation for both the potential U and the concentration c is chosen. The unknowns U and c are placed in the vertices of the triangles. There are in total N nodes and therefore $N(n+1)$ unknowns, for an electrolyte with n species. The integral of the time dependent equation (6) on each element, yields:

$$\int_T \frac{\partial c}{\partial t}\, dS = \Phi_{T,d} + \Phi_{T,c} + \Phi_{T,m}$$

$$= \int_T \overline{\nabla}.\left(D\overline{\nabla}c\right) dS + \int_T -\overline{v}.\overline{\nabla}c\, dS + \int_T K\overline{\nabla}.\left(c\overline{\nabla}U\right) dS \,, \tag{7}$$

with $\Phi_{T,d}$, $\Phi_{T,c}$ and $\Phi_{T,m}$ the "fluctuations" due to diffusion, convection and migration respectively. These fluctuations give rise to changes in the concentrations in the nodes. The time dependent left-hand side of equation (7) is replaced by a first order approximation. Suppose that the fluctuation on an element, given by the right-hand side of (7), is only distributed to its own vertices. This is expressed in the following way:

$$\Phi_{T,d} = \sum_{j=1}^{3} \alpha_j^{T,d}\Phi_{T,d}$$

$$\Phi_{T,c} = \sum_{j=1}^{3} \alpha_j^{T,c}\Phi_{T,c} \tag{8}$$

$$\Phi_{T,m} = \sum_{j=1}^{3} \alpha_j^{T,m}\Phi_{T,m}$$

with the following relation between the 3 α_j^T of each element:

$$\sum_{j=1}^{3}\alpha_j^{T,d} = \sum_{j=1}^{3}\alpha_j^{T,c} = \sum_{j=1}^{3}\alpha_j^{T,m} = 1\,, \forall\, T \in \Omega\,.$$

Combination of equations (7) and (8) results in:

$$c_i^{n+1} = c_i^n + \frac{\Delta t}{S_i}\sum_T \left\{\alpha_i^{T,d}\Phi_{T,d} + \alpha_i^{T,c}\Phi_{T,c} + \alpha_i^{T,m}\Phi_{T,m}\right\}.$$

The α_i^T are called the distribution coefficients; they determine in which way the fluctuation is distributed among the nodes. c_i^{n+1} is the concentration in node i at time step n+1, Δt the time step and S_i the dual mesh of node i. The dual mesh is one third of the area of all triangles who have the vertex i in common.

The idea behind MDUM is to chose the α_i^T in such a way that the fluctuation on each element is distributed among its own nodes with respect to the physical properties of the fluctuation sources. In other words the α_i^T are different for diffusion, convection and migration.

MDUM for the Diffusion Term

Since diffusion is isotropic and $\overline{\nabla}c$ is constant on an element, it is obvious to take the fluctuation coefficients α_i^T equal to 1/3. This means that the fluctuation on an element due to diffusion is equally distributed to its 3 vertices. The change in concentration can then be computed for each node in the domain. This results in the following system of equations:

$$\{\Delta C, d\} = D\,\overline{\overline{A}}\left\{C^n\right\}\,, \tag{9}$$

with:

$\{\Delta C, d\}$: the vector of the fluctuation in each node due to diffusion,

$\overline{\overline{A}}$: the matrix for diffusion,

$\{C^n\}$: the vector of the unknown concentration at time step n.

MDUM for the Convection Term

The velocity field causes a change in concentration in the direction of this field. The fluctuation coefficients α_i^T are therefore strongly depending on the direction of the velocity. Figure 1 gives an idea of the fluctuation due to convection. For the element on the left side (2 inflow sides) points 1 and 2 are updated. The element with the single inflow side distributes its total fluctuation to point 3.

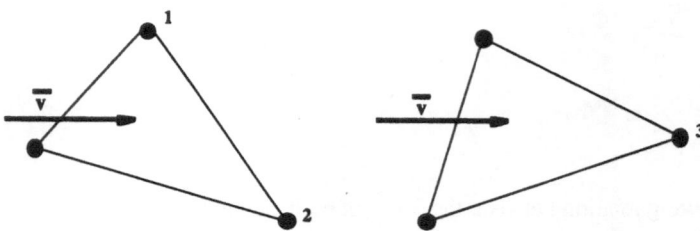

Figure 1. Fluctuation due to convection. Case of a double inflow side (left) and a single one (right).

More detailed information on the determination of the α_i^T can be found. As in the case of diffusion the following equation is obtained:

$$\{\Delta C, c\} = \overline{\overline{B}} \{C^n\}, \tag{10}$$

with:

$\{\Delta C, c\}$: the vector of the fluctuation in each node due to convection,

$\overline{\overline{B}}$: the matrix for convection depending on the local velocity field.

MDUM for the Migration Term

The case of migration is more complicated because of the appearance of both potential and concentration. It was decided to take all α_i^T equal to 1/3. In doing so a vector expression is obtained for the fluctuation due to migration in which the matrix depends on the concentration:

$$\{\Delta C, m\} = K \overline{\overline{E}}(c^n) \{U^n\}, \tag{11}$$

with:

$\{\Delta C, m\}$: the vector of the fluctuation in each node due to migration,

$\overline{\overline{E}}(c^n)$: the matrix for migration (depending on c),

$\{U^n\}$: the vector of the potential at time step n.

The MDUM equivalent for the transport of mass, obtained after putting equations (9) to (11) together, is then given by:

$$\left\{C^{n+1}\right\} = \left\{C^n\right\} + \frac{\Delta t}{S^*}\left[D\overline{\overline{A}}\left\{C^n\right\} + \overline{\overline{B}}\left\{C^n\right\} + K\overline{\overline{E}}(c^n)\left\{U^n\right\}\right] . \tag{12}$$

A coherent scheme for the concentration in the nodes is now available. This scheme contains the unknown potentials which are to be derived from the equation expressing conservation of charge.

MDUM for the Conservation of Charge

The MDUM equivalent for the equation describing the transport of mass for ion i - given by (4) and with $R_i = 0$ - is:

$$\left\{C_i^{n+1}\right\} = \left\{C_i^n\right\} + \frac{\Delta t}{S^*}\left[\left(D_i\overline{\overline{A}} + \overline{\overline{B}}\right)\left\{C_i^n\right\} + K_i\overline{\overline{E}}(c_i^n)\left\{U^n\right\}\right] , \tag{13}$$

At each time step the concentrations c_i in the nodes must be such that conservation of charge, given by equation (3), is obeyed. To obtain this, one replaces in (13) the vector for the potential at time step n $\left\{U^n\right\}$ by $\left\{U^{n+1}\right\}$ which is unknown at the moment. Equation (3) has to be discretised to obtain N new equations for the determination of the new potential. This yields the following expression:

$$\left\{\sum_i z_i C_i^{n+1}\right\} = \left\{\sum_i z_i C_i^n\right\} + \left\{\sum_i z_i \Delta C_i^n\right\} = 0. \tag{14}$$

Equations (13) and (14) give the following expression for the potential:

$$\left[\sum_i z_i \frac{\Delta t}{S^*} K_i \overline{\overline{E}}(c_i^n)\right]\left\{U^{n+1}\right\} = \\ -\sum_i z_i \frac{\Delta t}{S^*}\left(D_i\overline{\overline{A}} + \overline{\overline{B}}\right)\left\{C_i^n\right\} - \sum_i z_i\left\{C_i^n\right\} \tag{15}$$

which completes the set of equations.

The problem to be solved is now described by means of equations (13) and (15). Starting from an initial concentration field for the n ions that obeys the conservation of charge one can find the potential distribution with equation (15). With this potential and the given concentrations the new concentrations can be computed using equation (13). This procedure is repeated until the steady state solution is obtained.

APPLICATIONS

With the developed code the concentration and the potential in electrochemical systems with multiple ions can be computed. In this paper the geometry consists of a rectangular cell with 2 insulating walls and 2 electrodes with imposed current. The electrolyte contained 3 ions, one reacting ion (ion 1) and two non-reacting ions (ions 2 and 3). The boundary conditions on the insulators are given by:

$$\overline{\nabla}c_i.\overline{l}_n = 0 \ , \ \forall i$$
$$\overline{\nabla}U.\overline{l}_n = 0 \tag{16}$$

with $\overline{l}_n$ the normal to the boundary.
On the electrodes the boundary conditions for the non-reacting ions are given by:

$$\overline{N}_2.\overline{l}_n = 0$$
$$\overline{N}_3.\overline{l}_n = 0 \tag{17}$$

The flux of the reacting ion is given by:

$$\overline{N}_1.\overline{l}_n = \frac{\overline{J}_{imposed}.\overline{l}_n}{z_1 F} \tag{18}$$

The constants for the involved ions are:

$$D_1 = D_2 = D_3 = K_1 = K_2 = z_1 = z_2 = 1,$$

$$K_3 = z_3 = -1 \ . \tag{19}$$

These constants have no physical meaning. However, in the one dimensional case, they lead to an analytical solution which is the best tool to verify the numerical values.

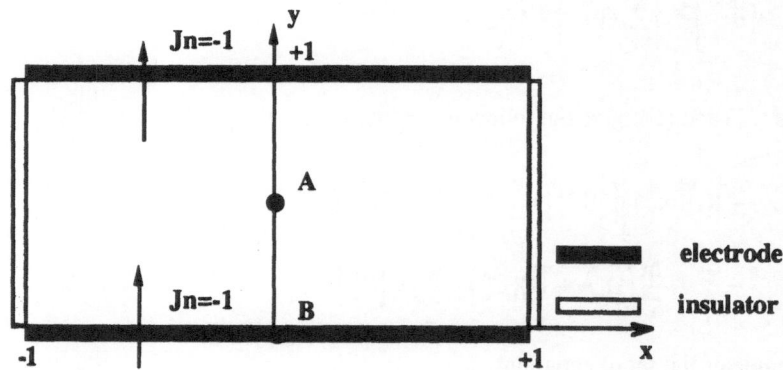

Figure 2. The geometry for the one dimensional problem.

The One Dimensional Problem - Diffusion and Migration

Considering the geometry of figure 2, the current density on the electrodes is imposed as well as the concentration in point A ($c_1 = c_2 = 1$, $c_3 = 2$). A fixed reference potential of zero value is placed in point B.

With the given boundary conditions and considering only diffusion and migration, equations (4) and (5) result in a one dimensional set of equations with solution[5]:

$$c_1(y) = \frac{2y+7}{4} - \frac{8}{2y+7} \quad , \quad c_2(y) = \frac{8}{2y+7} \quad ,$$

$$c_3(y) = \frac{2y+7}{4} \qquad , \quad U(y) = \ln\left(1 + \frac{2y}{7}\right) . \tag{20}$$

The test has been done on a grid with only 22 nodes. Because of the one dimensional properties of the domain only values for x equal to 0 and 1 were taken. The results of the one dimensional case are given in table 1. They are in very good agreement with the theoretical values.

Table 1. Results for the one dimensional case. Exact and numerical values are listed.

y	c1,exact	c1,num.	c2,exact	c2,num.	c3,exact	c3,num.	U,exact	U,num.
0	0.6071	0.6073	1.1429	1.1427	1.7500	1.7500	0.0000	0.0000
0.2	0.7689	0.7691	1.0811	1.0808	1.8500	1.8500	0.0556	0.0556
0.4	0.9244	0.9245	1.0256	1.0255	1.9500	1.9500	0.1082	0.1082
0.6	1.0744	1.0746	0.9756	0.9754	2.0500	2.0500	0.1582	0.1582
0.8	1.2198	1.2199	0.9302	0.9301	2.1500	2.1500	0.2059	0.2058
1.0	1.3611	1.3612	0.8889	0.8888	2.2500	2.2500	0.2513	0.2513

The Two Dimensional Problem

Figure 3 presents the geometry and boundary conditions for the two dimensional case. The test was performed on a regular grid with 153 nodes (17 in the x-direction, 9 in the y-direction). With the obtained numerical values, isolines were drawn for the potential U, and the concentrations c1 and c2.

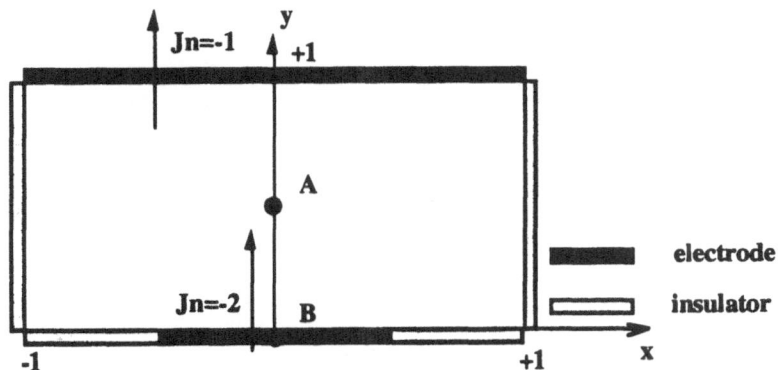

Figure 3. The geometry for the two dimensional problem.

The results in the diffusion-migration case are given in figure 4. The symmetry around the y-axis is visible. It can be clearly seen that the concentration of the reacting ion (ion 1 with charge +1) is small at the lower electrode (U = 0 Volt) and high at the upper electrode (U = +0.35 Volt). This potential is higher than in the one dimensional problem (U = +0.25 Volt). For the non-reacting positive ion (ion 2), the concentration increases from top to bottom. These results are in full agreement with the expectations.

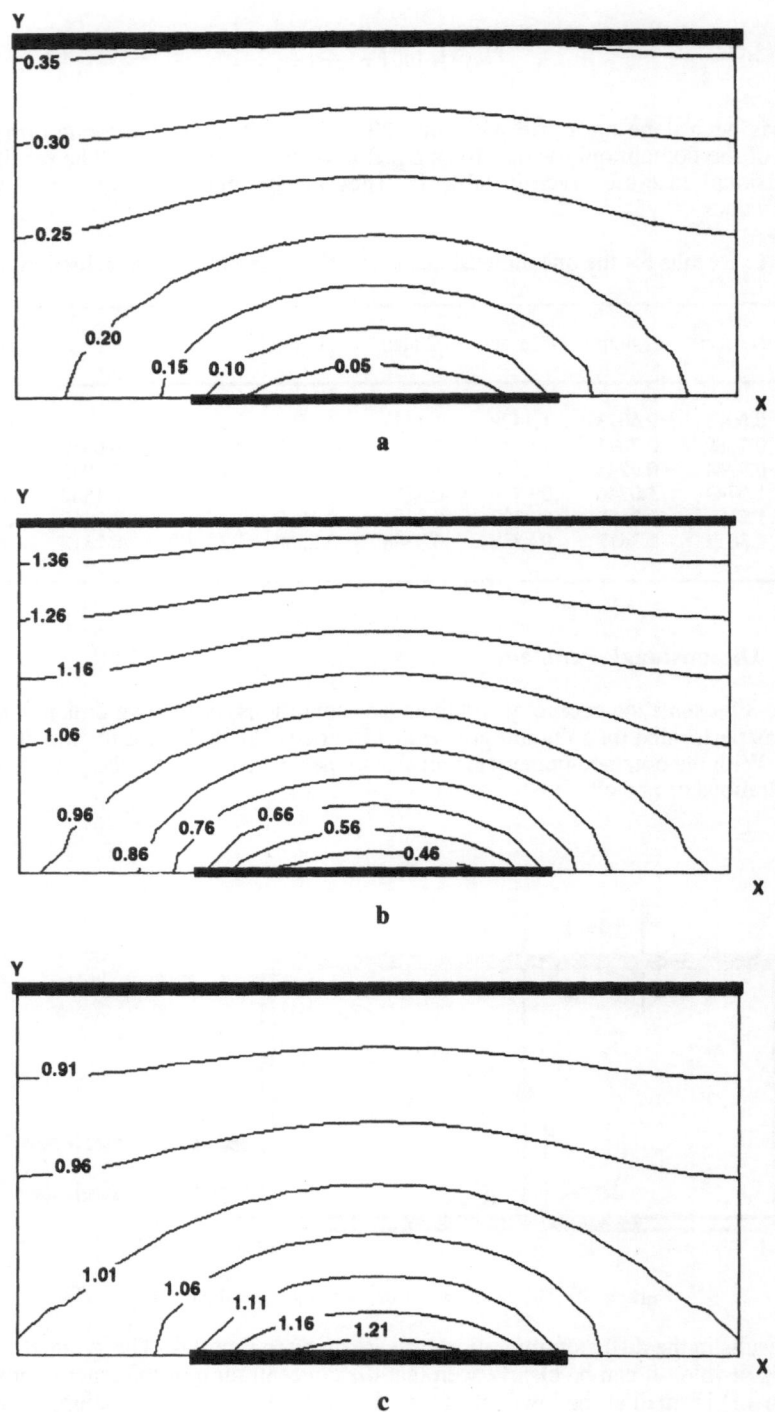

Figure 4. Two dimensional problem with diffusion and migration.
Isolines for the potential (a), concentration c1 (b) and concentration c2 (c) are shown.

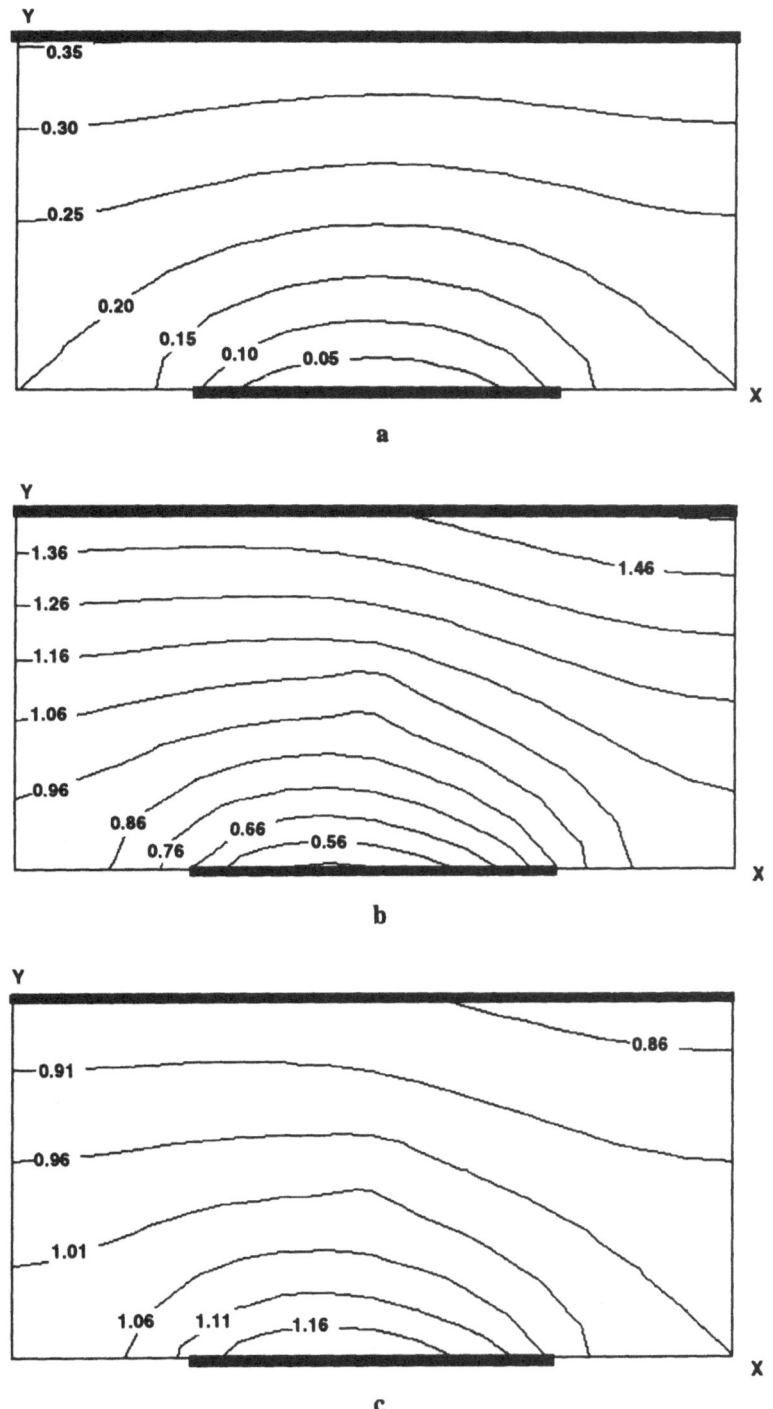

Figure 5. Two dimensional problem with diffusion, convection and migration.
Isolines for the potential (a), concentration c1 (b) and concentration c2 (c) are shown.

The influence of a velocity field along the x-axis ($\bar{v} = -\bar{1}_x$) has been investigated. Results are shown in figure 5. Almost the same values as in the above case have been found. The curvature of the concentration isolines can be explained by examining the sign of the dot product $\bar{v}.\bar{\nabla}c_i$ which appears in equation (4).

CONCLUSIONS

A numerical method has been proposed for the calculation of concentration, potential and current distributions controlled by diffusion, convection and migration. The method can deal with multiple ions and non-linear boundary conditions. Some simple geometries have been investigated. In cases where an analytical solution was available, the results were very good. In the other cases the results were in accordance with the physical expectations. It is believed that future extension to three-dimensional and axisymmetrical problems will not be too difficult. The next step in the development of the numerical model will implement more realistic problems rather than academic ones.

ACKNOWLEDGEMENT

This research was sponsored by the European Community under Brite-Euram project nr BE-5187, contract number BRE2-CT92-0170.

REFERENCES

1. J. Newman. "Electrochemical Systems", Prentice Hall. Inc., Englewood Cliffs, N.J. (1973).
2. J. Deconinck, Stroomverdelingen en ontwerp van elektrochemische reactoren, *ATB Metallurgie XXIX*, n° 3-4, (1989).
3. F. Balanck, J. Deconinck, Het wel en wee van de concentratie-overpotentiaal, Internal report, (1991).
4. R. Struijs, H. Deconinck, P.L. Roe, Fluctuation splitting schemes for the 2D Euler equations, von Karman Institute for Fluid Dynamics, *Lecture Series* 1991-01.
5. H. Deconinck, R. Struijs, G. Bourgois, H. Paillere, P.L. Roe, Multidimensional upwind methods for unstructured grids, *Paper n°4 AGARD* R-787, (1992).
6. A. Katagiri, Calculation of steady-state distributions of concentrations and potential controlled by diffusion and migration of ions, *J. Appl. Electrochem.* 21:487-495 , (1991) .

TOWARDS A CLEANER ENVIRONMENT USING ELECTROCHEMICAL TECHNIQUES

G. Kreysa and K. Jüttner

Karl-Winnacker-Institut der DECHEMA e.V.,
D-60061 Frankfurt am Main, Postfach 15 01 04

ABSTRACT

Electrochemical processes can contribute considerably to environmental protection by means of waste purification processes and production integrated waste minimization. As examples of waste purification techniques, electrochemical processes for gas purification and removal of toxic metal ions from waste water will be described. For production integrated processes, the examples given are fluidized bed electrolysis for metal recovery in cellulose acetate production, the membrane process for alkali chloride electrolysis, and the electroreduction of dichloroacetic acid.

INTRODUCTION

With respect to environmental protection, the inherent advantage of electrochemical processes lies in the fact that the removal of pollutant species by oxidation/reduction processes can be carried out directly with electrons as the reagent in an electrochemical cell without the addition of waste-producing redox chemicals. Considering electrochemical environmental protection measures in general, a distinction has to be made between two alternative concepts:

(i) **Electrochemical purification processes for treating waste**
    This includes all techniques where toxic compounds are removed from gases, liquids or even solids.

(ii) **Process-integrated environmental protection**
    This means substitution of waste-producing processes by cleaner electrochemical technologies producing little or no waste.

Besides these process-oriented benefits, electrochemistry provides considerable advantages in the field of electroanalytical techniques. Such techniques are suitable for automatic control and continuous monitoring of pollutants. These electroanalytical techniques, however, will not be considered here.

Electrochemical Engineering and Energy, Edited by
F. Lapicque *et al.*, Plenum Press, New York, 1995

ELECTROCHEMICAL PURIFICATION TECHNIQUES

Gas purification

Many gaseous pollutants, such as sulphur dioxide, nitrous oxides or chlorine, permit electrochemical conversion in an aqueous environment. For example, chlorine can be reduced to chloride, nitrous oxides can be oxidized to nitric acid and sulphur dioxide can be oxidized to sulphuric acid. An increasing demand for off-gas purification, particularly for smaller scale power plants, heating combustion units and chemical plants, thus encouraged the development of new concepts of gas purification based on electrochemical techniques. From the viewpoint of process engineering, alternative ways of carrying out an electrochemical gas purification process have to be considered [1].

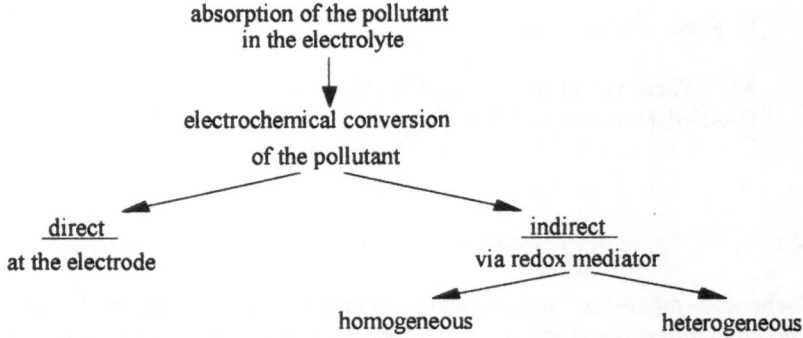

As shown schematically, the initial step consists of the absorption of the pollutant species into a liquid phase. Due to the low solubility of many gases in aqueous solutions, physical absorption must be supported by a reaction in the liquid phase which permanently converts the primarily dissolved species into a more soluble product, thus shifting the gas liquid equilibrium towards the liquid phase. In the simplest case this can be done directly by an electrochemical reaction at the electrode of an electrochemical cell.[2-4] The second possibility consists of an indirect process involving the use of a redox mediator to oxidize or reduce the dissolved species. The redox mediator, which may be present in a significantly higher concentration than the dissolved gas in the purging solution, has to be regenerated electrochemically in a consecutive step.

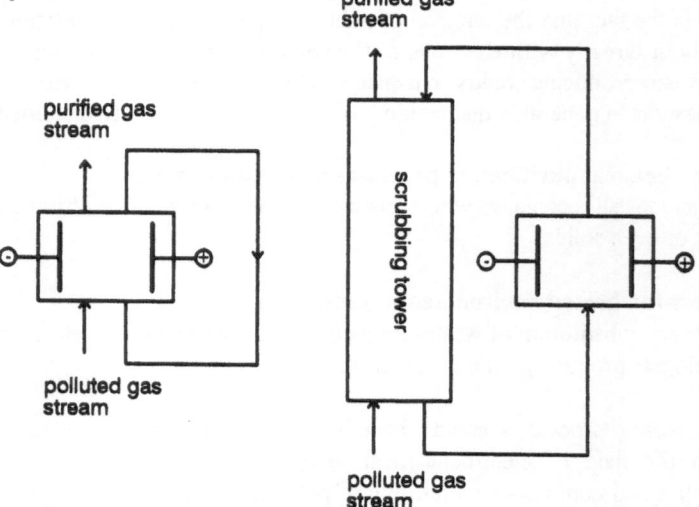

Fig. 1 Alternative concepts of electrochemical gas purification by inner cell and outer cell processes

Not only homogeneously dissolved reagents may serve as redox mediator systems, but also heterogeneous mediators, such as metal oxides, can be applied, which means that the separation of reaction products and the mediator system can be avoided. Both the electrochemical conversion of the pollutant and the electrochemical regeneration of the redox mediator can be achieved either by an inner cell or an outer cell process[5], as illustrated in Fig. 1.

For the direct electrochemical conversion of dissolved gases by an inner cell process a specially designed absorption column has been developed, consisting of a three-dimensional packed bed electrode of conducting particles in contact with a cylindrical feeder electrode and a counter electrode separated by an ion exchange membrane [2-4].

Fig. 2 shows schematically the construction of an electrochemical absorption column. Such a column, which usually operates in counter current flow conditions with respect to the gas phase and the liquid phase, exhibits a high space-time yield and has been successfully applied to the absorption of sulphur dioxide and chlorine [3].

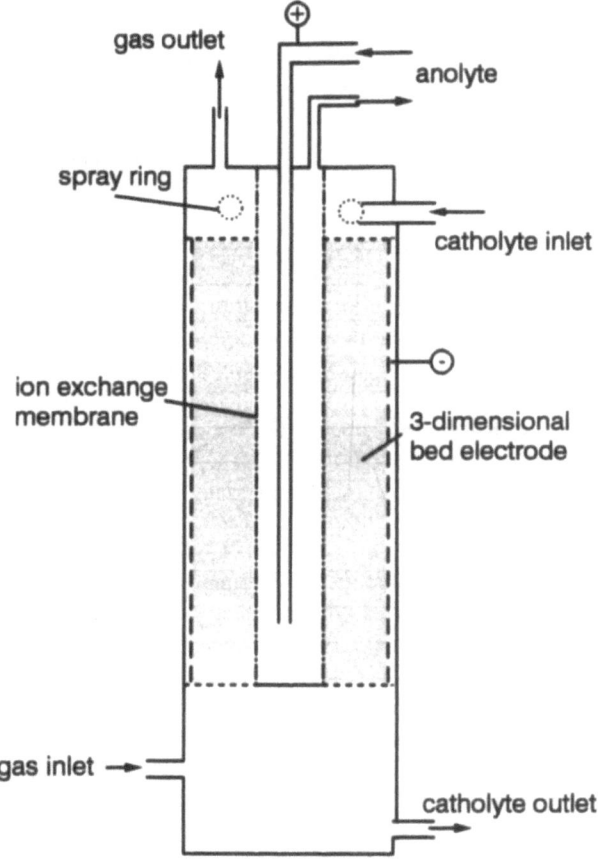

Fig. 2   Scheme of an electrochemical absorption column with 3-D particle bed electrode

Fig. 3 illustrates the efficiency of such a device by two examples, the oxidation of SO_2 to sulphuric acid and the reduction of Cl_2 to hydrochloric acid. Initially the absorption column works without a current and the outlet concentration of the gas approaches the inlet concentration asymptotically. When the current is switched on, the outlet gas concentration immediately drops and maintains a lower level. In these experiments the packed bed consisted of graphite spheres of 0.5 or 1.0 cm diameter.[3]

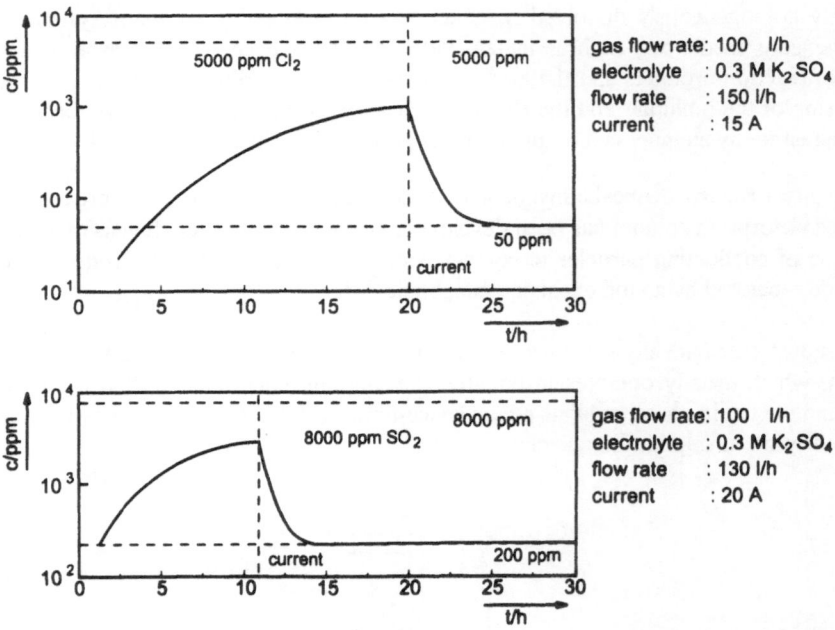

Fig. 3   Gas outlet concentrations of an electrochemical absorption column with graphite spheres

Examples of indirect electrochemical gas purification with homogeneous redox mediators are the modified "Mark 13 A" process developed at the Joint European Research Centre at Ispra [6], which uses bromine for the indirect oxidation of SO_2, and the so-called "Peracidox" process developed by Lurgi [7] in which peroxodisulphate is employed as the redox mediator for the oxidation of SO_2. In both processes the electrochemical regeneration of the redox mediator is performed by an outer cell process.

Examples of gas purification processes with heterogeneous redox mediators are the lead dioxide catalysed oxidation of SO_2, suggested by Strafelda [8], and the copper catalysed process for the combined electrochemical and catalytic removal of SO_2 with Cu_2O/Cu^{2+} as catalyst [9].

Since flue gases not only contain SO_2, but also NO_x to a certain extent, the development of a new process to remove both components simultaneously has been the subject of a recent joint European BRITE/EURAM project with partners from industry and fundamental research [10]. As a result two different processes have been developed and tested under semi-technical conditions. These are the indirect **Cerium (IV)-Assisted Process** as a typical outer cell process for the oxidation of SO_2 and NO_x to sulphuric acid and nitric acid by using Ce^{4+}/Ce^{3+} as the redox mediator system, and secondly the **Lead Dioxide Dithionite Process (LDP)**, combining an outer cell with dithionite as the homogeneous redox mediator for the chemical reduction of NO_x and an inner cell process with lead dioxide as the heterogeneous electrocatalyst for the electrochemical oxidation of SO_2.[11,12]

A simplified scheme of the **LDP** process is shown in Fig. 4. In the first step, the inlet gas mixture enters the absorption column (A) where NO, as the main component of nitrous oxide flue gases, is chemically reduced by the dithionite, $S_2O_4^{2-}$. The second component, SO_2, passes through the absorption column (A) without reaction and enters the anode compartment (B) of the electrochemical cell, where it is electrochemically oxidized to sulphuric acid at the anode of a lead dioxide packed bed electrode. At the cathode (C) of the electrochemical cell the regeneration of dithionite by reduction of SO_3^{2-} can be carried out simultaneously and the dithionite is recycled into the absorption column (A). Concentrated sulphuric acid is formed in the anodic process as a valuable by-product.[10-12]

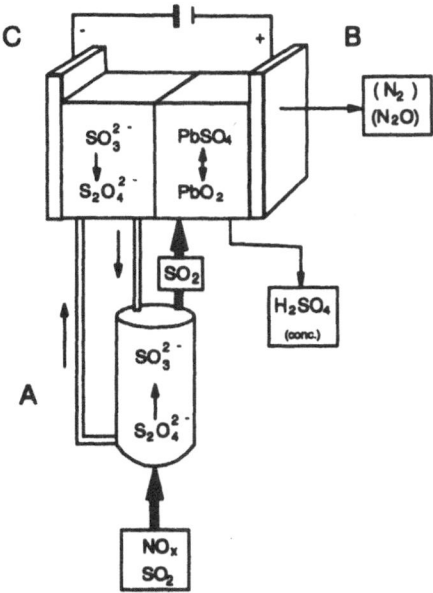

Fig. 4 The Lead dioxide Dithionite Process LDP for simultaneous removal of SO_2 and NO_x

Microkinetic studies of NO_x absorption show that the degree of NO conversion with dithionite is below 35% but, by adding a complexing agent, such as Fe(II)EDTA, the degree of conversion can be significantly improved as demonstrated by the experiment in Fig. 5.[12]

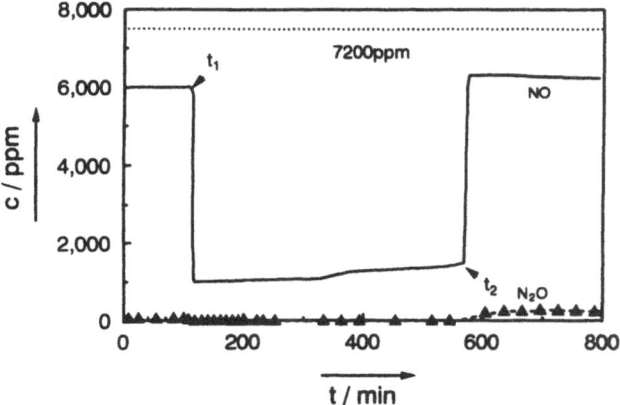

Fig. 5 Effect of Fe(II)EDTA (1mM) on the absorption of NO with dithionite (0.16 M $S_2O_4^{2-}$) ; pH 6.3; gas flow rate 3.9 l min^{-1}, NO inlet concentration 7200 ppm, addition of Fe(II)EDTA at t_1

On the addition of Fe(II)EDTA at t_1 in Fig. 5, the partial pressure of NO at the outlet is drastically decreased and remains at a lower level as long as dithionite is present in the absorption column. Analysis of the gas phase composition indicated that, under these conditions, the formation of N_2 is unfortunately negligibly small. Solution analysis by ion chromatography indicates that, in addition to SO_3^{2-}, several reaction products, such as NH_4^+ and amido-sulphonic acid $H_2N \cdot SO_3H$ can be identified, as shown in Fig. 6. Therefore, an optimization of this process requires further investigations into the complex chemistry of the nitrogen oxides. The regeneration of dithionite by cathodic reduction of the reaction product SO_3^{2-} has been performed at lead electrodes with a current efficiency of 0.8 in solutions of $4 < pH < 7$.[12]

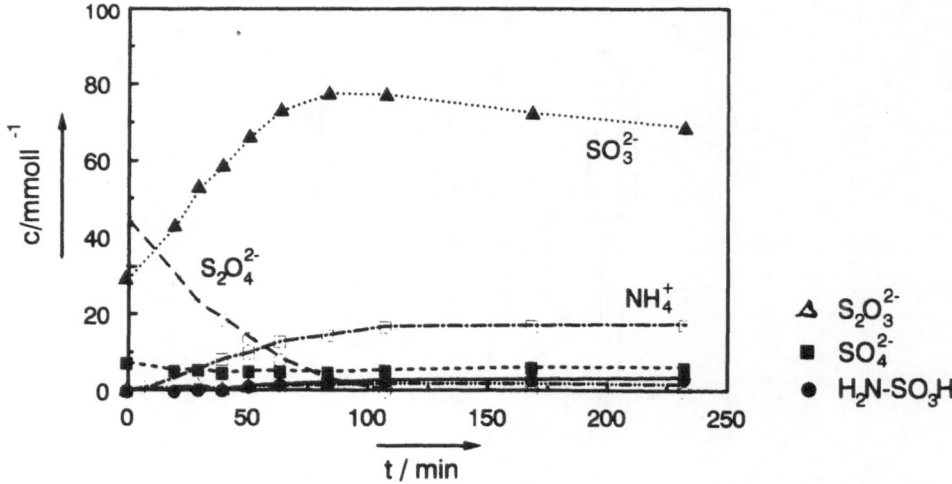

Fig. 6 Solution analysis of reaction products formed during NO absorption with dithionite in the presence of Fe(II)EDTA; (experimental conditions see Fig. 5)

Microkinetic studies of the anodic oxidation of sulphur dioxide on a lead rotating disk electrode (RDE) showed that a distinction has to be made between a direct and an indirect reaction mechanism. The direct oxidation takes place at the PbO_2 on the lead surface, which has good electronic conductivity, according to the overall reaction [12,13]

$$SO_3^{2-} + H_2O \xrightarrow{k_c} SO_4^{2-} + 2H^+ + 2e^- \qquad (1)$$

whereas the indirect process involving the heterogeneous chemical reduction of PbO_2 follows the reaction sequence

$$SO_3^{2-} + PbO_2 + 2H^+ \xrightarrow{k_1} PbSO_4 + H_2O \qquad (2)$$

$$PbSO_4 + 2H_2O \xleftrightarrow{k_2} PbO_2 + SO_4^{2-} + 4H^+ + 2e^- \qquad (3)$$

Correspondingly, the overall current density is a sum of two contributions

$$i = 2Fk_c\,c_s^*(1-\theta) + 2Fk_1^\circ\,c_s^*(1-\theta)\exp\left[\frac{\alpha_1 F}{RT}(E-E_o^1)\right] \qquad (4)$$

where c_s^* is the concentration of SO_3^{2-} at the electrode surface, θ is the degree of coverage of $PbSO_4$ and $E_o^1 = -0.108$ V is the standard potential of reaction (1). If reaction (3) is considered to be fast and in equilibrium, the degree of coverage θ can be expressed in terms of a Langmuir absorption expression. The concentration c_s^* can be replaced by the bulk concentration c_s and the diffusion limiting current density i_m according to

$$c_s^* = c_s\left(1 - \frac{i}{i_m}\right) \qquad (5)$$

and
$$i_m = 0.6 \, \nu_e F c_s D^{2/3} \nu^{-1/6} \omega^{1/2}$$
(6)

for the RDE. The overall current density i then becomes a function of the rate constants k_c, k_1^0, the equilibrium constant K_2 of reaction (3), the diffusion coefficient D of SO_3^{2-} and the charge transfer coefficient α_1. Fig. 7 shows an optimum fit of the experimental current density potential curves obtained at different SO_3^{2-} concentrations.[13]

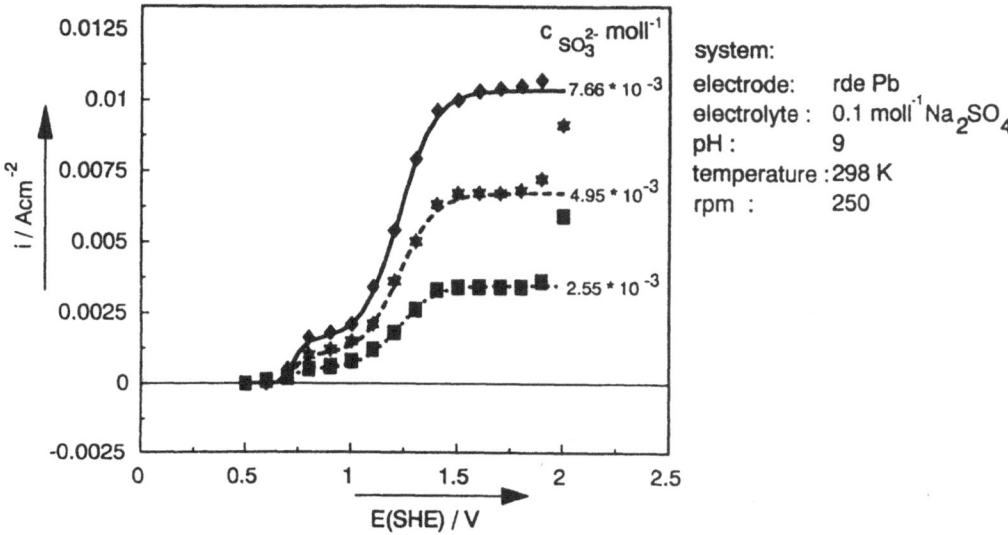

Fig. 7 Current density potential curves of anodic SO_3^{2-} oxidation on lead rotating disk electrodes in 0.1 M Na_2SO_4 at different SO_3^{2-} concentrations, pH 9, T = 298 K, $\omega = 26.2 \, s^{-1}$, experimental points and simulation (———) using eq(4).

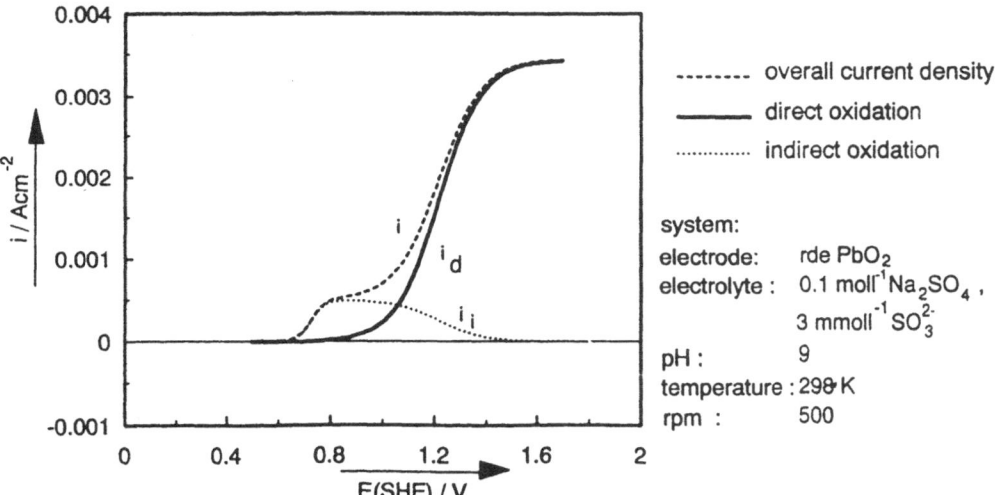

Fig. 8 Partial current densities corresponding to the direct and indirect oxidation of SO_3^{2-}

261

The partial current densities corresponding to the direct and the indirect reaction mechanism are shown in Fig. 8. At low electrode potentials the indirect mechanism predominates, whereas at higher potentials sulphite oxidation mainly takes place by the direct mechanism. On the basis of these results an electrochemical absorption column of 1.1 m height and 12 cm diameter was designed for the oxidation of SO_2. Parameter studies with respect to electrolyte concentration ($< 40\%$ H_2SO_4), liquid flow rate (< 0.25 m^3 h^{-1}), gas flow rate (< 1.2 m^3 h^{-1}) and SO_2 inlet concentration (< 4000 ppm) were carried out. The relevant technical process parameters are summarized in Table 1 together with the data for the NO conversion obtained in similar laboratory scale experiments.[10-13]

Table 1 Process parameters for the Lead dioxide Dithionite Process[10-13]

Process Parameter	SO_2	NO_x
degree of conversion	X = 94 - 97%	X = 93% - 98%
current effiency	Φ ≥ 0.88	Φ = 0.78
outlet concentration	≤ 150 ppm	≤ 150 ppm
gas residence time	τ = 2.7 s	τ = 7.1 s
normalized space velocity for X = 0.9	≥ 208 l/lh	≥ 200 l/lh
specific energy consumption (inlet concentration)	E = 0.0127 kWh/m^3 (c = 1750 ppm)	E = 0.018 kWh/m^3 (c = 500 ppm)

Waste water purification[14]

Waste water from industrial branches, such as electroplating, photographic development, printed circuit or battery production, requires special treatment for the removal of toxic metal ions. The maximum permissible concentrations of metals in effluents can no longer be obtained by the conventional technique of neutralization with metal hydroxide precipitation, Table 2.

Table 2 Effluent limits and metal concentrations obtained by hydroxide precipitation[15]

Metal	Effluent limits in ppm				Metal concentration after precipitation at pH 8 in ppm
	D ATV A115 Jan. 1983	D Abwasser-VWV Jan. 1990	USA 1983	CH 1983	
Pb	2	0.5	0.5	0.5	21
Cd	0.5	0.2	0.3	0.1	1500
Cu	2	0.5	0.5	0.5	1
Ni	3	0.5	0.5	2	340
Hg	0.05	0.05	-	0.01	-
Ag	2	0.1	-	0.1	-
Zn	5	2	0.5	2	2.6
Sn	5	2	-	2	-

This situation and the increasing cost of disposing toxic metal containing sludges has led to the development of new and cost efficient processes for the treatment of waste water. As an alternative to hydroxide precipitation, ion exchange techniques are gaining increasing importance.[16] However, these methods are often too costly for many types of effluents and regenerable ion exchange resins are not available for all metals. The precipitated heavy metal ions are replaced by alkali or alkaline earth metal ions and in many cases the resulting salt content prevents recycling of the water to the process. Since most of the metal ions can be removed by electrochemical deposition, electrochemical processes have been developed, some of which have been commercialized and are now being used in industry.[17-21] Since the concentrations of metal ions in waste water are usually small, the electrodeposition process is diffusion controlled. The electricity costs of removing such small quantities are quite low. More important are the specific investment costs which are determined by the space-time yield

$$\rho = a_e \frac{M \, \Phi^e \, i}{v_e \, F} \tag{7}$$

where i is the microkinetic current density, a_e the specific electrode area (A/V), Φ^e the current efficiency, M the molecular weight and v_e the charge of the metal ion to be deposited. Introducing the diffusion limiting current density $i_m = v_e \cdot F \cdot k_m \cdot c$ yields the key formula for the design of an electrochemical reactor for waste water purification

$$\rho = a_e \, M \, \Phi^e \, k_m \, c \tag{8}$$

showing that for a given metal ion concentration c a high mass transfer coefficient k_m and a large specific electrode area a_e are required. Efficient cell design has been directed towards optimizing this space-time yield, which means cell types with either enlarged specific electrode surface and/or improved mass transfer. A number of industrial cells which have been described in the literature [x] are listed in Table 3 and characterized with respect to these criteria.

Table 3 Different cell types developed for the purification of effluents

Cell type	References	improved mass transfer	enlarged electrode surface	Remarks
Pump cell	24,25	●	o	moving two-
Eco-Cell ®	26-30	●	●	dimensional
Beat rod cell	29-30	●	●	electrodes
Vibrating cells	31	●	o	
Multi-cathode cell	32	●	●	"semi-three-
Swiss-roll cell	33	●	●	dimensional"
ESE-Cell ®	34	o	●	structures
Porous flow-through cell	35-38	●	●	
Packed bed cell	39-41	●	●	
Fluidised bed cell	42-46	●	●	three-dimensional
Rolling tube cell	47-48	●	●	electrodes
Circulating bed	49	●	●	
Electrocatalytic reactor	50	●	●	currentless metal deposition

effects: o small   ● medium   ● strong

The space-time yield ρ is dependent on the inlet concentration and the degree of conversion and does not allow a comparison of reactor performances independent of the waste water properties. A normalized space velocity $\rho_s{}^n$ has therefor been suggested for the characterization of waste water cells [22,23].

The packed bed cells using graphite particles as three dimensional electrodes have been successfully commercialized, particularly for the treatment of waste water. According to the theory of three dimensional electrodes, the penetration depth of the current is limited due to ohmic losses in the electrolyte. The local current density in the bed approaches zero with increasing distance from the counter electrode. For a diffusion controlled reaction, the optimum penetration depth of the limiting current density can be calculated by the following formula: [20,21,51-55]

$$h_p = \sqrt{\frac{2\,\varepsilon\,\kappa\,\Delta\eta}{a_e\,k_m\,v_e\,F\,c}} \qquad (9)$$

where κ is the solution conductivity, ε the voidage of the three-dimensional volume structure and $\Delta\eta$ the range of the overpotential where metal deposition proceeds under limiting diffusion control. According to this equation, the penetration depth increases with decreasing concentration along the flow direction in a packed bed cell. This led to the design concept of the enViro cell which is characterized by an increasing thickness of the electrode in the flow direction.[39-41] A schematic representation of the enViro cell concept is shown in Fig. 9.

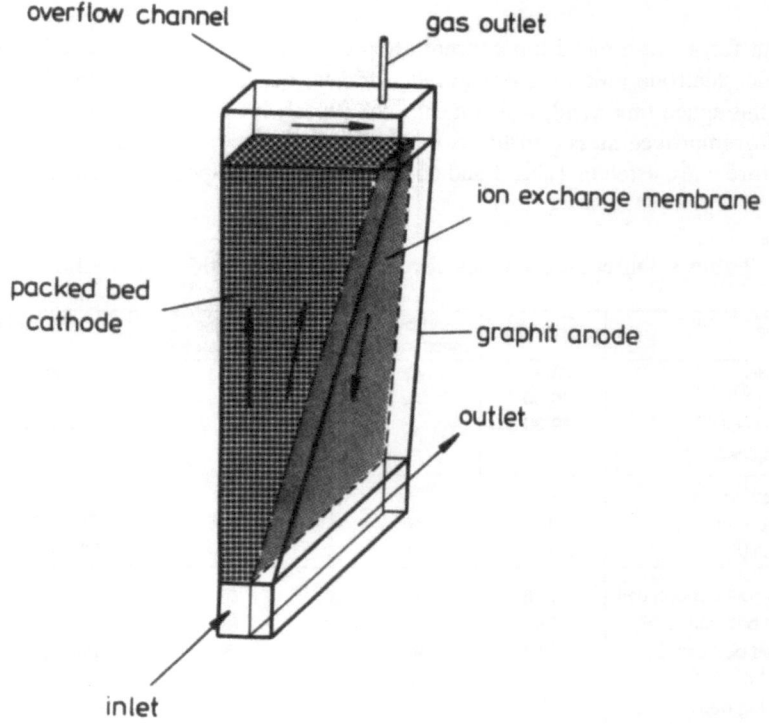

Fig. 9   Schematic representation of the enViro cell

With residence times of only a few minutes, the metal ion concentration can be reduced by up to a factor of 1/1000. The enViro cell has found wide range of industrial applications, some of which are summarized in Table 4 together with characteristic operational data.

Table 4 Industrial applications of the enViro cell and characteristic operational data[15]

Application field	Metal	Troughput m^3/h	Inlet conc ppm	Output conc. ppm	Energy consumption kWh/m^3	Anode area m^2
Production of measuring instruments	Hg	0.3	300	0.05	1.2	1
Film processing	Ag	0.2	15	1.0	0.15	1
Salt production	Pb	0.5	2	0.1	0.07	1
Electroplating	Cd	0.2	20	0.1	0.18	1
Battery production	Hg/Cd	0.08	500	0.01	1.7	3
Cellulose acetate production	Cu	20	20	1.9	0.08	40
Pickling (recycling of solution)	Cu	3	150	50	0.19	5
Dyestuff production	Cu	6	400	2.0	4.0	90
Dyestuff production	Hg	2	4	0.05	2.5	15

PROCESS INTEGRATED ENVIRONMENTAL PROTECTION

Fluidized bed electrolysis

Some of the cells in Table 4 have been successfully applied to plating baths and re-use of rinsing water in the metal plating industry. For such purposes packed bed electrolysis is not suitable since metal ion concentrations are too high. Another concept for metal deposition from dilute solutions is fluidized bed electrolysis, which is suitable for metal ion concentrations up to 5-10 g/l, as designed by Goodridge and Fleischman. [43]

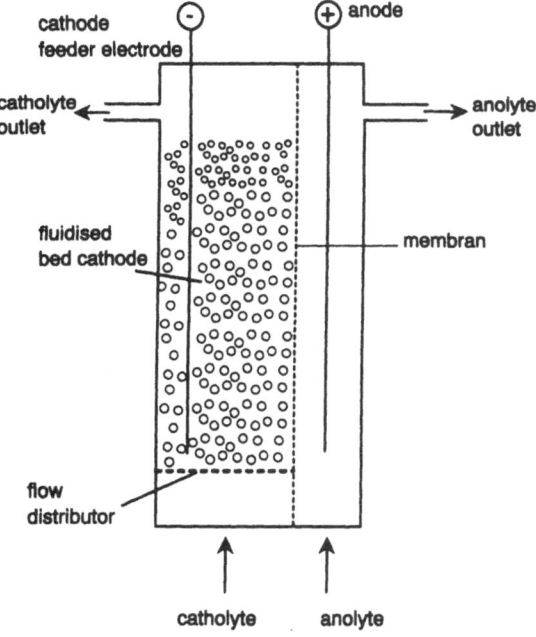

Fig. 10 Sketch of the fluidized bed electrolysis cell

A scheme of the fluidized bed cell is shown in Fig. 10. The electrolyte is flowing from bottom to top at a flow rate above the fluidization velocity. The particles are charged by the feeder electrode and metal ions are discharged and deposited on the particles. On increasing weight the heavier particles are removed at the bottom of the cell and are replaced by fresh ones. To meet the fluidization conditions, the flow velocity of the solution must be relatively high. Therefore a limited concentration decrease per pass can be achieved. This problem can be overcome by continuos operation with recirculation of the solution and cascade arrangements of cells. Fluidized bed electrolysis has been examined for different applications in many laboratories[42,56-60] and has also been applied on an industrial scale.[44-46] A typical application is the removal of copper from a process liquid of the cellulose acetate production for membrane fabrication of artificial kidneys. The process scheme of a cascade arrangement and the technical operation data are shown in Fig. 11.

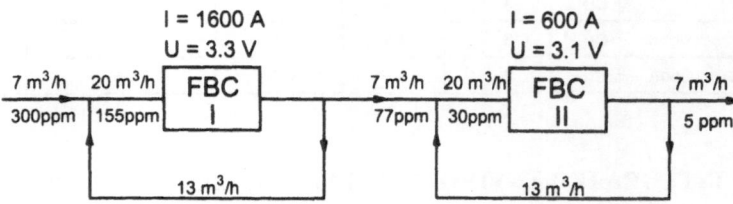

Fig. 11 Flow sheet of copper removal by fluidized bed electrolysis[14]

Chloralkali membrane electrolysis[61-63]

The conventional process for chloralkali electrolysis is the amalgam process. This technology is associated with the emission of mercury vapour into the air and mercury containing waste waters. With the development of fluorinated ion exchange membranes a new process became available which represents a much cleaner technology. In recent years new plants for chloralkali electrolysis were mainly constructed using the new membrane technology. A scheme of a membrane electrolysis is shown in Fig. 12.[61]

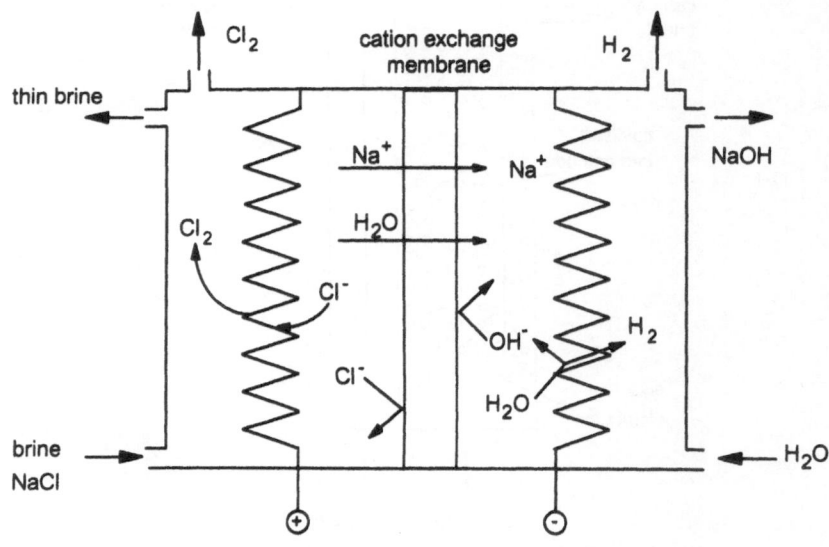

Fig. 12 The membrane process for alkali chloride electrolysis

This membrane technology also allows the application of the zero gap cell concept which is characterized by electrodes which are pressed directly onto the diaphragm in order to minimize the iR-drop due to gas bubbles. Such a version of membrane technology has been commercialized by AZEC-cell in Japan. Another interesting improvement of the process has been achieved by ICI with the FM 21 electrolyser which is a plate and frame cell bloc of only 40 cm in height and a width up to 2 m. This low height of the electrolyser avoids gas accumulation and contributes to energy savings. The progress in decreasing the specific energy consumption for chloralkali electrolysis by the invention of these new processes is summarized in Fig. 13.

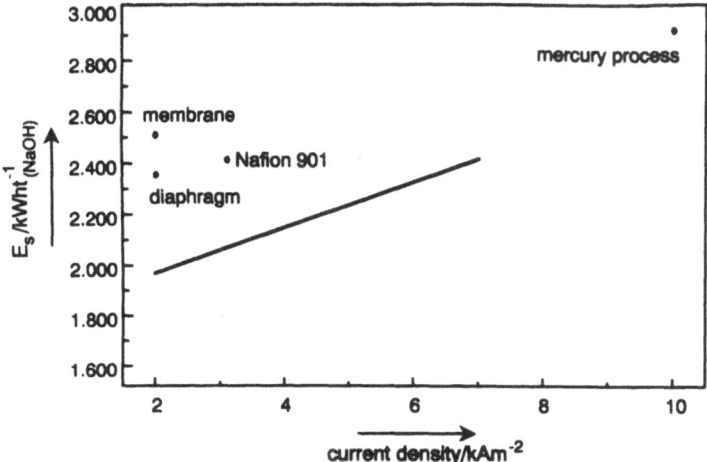

Fig. 13 Specific energy of different process for alkali chloride electrolysis[61]

Monochloroacetic acid production[64]

Monochloroacetic acid is produced by chlorination of acetic acid. In this process dichloroacetic acid is formed as an unwanted by-product by further chlorination of the main reaction product. In the conventional process dichloroacetic acid is reconverted into monochloroacetic acid by catalytic hydrogenation. The process scheme and the main reaction steps are shown in Fig. 14.

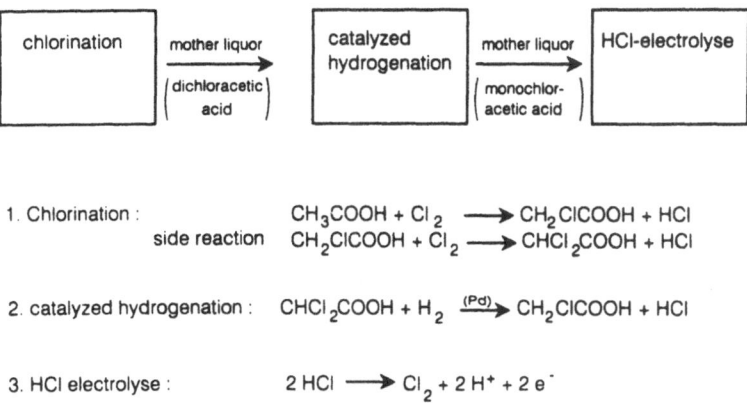

Fig. 14 Process scheme of the conventional monochloroacetic acid production

However, a more elegant way of recovering monochloroacetic acid from dichloroacetic acid is to reverse the direction of the side reaction. Reversal of a spontaneous reaction is only possible by way of electrolysis. The reaction scheme and process data are shown in Fig. 15.

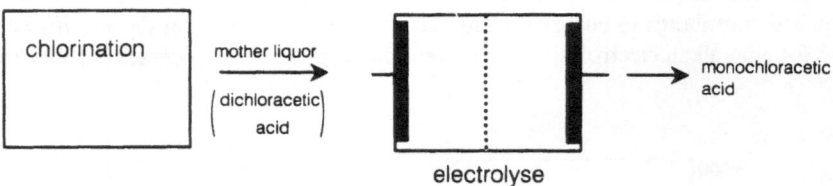

Process data : electrodes: impregnated graphite, electrolyte : 50%final liquor+10-50ppm coppernitrate, current density : 8000 A/m^{-2}

1. Chlorination :

$$CH_3COOH + Cl_2 \longrightarrow CH_2ClCOOH + HCl$$

side reaction $CH_2ClCOOH + Cl_2 \longrightarrow CHCl_2COOH + HCl$

2. Electrochemical reconversion

cathode : $CHCl_2COOH + 2\,e^- + 2\,H^+ \longrightarrow CH_2ClCOOH + HCl$

anode : $2\,HCl \longrightarrow Cl_2 + 2\,H^+ + 2\,e^-$

cell : $CHCl_2COOH + HCl \longrightarrow CH_2ClCOOH + Cl_2$

Fig. 15 Scheme of the new electrochemical process of monochloroacetic acid production

In laboratory experiments this process has caused some serious corrosion problems of different electrode materials like magnetite, graphite and lead. But it has been found that the reaction can be catalysed at graphite electrodes in the presence of traces of different metal ions (Pb^{2+}, Cu^{2+}, Au^{3+}). By means of such metal ion catalysis the current density at graphite electrodes can be enhanced by roughly a factor of 10 without any significant corrosion of the graphite electrode.

This process is a very good example of how electrochemical processes can contribute to improved production integrated environmental protection. This example shows that only electrochemical processes are able to reverse the production of unwanted follow-up products.

REFERENCES

1. G. Kreysa,and W. Kochanek. Möglichkeiten der elektrochemischen Abgasreinigung,*Chem. Ind.* 36:45 (1984).

2. G. Kreysa, and H.-J. Külps, Ein elektrochemisches Absorptionsverfahren zur Gasreinigung, *Chem. Ing. Tech.* 55:58 (1983).

3. G. Kreysa. and H.-J. Külps. A new electrochemical gas purification process, *Ger. Chem. Eng.* 6:352 (1983).

4. G. Kreysa, H.-J. Külps. and C. Woebcken. Elektrochemische Gasreinigung, *Dechema-Monographien* 94:199 (1983).

5. G. Kreysa. Elektrochemische Aspekte der Rauchgas-Reinigung. *Dechema-Monographien* 109:9 (1987).

6. D. van Velzen. H. Langenkamp. and A. Moryoussef, HBr electrolysis in the ISPRA Mark 13A flue gas desulphurization process: electrolysis in a DEM cell, *J. Appl. Electrochem.* 20:60 (1990).

7. Lurgi Schnellinformation C 1217/12.76 (1976). U.K. Pat. 930,548.

8. F. Strafelda, and J. Krofta, Czcheh. Pat. 153372 (1974).

9. G. Kreysa, J. M. Bisang, W. Kochanek, and G. Linzbach, Fundamental studies on a new concept of flue gas desulphurization, *J. Appl. Electrochem.* 15:639 (1985).

10. Final technical report: Heterogeneously and homogeneously catalysed electrochemical gas purification for SO_2 and NO_x removal, Brite-Euram project No P-2026;
    Poject partners: DECHEMA e.V. (D), enViro-cell GmbH (D), LSGC-CNRS, Nancy (F),
    SOCREMATIC SA (F), 1988-1992.

11. K. Jüttner, and G. Kreysa, New electrochemical concepts of gas purification, *Bull. Electrochem.* 8:1 (1992).

12. K. Jüttner, G. Kreysa, K.-H- Kleifges, and R. Rottmann, Elektrochemisches Abgasreinigungsverfahren zur simultanen Entfernung von SO_2 und NO_x, *Chem.Ing.Tech.* 66:82 (1994).

13. R. Rottmann, "Entwicklung und Untersuchung verschiedener Konzepte zur elektrochemischen SO_2-Entfernung aus Rauchgasen," Thesis, Fortschrittsber. VDI, Reihe 15, VDI-Verlag, Düsseldorf 1993.

14. G. Kreysa, Reactor design for electrochemical water treatment, *in:*" Process technologies for water treatment," ed., S. Stucki, Plenum Publishing Corp., New York, London, Washington, Boston (1988).

15. K.J. Müller, and G. Kreysa, Festbettelektrolyse - Design-Konzepte und ihre industrielle Anwendung, *Dechema Monographien* 98:367 (1985).

16. R. Kammel, and H.-W. Lieber, Möglichkeiten zur Behandlung galvanischer Abwässer unter Vermeidung von Sonderabfällen, Teil III, *Galvanotechnik* 68:413 (1977).

17. A.T. Kuhn, Electrochemical techniques for effluent treatment, *Chem. Ind.* 946 (1971).

18. R. Kammel, and H.-W. Lieber, cf 16., Teil VI, *Galvanotechnik* 68:883 (1977).

19. R. Kammel, and H.-W. Lieber, cf 16., Teil VII *Galvanotechnik* 69:317, Teil VIII 69:624 (1978).

20. G. Kreysa, Festbettelektrolyse - ein Verfahren zur Reinigung metallhaltiger Abwässer, *Chem.Ing.Tech.* 50:332 (1978).

21. G. Kreysa, Reinigung und Recycling metallhaltiger Abwässer durch Festbettelektrolyse, *Metalloberfläche* 34:494 (1980).

22. G. Kreysa, Performance criteria and nomenclature in electrochemical engineering, *J. Appl. Electrochem.* 15:175 (1985).

23. G. Kreysa, Normalized space velocity - a new figure of merit for waste water electrolysis cells, *Electrochim. Acta* 26:1693 (1981).

24. R.E.W. Janson, and R.J. Marshall, *Chem. Engineer* 315:769 (1976).

25. R.E.W. Janson, and N.R. Tomov, *Chem. Engineer* 327:867 (1977).

26. D.R. Gabe, The rotating cylinder electrode, *J. Appl. Electrochem.* 4:91 (1974).

27. F.S. Holland, The development of the Eco-Cell process, *Chem. Ind.* 453 (1978).

28. L.J. Ricci,*Chem. Eng.* 29 (1975).

29. R. Kammel, and H.-W. Lieber, cf 16, Teil IV, *Galvanotechnik* 68:710 (1977).

30. W. Götzelmann, cf 16, Teil V, Galvanotechnik 68:789 (1977).

31. D. Bruhn, W. Dietz, K.-J. Müller, and C. Reynvaan, EPA 86109265.8 (1986).

32. A. Storck, P.M. Robertson, and N. Ibl, Mass transfer study of three-dimensional electrodes composed of stacks of nets, *Electrochim. Acta* 24:373 (1979).

33. P.M. Robertson, F. Schwager, and N. Ibl, A new cell for electrochemical processes, *J. Electroanal. Chem.* 65:883 (1975).

34. K.B. Keating, and J.M. Williams, *Res. Rec. Conserv.* 2:39 (1976).

35. D.N. Bennion, and J. Newman, Electrochemical removal of copper ions from very dilute solutions, *J. Appl. Electrochem.* 2:113 (1972).

36. G.A. Carlson, E.E. Estep, and D. Jacqueau, Eine poröse Kathodenzelle für die Laugenreinigung, *Chem.Ing.Tech.* 45:217 (1973).

37. R.S. Wenger, and D.N. Bennion, Electrochemical concentrating and purifying from dilute copper solutions, *J. Appl. Electrochem.* 6:385 (1976).

38. J. van Zee, and J. Newman, Electrochemical removal of silver ions from photographic fixing solutions using a porous flow through electrode, *J. Electrochem. Soc.* 124:706 (1977).

39. G. Kreysa, Elektrochemie mit dreidimensionalen Elektroden, *Chem. Ing. Tech.* 55:23 (1983).

40. G. Kreysa, and C. Reynvaan, Optimal design of packed bed cells for high conversion, *J. Appl. Electrochem.* 12:241 (1982).

41. G. Kreysa, Elektrochemische Zelle, DE 26 22 497 (1976).

42. J.R. Backhurst, J.M. Coulson, F. Goodridge, R.E. Plimley, and M. Fleischmann, A preliminary investigation of fluidized bed electrodes, *J. Electrochem. Soc.* 116:1600 (1969).

43. J.R. Backhurst, M. Fleischmann, F. Goodridge, and R.E. Plimley, GB Pat. 1 194 181 (1970).

44. H. Scharf, DE 22 27 084 (1972).

45. C. Raats, H. Boon, and W. Eveleens, *Erzmetall* 30:365 (1977).

46. G. v. Heiden, C. Raats, and H. Boon, Fluidised bed electrolysis removal or recovery of metals from dilute solutions, *Chem. Ind.* 465 (1978).

47. R. Kammel, and H.-W. Lieber, cf 16. Teil IX, *Galvanotechnik* 69:687 (1978).

48. W. Götzelmann, Die elektrolytische Metallrückgewinnung - Wirtschaftlichkeit und praktischer Einsatz *Galvanotechnik* 70:596 (1979).

49. K. Scott, A consideration of circulating bed electrodes for recovery of metal from dilute solutions, *J. Appl. Electrochem.* 18:504 (1988).

50. T.Bachmann, I. Vermes, and E. Heitz, Abbau umweltbelastender halogenierter Kohlenwasserstoffe durch Metalle, *Dechema-Monographien* 124:221 (1991).

51. J. Newman, and C.W. Tobias, Theoretical analysis of current distribution in porous electrodes, *J. Electrochem. Soc.* 109:1183 (1962).

52. A.K.P. Chu, M. Fleischmann, and G.J. Hills, Packed bed electrodes I. The electrochemical extraction of copper ions from dilute aqueous solutions, *J. Appl. Electrochem.* 4:323 (1974).

53. R. Alkire, and P.K. Ng, Two-dimensional current distribution within a packed-bed electrochemical flow reactor, *J. Electrochem. Soc.* 121:95 (1974).

54. D.N. Bennion, Electrochemical removal of copper ions from very dilute solutions, *J. Appl. Electrochem.* 2:113 (1972).

55. J.A. Traiham, and J. Newman, A flow through porous electrode model: application to metal-ion removal from dilute streams, *J. Electrochem. Soc.* 124:1528 (1977).

56. D.S. Flett, Methods, apparatus: new product research, process development and design, *Chem. Ind.* 13:300 (1971).

57. D.S. Flett, The fluidized-bed electrode in extractive metallurgy, *Chem. Ind.* 983 (1972).

58. H.-D. Steppke, and R. Kammel, *Erzmetall* 26:533 (1973).

59. A.J. Monhemius, and P.L.N. Costa, *Hydrometallurgy* 1:183 (1975).

60. G. Kreysa, Theoretische Grundlagen, technischer Stand und Anwendungsmöglichkeiten von Fest- und Wirbelbettelektroden, *Erzmetall* 28:440 (1975).

61. G. Kreysa, Aktuelle Entwicklungslinien der elektrochemischen Prozeßtechnik, *Chem. Ing. Tech.* 55:267 (1983).

62. D. Bergner, *Chem.-Ztg.* 104:215 (1980).

63. D. Bergner, Membrane cells for chlor-alkali electrolysis, *J. Appl. Electrochem.* 12:631 (1982).

64. S. Dapperheld, Elektrokatalytisches Verfahren zur Aufbereitung von Mutterlaugen aus Chloressigsäure-Produktion, *Dechema-Monographien* 112:317 (1989).

INDEX